A Place Against Time

Studies in Environmental Anthropology
edited by Roy Ellen, University of Kent at Canterbury, UK

This series is a vehicle for publishing up-to-date monograph studies on particular issues in particular places which are sensitive to both socio-cultural and ecological factors (i.e. sea level rise and rain forest depletion). Emphasis will be placed on the perception of the environment, indigenous knowledge and the ethnography of environmental issues. While basically anthropological, the series will consider works from authors in adjacent fields.

Volume 1
A Place Against Time
Land and Environment in the Papua New Guinea Highlands
Paul Sillitoe

A Place Against Time

Land and Environment in the Papua New Guinea Highlands

Paul Sillitoe
University of Durham, UK

LONDON AND NEW YORK

Copyright © 1996 by OPA (Overseas Publishers Association) Amsterdam B.V.
Published in The Netherlands by Harwood Academic Publishers GmbH.

All rights reserved.

No part of this book may be reproduced or utilized in any form or by any means, electronic or mechanical, including photocopying and recording, or by any information storage or retrieval system, without permission in writing from the publisher.

First published 1996 by Harwood Academic Publishers

This edition published 2013 by Routledge
2 Park Square, Milton Park, Abingdon, Oxfordshire OX14 4RN
711 Third Avenue, New York, NY 10017

First issued in paperback 2016

Routledge is an imprint of the Taylor & Francis Group, an informa business

British Library Cataloguing in Publication Data
Sillitoe, Paul
  A place against time: land and environment in the Papua
New Guinea highlands. — (Studies in environmental
anthropology; v. 1)
1. Land use — Papua New Guinea 2. Human ecology — Papua New
Guinea
I. Title
333.7'3'09956

ISBN 13: 978-1-138-97849-2 (pbk)
ISBN 13: 978-3-7186-5925-8 (hbk)

FRONT COVER: Photograph by the author

For two major elements in my environment: Hugh and Ralph.

"Often again it profits to burn the barren fields, firing their light stubble with crackling flame: It is whether the earth conceives a mysterious strength and sustenance thereby, or whether the fire burns out her bad humours and sweats away the unwanted moisture or whether the heat opens more of the ducts and hidden pores by which her juices are conveyed to the fresh vegetation — or rather hardens and binds her gaping veins against fine rain and consuming sun's fierce potency and the piercing cold of the north wind."

Virgil *The Georgics*

CONTENTS

List of maps	x
List of figures	xi
List of tables	xiii
List of plates	xvii
Preface	xxi

Section I   INTRODUCTION
Chapter 1	Environmental ethnoscience: land use, soils and society	3
Chapter 2	Rotating land not crops: shifting cultivation and sweet potato	25

Section II   THE CLIMATE FACTOR
Chapter 3	Ethnometeorology: the climate	53
Chapter 4	Coping with climatic variations: the spectre of famine	79

Section III   THE LAND RESOURCES FACTOR
Chapter 5	Ethnogeoscience: topography and geology	105
Chapter 6	Living with land loss: the costs of erosion	139

Section IV   THE BIOTIC FACTOR
Chapter 7	Ethnobotany: the plants and vegetation communities	167
Chapter 8	Contending with forest and fallow: demons to regrowth	201
Chapter 9	Into the soil: nutrient cycling and decomposition	229

Section V   THE SOIL
Chapter 10	Ethnopedology: the soils	265
Chapter 11	Keeping up with soil status: the implications of variability	297
Chapter 12	Out of the soil: fertility under cultivation	339

Section VI   CONCLUSION
Chapter 13	Mounds of soil: grounds for sound land resource management	375
Chapter 14	Racing with time: degradation or conservation?	393

APPENDIX
Glossary of Wola environmental terms	405

REFERENCES   411

INDEX   425

LIST OF MAPS

1.1	The island of New Guinea (locating some of the places mentioned in this book).	3
1.2	The Wola region.	16
3.1	The effect of topography on wind direction across the Wola region.	68
5.1	Topography of the Wola region.	109
5.2	Local place names in one part of the Was valley.	111
5.3	Landform units of the Wola region.	116
5.4	Geology of the Wola region (with cross-section).	124
5.5	Orogeny in a localised region of the Was valley.	126
5.6	The distribution of recent (<50,000 yrs B.P.) volcanic ash deposits across the Highlands.	131
7.1	Vegetation of the Kerewa-Giluwe region.	180
7.2	Vegetation of the Was valley study area	182

LIST OF FIGURES

2.1	Changes in exchangeable bases in an Ultisol and an Alfisol.	31
2.2	Percent changes in topsoil organic carbon content with different crop-fallow ratios.	33
2.3	Diagram of forest-soil nutrient cycle.	35
2.4	The yields of sweet potato from gardens cultivated semi-continuously for different periods of time.	44
2.5	Sweet potato (*Ipomoea batatas*).	45
2.6	Environmental factors affecting sweet potato dry matter production.	47
2.7	Development of sweet potato roots in relation to cambium activity and stele cell lignification.	49
3.1	Mean monthly rainfall (mm) for six locations in the Wola region.	59
3.2	Monthly wind roses (direction and speed) for Haelaelinja.	67
3.3	Mean, maximum and minimum monthly temperatures.	70
4.1	Responses of sweet potato to different temperatures.	83
4.2	Effects of waterlogging on sweet potato.	84
5.1	Topographical cross-section with some Wola terms for features.	110
6.1	Soil loss according to slope on bare soil where vegetation cover 0%.	153
6.2	Soil loss according to slope where crop vegetation cover <50%.	153
6.3	Soil loss according to slope where crop/fallow vegetation cover >50%.	154
7.1	Cane grass (*Miscanthus*).	195
8.1	Southern beech (*Nothofagus*).	202
8.2	Casuarina (*Casuarina oligodon*).	226
9.1	Mineral cycling under New Guinea montane rain forest.	230
9.2	Mineral cycling under secondary woodland.	231
9.3	Mineral cycling under secondary cane grassland.	232
9.4	Mineral cycling under secondary coarse grasses.	233
9.5	The mass of earthworms in the soil compared with sweet potato tuber yield.	258

12.1	Comparison of garden type with time under cultivation.	340
12.2	The relationship between topsoil pH and time sites abandoned.	348
12.3	The relationship between topsoil phosphorus and site cultivation status.	350
12.4	The relationship between topsoil exchangeable potassium and site cultivation status.	352
12.5	The relationship between topsoil carbon content and site cultivation status.	355
12.6	The relationship between topsoil nitrogen content and site cultivation status.	356
13.1	Mean soil bulk densities in gardens cultivated for different periods of time, according to time for which sweet potato mounds planted.	380
13.2	Sweet potato yields, Aiyura soil exhaustion experiment.	387
13.3	Coarse fallow grass (*Ischaemum polystachyum*).	389

LIST OF TABLES

2.1	The elemental composition of, and nutrients added in, burned Peruvian secondary forest.	27
2.2	Semi-permanent cultivation within the context of a shifting regime.	39
3.1	The daily gross weather pattern each month (percent days/month).	55
3.2	Cloud cover data from two locations: mean daily total cloud cover, and percentage occurrence of cloud types each month.	63
3.3	Humidity data for two locations: mean monthly wet and dry bulb temperatures (°C at 09.00 hrs, 15.00 hrs), and relative humidity (%).	71
3.4	Solar radiation data: mean daily total radiation (MJ/m^2).	72
4.1	List of plant foods resorted to in times of famine.	80
5.1	The areas and relative proportions of the Kerewa-Giluwe region covered by different landforms.	120
5.2	Geology of the Wola region: detailed lithology, local formation names and their geological periods.	125
6.1	Probability of rainfall intensities (average no. days per month with rainfall within specified classes).	143
6.2	Kinetic energy and erosivity calculations for rainfall across the Wola region.	144
6.3	Kinetic energy by rainfall intensity class across the Wola region.	146
6.4	Cropping cycle in the New Guinea Highlands.	159
7.1	The areas and relative proportions of the Kerewa-Giluwe region covered by different vegetation.	181
8.1	The vegetation characterising different stages of regrowth after garden abandonment.	218–219
8.2	Analysis of soil samples, collected in the Was Valley under casuarina trees compared with samples collected under adjacent open fallow land.	225
9.1	The average nutrient status of comminuted litter layer under different vegetation communities.	237
9.2	Measures of topsoil chemical fertility compared with site vegetation cover.	238

9.3	The contribution of different floristic life-forms to the biomass of different vegetation communities.	241
9.4	The percentage distribution of nutrient elements between different fractions of above-ground vegetation under various successions.	244
9.5	The estimated release of nutrients following the clearance of different vegetation successions for cultivation (kg ha^{-1}).	245
9.6	Common crop pathogens and diseases of the Southern Highlands Province.	250
9.7	Soil fauna population densities: organisms observed during (1) the digging of 85 soil profile pits, (2) the sieving of soil from 31 sweet potato mounds and (3) the excavating of a one cubic meter pit.	260–261
10.1	A selection of horizon sequences named and identified by the Wola.	274–275
10.2	The soils of the Wola region according to various classifications, including the indigenous scheme.	277
10.3	The nutrient status of *haen hok* rendzina soil.	279
10.4	The average nutrient status of *hundbiy* clay soils.	282
10.5	The average nutrient status of *tiyptiyp* ashy clay soils.	288
10.6	The average nutrient status of *iyb dor tilai* alluvial soils.	290
10.7	The average nutrient status of *iyb muw* sandy soils.	291
10.8	The average nutrient status of *tongom* gley soils.	293
10.9	The average nutrient status of *iyb uw damiy* peat soils.	294
10.10	The nutrient status of *waip* peat soil.	296
11.1	The soil resource classification.	304
11.2	The distribution of local assessment and land use classes between the soil profile and site components of the soil resource classes.	305
11.3	Wola profile classification compared with soil profile clusters.	307
11.4	Indigenous profile classification compared with local assessment of soil agricultural potential.	308
11.5	Mean garden site and soil factors compared with gardeners' social status.	318–319
11.6	Mean garden site and soil factors compared with gardeners' ages.	324–325
11.7	Mean garden site and soil factors compared with gardeners' kin group affiliation.	328–329

List of Tables

12.1	Mean garden site and soil factors compared with garden type and times cultivated.	342–343
12.2	Measures of topsoil chemical fertility compared with site land use status.	346
12.3	A comparison of local assessment of cultivation potential with mean site soil chemical fertility.	360
12.4	A comparison of soil survey profile classes with mean site soil chemical fertility.	361
12.5	The average nutrient status of ash and topsoil samples from newly cultivated garden sites cleared from different vegetational communities.	363
13.1	Chemical composition and nutrients supplied by vegetation successions used as compost in the Wola region.	390
13.2	The average nutrient status of soils on abandoned house sites.	392

LIST OF PLATES

1.1	The rugged terrain of Wolaland: looking along the Was river valley.	15
1.2	The fast flowing Was river at Kaerep with adjacent garden and house.	17
1.3	An exposure showing the region's deep blocky structured clay subsoil overlain by a friable dark topsoil, supporting *Miscanthus* cane.	18
1.4	A profile of one of the major soils of Wolaland: a Tropept, called locally *hundbiy*.	19
2.1	Burning small pieces of vegetation remaining on a garden site before starting to plant it.	40
2.2	Heaping up dried fallow grasses to compost a sweet potato mound.	41
2.3	Breaking up soil to build a mound.	42
2.4	Heaping up turned soil into a mound for sweet potato propogation.	43
3.1	A deciduous *hok* tree (*Sterculia*) that has shed its leaves in the *bulhenjip* season.	58
3.2	Heavy rain turns *howma* ceremonial clearing into a muddy expanse.	60
3.3	Wood smoke seeps through the thatch of a men's house *enza* to drift into the sky and become cloud.	64
3.4	Cloud fills valleys in the early morning, "feeding" on the night's fire smoke.	64
4.1	One of the up-welling subterranean pools associated with the injured female spirit Saliyn's flight, at Iybwimb.	92
4.2	A man decorated in full finery like the *moray nais* at the close of the large *iyshpondamahenday* sequence.	93
4.3	At Ngoltimaenda, a man, Pelsuwol Njep, points to the two ritual stones Hab (left) and Homok (right).	96
4.4	Pelsuwol Kwimb holds a hammerstone used to crack the leg bones of marsupials against Hab and Homok.	97
5.1	Turbulent mountain rivers like the Was flow along valley bottoms.	107
5.2	Looking down on a vine bridge suspended across a narrow gorge in the Was river's course.	108

List of Plates

5.3	A limestone pinnacle in the Ak valley, typical of karst country.	117
5.4	A doline in limestone terrain (the sinkhole is near the many-crowned pandan at the centre).	118
5.5	Stratification evident in a limestone exposure at Bombok (solution weathering is evident in the calcareous surface deposit).	121
5.6	Peering down a cleft in the limestone into a pothole.	122
5.7	A *mundit* landslide that has torn a gaping hole in a hillside, sweeping away trees in its path.	136
6.1	Evidence of surface wash erosion from and between sweet potato mounds where soil is exposed by recent cultivation.	140
6.2	Incipient rill developing on soil exposed after pulling up remaining crops and weeds prior to recultivating a garden.	141
6.3	The dense ground cover of established sweet potato vines affords excellent protection of the soil.	156
6.4	A pig rooting over topsoil, moving it downslope and exposing it to rain erosion.	157
7.1	The *iysh* family: a Southern beech tree (*Nothofagus*).	168
7.2	The *henk* and *gaimb* families: a tree fern (*Cyathea*) growing in cane grass (*Miscanthus*).	169
7.3	Plants of no family: a *aenk* screw pine (*Pandanus* sp.) growing in *gaimbs* grass (*Miscanthus*).	170
7.4	A homestead: a bamboo (*Nastus*) overhanging a women's house.	171
7.5	The *den* family: a garden site overgrown with herbaceous invaders (*Arthraxon, Polygonum, Viola, Crassocephalum* etc.), some sweet potato (*Ipomoea*) and Highland pitpit (*Setaria*) among them.	172
7.6	Some cultivated food plants: a mixed vegetable garden (em gemb) with taro (*Colocasia*), sugar cane (*Saccharum*) and pumpkin (*Cucurbita*), among other crops.	173
7.7	The *iyshabuw* lower montane rainforest, seen from a garden, Southern beech (*Nothofagus*) predominating (including the gaunt, moss-clad trees).	185
7.8	An expanse of *pa* wetland, swamp grass (*Leersia*) predominating, giving way to a cane grass (*Miscanthus*) community.	186
7.9	A steep-sided limestone pyramid clothed in *haenbora* rocky vegetation of ferns (*Pteridium*), orchids (*Spathoglottis*) and kunai grass (*Imperata*).	188
7.10	Moss-clad stunted trees of high altitude cloud forest adjacent to the *yom* alpine community.	189

List of Plates

7.11	Sweet potato (*Ipomoea*) mounds in newly cultivated *em* garden, with some sugar cane (*Saacharum*) and Highland pitpit (*Setaria*).	190
7.12	Wenja Muwiy stands in an established sweet potato garden with a quadrat frame used in the survey of vegetation community composition.	191
7.13	A sweet potato garden in the later *puw* yield stage with a dense cover of invading weeds (*Bidens, Polygonum, Paspalum* etc.).	192
7.14	The early *mokombai* phase of recent regrowth with early herbaceous invaders (*Arthhraxon, Bidens, Polygonum, Viola* etc.) well established.	193
7.15	The later *bol* phase of recent regrowth with coarse grasses (*Ischaemum* predominating) well developed (the smoke in the background comes from fires lit in recultivating such an adjacent fallow area).	194
7.16	An expanse of *gaimb* cane (*Miscanthus*) grassland adjacent to the Uwt stream in the Nembi valley.	198
7.17	The *obael* phase of secondary regrowth, soft-wooded trees, pandans and cane grass.	199
8.1	The montane beech forest home of forest demons.	205
8.2	A man camouflaged like a forest demon, rushing to and fro to keep spectators back at a dance.	206
8.3	The decoration worn during the *hwiybtowgow* curing rite.	210
8.4	The *komay* leaf wall erected for the *hwiybtowgow* curing rite.	211
8.5	Two men with aroid leaf parcels in their mouths stand behind the leaf wall in the *hwiybtowgow* rite.	212
8.6	A fence line planted with palm lilies (*Cordyline*), which may serve as boundary markers years later if the site is recleared.	223
8.7	A casuarina tree showing characteristic wispy cassowary-feather-like foliage (*Casuarina oligodon*).	227
13.1	A newly planted sweet potato mound (*mond*).	376
13.2	Using a digging stick to uproot coarse grass in a fallowed garden.	377
13.3	Burning sun-dried fallow grass vegetation before mounding commences.	378
13.4	Heaping soil into an *aend saway* square to accommodate compost for the centre of a mound.	379
13.5	Starting to throw turned soil over compost in mound centre.	381
13.6	A mound starts to take shape as the compost is buried.	383
13.7	Scooping up soil onto a mound to produce a friable bed.	384
13.8	A sweet potato mound nears completion.	385

14.1	A mosaic of gardens, fallow grassland and cane regrowth with cloud shrouded primary forest on ridge crest.	394
14.2	A pig, item of wealth and agent of soil erosion, standing in an area of soil that it has rooted over.	396
14.3	The prelude to recultivation of a garden: pulling up coarse fallow grasses.	399

PREFACE

An abiding memory of my early days among the Wola, while still awe-struck at the realisation that I was with them doing fieldwork in the Papua New Guinea highlands, concerns a survey of garden sites that I decided to conduct to ascertain something about land holdings in the region where I was to live and work. We were standing in a garden at a place called Bombok in the ruggedly beautiful Was river valley (see Map 5.2), and I was dumbfounded when I asked how long ago those named as the tenants had cleared the site, expecting to hear that it was a year or so previously, to be told that they had not cleared it at all, but that their father Waebis, or maybe his father Wa, had established it. Both men had been dead for some time, Waebis for over a decade, and his father for considerably longer. The reputation of Waebis had already caught my imagination; he was a man of great renown who had ten wives, and was among the first to sight Hides' and O'Malley's patrol when, harbinger of the outside world, it penetrated Wolaland in the 1930s. Now I heard that he could apparently contravert what I thought I knew about subsistence agricultural regimes in tropical forested regions, featuring shifting cultivation.

Before arriving in the Southern Highlands of Papua New Guinea I was under the impression that the standard notion of shifting cultivation prevailed there, as the norm throughout the tropics.[1] During the year that I was preparing to leave for the field I recall attending sessions at which 'Skip' Rappaport, who was then visiting Cambridge, described Maring shifting cultivation, which complied with the classic model, having brief periods under cultivation followed by long intervals under natural fallow to allow recovery of soil fertility (Rappaport 1972). Nonetheless those who accompanied me to Bombok were adamant when questioned closely that the garden had been under continuous cultivation, with brief periods of grassy fallow, for so long as they could recall. And it was not the only one attributable in some way to Waebis' talents. They named several places where there were other long-term gardens, which they told me they call *em hul* (lit. garden bone — i.e. durable like bone). Nor was that all. They had further surprises in store for me. Some sites I learned later, far from experiencing a decline in staple crop yields, as the accepted model of low-input subsistence agriculture predicts, experience the reverse and improve with time under cultivation, with increases in yields. When I did some subsequent work on crop yields, it seemed to verify their assertions — the yield of crops from newly cleared garden areas was only some three-quarters of that from established ones and perhaps as low as 60% for staple sweet potato (Sillitoe 1983:225).

---

[1] The existence of long-term gardens had already been reported elsewhere in the highlands (e.g. for Simbu by Brookfield & Brown 1963; for Enga by Waddell 1972; for the highlands region generally by Kimber 1972), but I was quite unprepared for its extent among the Wola or the time scale involved on some sites.

The farming practices of the Wola seem to contradict widely accepted assumptions about traditional agriculture in the tropics. How can this be so? How are they able to continue with an agricultural system that features repeated cultivation of sites, with minimal or no fallow breaks and no outside amendments, without a catastrophic decline in productivity? How is it that the reverse sometimes occurs and crop yields improve? This book, a study of the natural environment informed by an indigenous perspective, addresses these and other natural resource questions. It is environmental ethnography, one which attempts to convey something about local people's viewpoints, in addition to those of environmental science (Inglis 1993). It focuses on the soil, a central agricultural resource, developing an understanding of natural resources through an analysis of local soil conditions and management practices. Since the pioneering work of the renowned Russian soil scientist Dokuchaev (1883), it has been recognised that:

> "Soils result from the combined activity of the following agencies: living and dead organisms (plants and animals), parent material, climate and relief."

This book, framed around this premise, gives accounts of each environmental agency and their contributions to the soil resources of the montane Wola region. It presents an investigation of natural resources by considering aspects of the environment as they contribute to the soil, establishing the general conditions that prevail for crop growth.

The soil initially suggests itself as an uninspiring subject of study to the uninitiated, equated in the minds of many urban dwellers, who are distanced from the land and food production, with dirt and muck. And an ethno-pedological investigation will strike some as doubly unpromising if not downright eccentric. But the soil is a remarkable medium (Hillel 1992). And peoples' understanding and management of it will be central to their existence. Life depends directly on the well-being and fertility of the soil. It is the origin and last resting place of everything that grows. It contains the inorganic nutrients plants use, along with water and atmospheric gases, to build the complex organic structures upon which the trophic pyramid of animal, including human life stands. It is a difficult subject to study because of the complexity of any soil system, its components inter-related in complicated and fascinating feedback relationships. An infinite variety of soils exist, like individual human beings no two soils are strictly identical, defying rigid classification. And the complex processes which produce soils from weathered rocks and decaying organic matter go on continuously, chemical changes releasing nutrient ions from mineral particles and organic substrates, and the growth of plants and activities of animals endlessly stirring up the soil until they return to it at death to contribute further soil forming materials. The wonder of the soil is conveyed by Wrightson and Newsham (1919) in their sturdy old book:

> "When we consider ... the complex character, the perpetual renovation ... in every fertile soil, no one can fail to be struck with wonder and admira-

tion. The soil is a graveyard in the widest sense, and yet it is the very mainspring of new life. If any object can demonstrate the possibility of resurrection, it is the soil upon which we walk."

The soil never stands still, it is constantly changing and evolving, rather like human society. Every contemporary soil exists against time. And human beings contribute to this natural change when they disturb the soil in cultivation, sometimes skilfully managing the soil to promote, or at least maintain, its fertility, like the people who feature in this book, other times plundering and degrading it.

Several strands weave through this book. One is an account of a montane New Guinea environment, framed around the soils resulting from the long-term interaction of Dokuchaev's various agencies of the environment. Another is the attempt made to convey how the local people perceive of their natural surroundings and contrive to manage them, relating the natural environment, particularly its soils, to their intriguing horticultural traditions. A further strand concerns their staple crop of sweet potato, which occupies a prominent place in this agricultural regime.

The study of the natural environment, and how it relates to people's subsistence arrangements, has a creditable history. While it has long been recognised that environmental determinism is wrongheaded — that a region's natural surroundings do not decide the kind of culture found there — it is nonetheless accepted that the environment sets certain parameters to what any human population can achieve within a given technology. A well-known approach to this subject, sometimes called cultural ecology, considers human beings to be adapted to particular niches, like all other animals, to ensure the efficient appropriation of a share of the energy captured by, and flowing through, an ecosystem (Rappaport 1972). But unlike other animals, human beings are not passive regarding nature's constraints but think about, manipulate and manage their environments. This is one of the triumphs of technology, and tragedies when misused (Sillitoe 1988). The romantic poets' notion of the environment imposed on the 'primitive', still strong in the popular imagination, is wrong. A central theme of this work is the relationship of Wola culture to the natural environment, how structured to cope with its constraints in a sustainable manner, effectively to exploit and manage its natural resources.

New Guinea is the large island perched bird-like above Australia in the South West Pacific (Map 1.1). It is geographically diverse, with environments ranging from coral fringed coasts through steamy lowland swamps to rugged highland ranges. It has large expanses of rainforest, together with swamp and grassland; an Australasian marsupial fauna and a sensational bird life (including the famous birds of paradise). It is also culturally diverse, with a range of agricultural systems, staple crops varying from sweet potato to bananas to taro to sago to yams; some writers suggest that it has half the world's languages. The natural environment, soils and agricultural system investigated in this book are located in the montane region of Papua New Guinea, in the Southern Highlands Province. The Wola occupy five valleys here, along which flow the Augu (locally called the Ak), Wage (Was), Nembi (Nemb), Lai and Mendi rivers (Map 1.2).[2] The ethnographic and environmental data presented here come largely from the west of this region, where I have lived in the settlement of Haelaelinja, situated within a sweeping bend of the Was river, and conducted fieldwork. Throughout I have endeavoured to work in Wola, a difficult verb-dominated tongue like all Non-Austronesian languages.[3] A companion to my previous study of their crops and cultivation (Sillitoe 1983), this work expands on environmental issues and their relation to the Wola farming regime.

A book inevitably owes something to the environment in which it is written, an author depending more upon his surroundings perhaps than many other organisms, demanding not only support but also inspiration. I have been particularly fortunate in my environment. My largest debt on the support side is to my family, which patiently endured up-rooting to the other end of the country for a stay at Wye in Kent, and my absences of several months to pursue fieldwork on the other side of the world. My wife Jackie has listened good-temperedly to my expounding, sometimes obsessively, on ideas which sometimes can have made little sense to her, for they did not always at the time make much sense to me as I struggled to understand them, but she was always generous in her support and encouragement. I also thank her and the boys for tolerating someone who could not always be prevailed upon to take on his full share of family responsibilities, his "mind in the midst of a sentence being like a ship at sea, knowing no rest or comfort till safely piloted into the harbour of a full stop".[4] I cannot expect this book to compensate for the forgone fishing, footballing and bike-riding trips, but perhaps when they are somewhat older the boys will have some sympathy with my endeavours to make sense of strange things.

My largest debt regarding inspiration is to my Wola friends, who have been a constant, if sometimes unwitting source of ideas with their unexpected remarks and practices, constantly challenging in a provocative manner what I thought I knew of natural resources and their management. I thank all of my Wola friends in the Southern Highlands Province for their assistance and hospitality while investigating their region's natural environment, observing their relations with it and learning about their subsistance farming activities. In particular I thank my two long-standing companions Maenget Pes and Wenja Neleb, who between periods working elsewhere, have faithfully assisted me in my endeavours, their reliability, resourcefulness and friendship contributing enormously to the success of my work. Also Ind Kuwliy and Wenja Muwiy, who have proved stalwart friends for many years on whose assistance

---

[2] See Sillitoe 1979 for a discussion of this delineation of the Wola region, founded on a common language; also Lederman 1986.
[3] See Sillitoe 1979 for a brief note on this language, the orthography used in writing it, and my approach to its translation, often featuring literal transcriptions.
[4] Thomas Hardy, *The Hand of Ethelberta*.

I could always depend, and more recently Mayka Sal and Mayka Haebay. Among others, Mayka Yaeliyp, Huwlael Kombap, Ind Haendaep and Mayka Waeriysha deserve mention for carrying equipment on my many journeys throughout their region to investigate plant communities and soils, sometimes in remote locations.

In my attempts to straddle different intellectual approaches and unite the perspective of Western environmental science with the cultural empathy of anthropology, I have been especially fortunate in the support that I have received from colleagues. In particular, I have to thank Robert Shiel of the Department of Agricultural and Environmental Science at the University of Newcastle upon Tyne for his good-humoured assistance and constant encouragement with this study; his wide knowledge of environmental science and ready advice have been invaluable. At Wye College I thank Ken Giller for his broad-minded approach and for stimulating discussions, Paul Burnham for helpful suggestions, George Cadisch for allowing me to draw upon his extensive knowledge of plant nutrition, and Sinclair Mantell for helpful advice on sweet potato horticulture. The senior common room of Eliot College at the University of Kent at Canterbury extended a fellowship to me during my stay in Kent, and I am grateful to Roy Ellen and his department for their warm welcome. My Durham colleague Todd Rae gave invaluable assistance in making technical information pertaining to this study available electronically on the World Wide Web. Also I thank T. Bayliss-Smith for constructive criticism of the manuscript. And Alan Clewer of Wye for thoughtful advice on statistical methods and interpretation, and the staff in computing services at both Wye College and Durham University for patiently helping me master the complexities of various computing packages. I am grateful to the following for facilitating the soil analyses central to this study: F. Fahmy (formerly of D.P.I. Labs. Konedobu, Papua New Guinea), M. Clarke (Geography Department, Manchester University), David Radcliffe (formerly of AFTSEMU in Mendi), Andrea Summerson and Jennifer Thompson (Anthropology Department, Durham University), Derek Coates (Geography Department, Durham University), and Carol Camsell (Agricultural and Environmental Science Department, Newcastle University). And for help identifying the plant material I thank in particular Bob Johns, Barbara Parris and Peter Edwards, among others at Kew herbarium.

The conduct of any research depends upon adequate financial support. I am very grateful to the Nuffield Foundation for electing me to a Fellowship at an opportune moment with respect to the development of my ideas, giving me valuable time to think and proceed with my work. Also the British Academy for personal research grants to pursue further fieldwork. This study is also inevitably informed by my earlier visits to the Southern Highlands of Papua New Guinea, for research of this kind is inevitably continuous; I thank elsewhere the many other bodies that have generously supported me over the years. The University of Durham has also generously supported my research, and I am grateful to my colleagues in the Anthropology Department for their tolerance and support.

SECTION I

INTRODUCTION

'*Suw ngo aiben? Ninau suw. Nau momon diy nau abon diy suw ngo waem aiben aiben bisor. Ninau suw aengora. Nau suw turiy homow konay uw maemb bor haekwa. Ninau isiy diy nguwp turiy homow uw maemb bor haem aiben aiben buw hakor.*'

'Whose land is this? My land. It was my grandfather's and my father's, this land has been passed on and on [down to me]. It's my land only. I'm happy to think of caring for my land. My sons too will be happy to care for it when it passes on and on [to them].'

Wenja Neleb

CHAPTER 1

ENVIRONMENTAL ETHNOSCIENCE: LAND USE, SOILS AND SOCIETY

Land is very important to New Guinea highlanders. It has a tremendous sentimental value for them. They frequently dispute, and sometimes even fight, over rights of access to particular areas. Living in a place reinforces social connectivity, it builds on kinship there. Association with the land establishes a strong relationship with others who live in the same locale. In the settlement of Haelaelinja, where I have lived and worked for some twenty years now, they refer to me for example as a *suw ora* (lit. land father), a guardian of the territory to which today's children are heirs. It signifies that I belong there, and have social obligations there. The Wola build on connection with place by taking part in wealth exchanges, maintaining social attachments there. These are cultural and spatial constructs on the biology and sociology of kin relationships.

The Wola perceive of themselves as the custodians of their land, protecting access to it and managing use of it responsibly, for themselves and their offspring, as reflected in Neleb's opening comments to this section. Land is not a resource with some market value, it is an inalienable property of great emotional worth. The Wola are perplexed when they hear about recent developments elsewhere, notably mineral exploitation, which bring landowners large sums of money and unimagined access to material goods but result in the destruction of their territories as they know them. The current visits to their region by helicopters on mineral prospecting trips has them particularly agitated. They may not be vocal conservationists anxious to protect particular

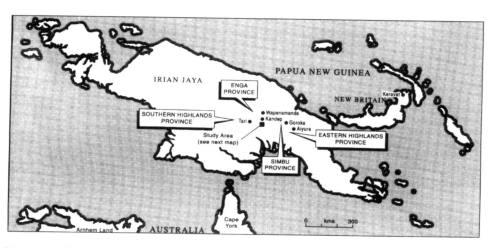

Map 1.1 The island of New Guinea (locating some of the places mentioned in this book)

natural communities, but loss of land worries them. They responsibly cultivate and sustainably manage their land, avoiding degradation where possible, whatever first appearances, in practicing sometimes a paradoxical 'stationary shifting' cultivation, which some might assume invites damage of land resources.

The cultivation regimes found in the rugged central highland regions of the mountainous interior of the Pacific island of New Guinea, like that practiced by the Wola, defy characterisation according to any farming systems classification. It is debatable whether they are variants of shifting cultivation, bush fallow, extensive/alternating cropping, land rotation or some other system.[1] On some plots, people practice classic shifting cultivation, clearing them for one, or possibly two crops, and then abandoning them to natural regrowth for many years. Other plots, the Wola 'bone gardens', they keep under more-or-less permanent cultivation for decades, with occasional, relatively brief periods of grassy fallow between some cultivations. And between these extremes, they cultivate a range of plots for varying periods of time. The result is a continuous spectrum of agricultural land use, comprising a single farming system. When they break new ground, clear an area of natural vegetation for a new garden, the farmers themselves often cannot say how long it will remain under cultivation.[2] They plant it and see how their crops fare. So long as yields are respectable, and the location of the garden convenient, they may continue to cultivate it indefinitely. All cultivations, whatever their productive life, are farmed using the same technology and procedures (Steensberg 1980). We have one farming regime here, not several discrete ones integrated into some complex system, confounding our misguided urge to pigeon-hole diverse tropical farming into a few simple classes.

Environment and Soils

This book is an investigation of the natural environment, focussing on soils and local knowledge, exploring as it goes this ambiguous and intriguing cultivation regime. It breaks new ground intellectually as an ethno-pedological study, which to-date is a little researched topic. In certain regards it is some-

---

[1] Perhaps this is only to be expected, global classification schemes are unlikely to accomodate the varied agricultural systems of the tropics, as we come to know them in detail. The classificatory endeavour is a will-o'-the-wisp pursuit. Nevertheless there is a considerable body of oft-quoted and influential literature on the classification of tropical farming systems, with a bewildering array of suggested terminologies; attempts have even been made to give numerical substance to suggested definitions. It says something perhaps about the urgent need for in-depth research into tropical farming systems that such classifications continue to hold sway. See Conklin 1957, Nye & Greenland 1960:5–6, and Sanchez 1976:348 on the bewildering array of terms used for shifting cultivation systems.
[2] Unless they intend to establish a stand of taro, a crop which they say rapidly reduces soil productivity; such a garden, usually on a wet site, will be a one-off cultivation. This contrasts with the use of wetland sites in the neighbouring Tari Basin, on which the Huli have continuously, for over a century, cultivated a wide range of crops following drainage (Wood 1984, 1991).

what unconventional, notably in attempting an ethno-environmental study that presents both our natural science and another culture's perspectives of the natural environment. In other regards the book is studiedly conventional, notably in its approach to soil science, being organised according to the classic factors of soil formation.

A region's natural environment sets parameters for crop growth; the temperature, the terrain, the plants and so on all contribute to the overall conditions with which any agricultural system contends. The soil, a natural resource critical to farming, is the product of the interaction of this range of environmental factors, it is where they meet. Soil scientists have long recognised that environmental factors and relationships contribute critically to any region's soils (Jenny 1941; FitzPatrick 1983). Those which they cite include:

- climatic factors
- topographic factors
- parent materials
- biotic factors
- time.

Some writers include a sixth factor, which is particularly relevant in an anthropological context:

- human activity.

The book is structured around these environmental factors, covering those resources central to crop production. It devotes sections to the first four, the other two occurring throughout it. The soils that result from the interaction of these factors feature in another section.

Early writers considered **climate** as the paramount environmental factor influencing soils, with their notion of zonal soils (classifying soils by climatic zones — e.g. Dokuchaev 1883). And the two most important aspects of climate from a soil perspective, are a region's rainfall and temperature. Precipitation is important because many of the physical, chemical and biological processes that go on in soil involve water: rainfall influences weathering conditions and leaching through the soil. And temperature prevails over the rate of chemical and biochemical processes. The climate consequently contributes to the rate of soil formation and to geomorphological processes, and to vegetation distribution too. Nevertheless, the soil-related-climate can vary in two adjacent locales under the same atmospheric conditions due to variations in topography, parent material, and so on (a depression may have a saturated soil and a neighbouring elevated place a freely drained one, each having different soils and vegetation communities). Other environmental factors are as important as the climate, it is the interaction between them that is significant.

The **topography** of a region relates to its local relief. It determines to a considerable extent the spatial distribution of soils across any landscape, through its influence on microclimate, surface geological processes, pedogenesis and so on, the effects of all of which are interrelated. Soils vary laterally with topography. The aspect of slopes affects microclimate, orientation influencing the moisture and temperature regime of any soil. Differences in elevation will also influence climatic conditions, notably temperatures. The steepness of slopes affects rates of surface-water runoff and erosion, in turn influencing soil properties. In mountainous terrain, soil conditions will vary up and down slopes, with soil eroded from some locales deposited as sediment in others. Topographic variations in aspect, slope, elevation and so on, in concert with other environmental factors, lead to the establishment of contrasting soil sequences. And tectonic processes such as volcanic eruptions and earthquake movements can have dramatic effects on soils, introducing new materials and rearranging old ones.

A soil's **parent material** comprises in the first instance the rock derived component from which its inorganic mineral fraction derives through weathering, and in the second the organic matter component that originates largely from plant remains through decomposition. Jenny (1941) refers to parent material as "the initial state of the soil system" by which he intends that it is the original material from which a soil derives by a series of interconnected physical, chemical and biological processes. The geological material may occur as consolidated rock, either crystalline like basalt or sedimentary like limestone, or it may occur as unconsoidated superficial deposits like volcanic ash or alluvium. The physical nature of the material determines the ease of its breakdown, permeability and surface area for instance affecting the weathering rates that prevail. Minerals also vary in their resistance. The mineralogical composition of parent materials is significant, notably the proportion of non-silicates to silicates and scope for clay sheet formation, because it not only influences soil evolution but also the physical properties and chemical make-up of the resulting soils.

The **biotic** factor covers all those organisms that may influence soil processes and development. It encompasses micro-organisms (such as bacteria, fungi and protozoa), through plants (anything from mosses up to trees, and including larger fungi), to animals (from invertebrates like insects and worms, through to vertebrates like lizards and rodents). This study, with its ethnographic focus, treats human-beings as a separate factor from other organisms. According to FitzPatrick (1983) "nearly every organism that lives on the surface of the earth or in the soil affects the development of soils in one way or another", making it difficult to distinguish the influence of some from others. The actual role of organisms in soil formation is difficult to assess because of close interactions between soil and vegetation, and the marked influence of climate on both. Vegetational effects can be attributed in some places to micro-climatological variations which have led to differences in plant communities. Whatever, the remains of organisms contribute all-important organic matter to soils, referred to under parent materials, having profound effects on soil structure and chemistry, in addition to mineral nutrient availabilities, as do the activities of organisms when alive, in particular facilitating decomposition and mixing of the soil.

All soils result from processes progressing over **time**, sometimes considerable periods. Soil development is a slow process, extending over thousands of years. The time involved varies for different soil features, from fairly rapid for properties dependent upon organic matter build-up, through to very slow for those dependent upon the weathering and translocation of primary mineral materials and alteration of clay products. It also varies between soil types and natural environments. Soils pass along a continuum as they evolve, divided into stages by soil scientists, featuring profiles of well-differentiated, characteristic horizons. The Inceptisols of the Wola region, for example, will develop fairly rapidly, in pedological terms, into Ultisols and ultimately into comparatively barren Oxisols, unless there is further volcanic activity resulting in soil rejuvination, which is possible for a region located close to an active plate margin, although it would be severely disruptive of the current environment. Geological events, like volcanoes erupting, or periodic changes in climate and vegetation, complicate the interpretation of contemporary soils, interrupting the course of soil formation. The existence of any soil, like a human life, may consequently be seen as a 'race against time'. The time factor runs throughout the entire book, for it conditions the other environmental factors. Any soil and accompanying ecosystem is dynamic and constantly changing; any contemporary natural environment exists as a place against time.

The agricultural manipulation of the natural environment by **human-beings** is also a struggle against time, as they try to control nature and battle to stop her reclaiming managed land. The time factor is central to the response of soils to farming too, relating to the age of gardens and changes in fertility status under cultivation. The ethnographic focus prompts the distinguishing of human-beings as a factor separate from other organisms, although in a broad sense, from a soil perspective, the human is only another animal, albeit a sometimes particularly intrusive one regarding soil processes, notably when cultivating soil to produce food or moving it when constructing buildings and roads. The relevant issues include how people think of, classify and appraise, their soils, how they manage them under cultivation, and the impact of their activities on soil conditions, notably what happens to soil processes, particularly productivity, under cultivation for varying periods of time, and so on. The activities of humans may either conserve, even improve soil productivity, where they practise sustainable farming, or diminish it, even degrade soil resources, by promoting erosion and such like, where agricultural practices are particularly inappropriate. The balance of evidence suggests that the Wola follow a remarkably conservational horticulture regime within the constraints of their technology.

Enviromental Ethnoscience

The place of a highland New Guinea population in its natural environment, like that of any other, depends critically on its awareness of its natural resources and its management of them. This suggests that setting an environmental study within ethnographic context, combining environmental science with ethnography, should prove particularly fruitful. And this study is as attentive to how the

Wola use their environment, as it is to how they think about it, with its interest in the agricultural potential and exploitation of natural resources. The uses to which people put their knowledge of their natural world is again the thrust of the ethnoscientific approach adopted here, not merely how they classify it (Sillitoe 1983). The dual meaning currently given to ethnoscience — with social anthropologists emphasising classification of natural phenomena in cultural context, and natural scientists reviewing indigenous uses of plant and other resources (frequently with an eye to finding something commercially exploitable) — strikes me as distorting and counter-productive.

This is not to subscribe to the simple-minded 'utilitarianist' stance. It is to criticise the emphasis that anthropological ethnoscience places on the classification of phenomena at the expense of exploring the understanding that people achieve of their natural world, in part using their taxonomic schemes, of their place in it and ability to manipulate it. The recent debate between so-called 'utilitarianists' and 'intellectualists', to give the unwieldy terms used to label the two extreme positions taken in it, is tangential to this viewpoint, reflecting the classificatory emphasis that has predominated to-date (Berlin 1991, 1992; Hays 1982, 1991; Hunn 1982). It is unquestionably wrongheaded to suggest that ethnobiological taxonomies derive wholly from the use of species by human-beings, facilitating ecological adaptation to habitats by arranging organisms according to their practical importance. The Wola name and classify many things in their natural world that serve no pragmatic end, though there is some correlation, which they themselves acknowledge, between the extensiveness of their taxonomies and the amount phenomena feature in their lives, as indicated in subsequent chapters.[3]

The definition of utility immediately poses problems too, prompting tautological arguments. Do people have to eat species, or use them directly in making things, for them to qualify as useful, or can those used as symbols or those featuring in myths count too? Furthermore, an apparently useless plant or animal may be important to a useful one, however defined, serving as food source, nesting site or whatever, and knowledge of it, maybe partly encapsulated in taxonomic status, thus indirectly serve on occasion a utilitarian purpose. This is an inevitable consequence of the interconnectedness of the natural world, manifest in ecology, where functional and cybernetic approaches firmly hold sway. It also relates to the point made above that people do not just think and classify in an unworldly vacuum, they act too and apply their knowledge of natural phenomena in their everyday lives, whether or not they have a direct use to them. The Amazonian environmentalist Ailton Krenak catches the essence of this view when he comments that the Brazilian Centre of Indian Research studies biology "not in the European sense, but how to handle it".[4]

---

[3] When asked why some plants have no names, are what they call *imbiy na wiy* (lit. name not have), my friends repeatedly referred to them having no *kongon* (lit. work), by which they intended 'use'.

[4] Rambali, P. 1993 *It's all true: in the cities and jungle of Brazil*. London: Heinemann.

The taxonomic schemes of people only reflect part of their perception of the natural world. They carry environmental knowledge, and transfer it between generations, in other ways too, although it is considerably more difficult for an outsider convincingly to gain some understanding of this other knowledge. Among the Wola, for example, it is not codified but diffuse and communicated piecemeal in everyday life. It is to a considerable extent knowledge gained through experience, passed on equally by example as by word, transmitted as, and when, daily events require. These people, like the heirs of other non-literate traditions, are not familiar with trying to express all that they know in words. They live rather than reflect on their environmental knowledge. It has a marked practical aspect to it. When asked, for instance, to comment on the soil at a particular site, individuals may inspect it, and even handle it, before passing judgement. If you then ask them to justify their assessment they will look somewhat bewildered. They will probably tell you to look at the soil for yourself, maybe even pass you a handful to feel, the implication being that you surely can just tell.

If you are a Wola you just know, you are not used to being asked how you know. The awareness that you have of the soil for instance, is an accumulation of experiences, built up over the years, cultivating the soil and hearing many comments from others on it.[5] The knowledge is passed on by informed experience and practical demonstration, more often shown than articulated, it is as much skill as concept. The author Jenny Diski catches the sense of what I am trying to convey in her novel *Rainforest*, where she writes:

> "He had come to understand, gradually, about these visitors from far-away cities, how they had another way of looking at things. They wanted names and descriptions; to look at things, but not touch or use. He had names for things, of course, but along with the word came pictures and memories of the part the thing played in his life and the life of his people. ... Things were known by names, but inside his head by their qualities. ... Like their cameras and tape recorders, they used words to carry bits of the forest away to strange and unimaginable places without the need of having the things themselves. He found this interesting, and supposed that the hunters and providers in those far places had other ways of doing things. It was necessary, he knew, to have a sense of the world one lived in and used, a feeling about it that was more than the names one spoke aloud to tell others that it was there."

The epistemological problems that attend attempts to document something about such indigenous understandings of the environment are truly awesome. Some years ago, under the structuralist agenda, an anthropologist might have felt confident enough to speculate that selected symbolic imagery, figurative speech and so on could be interpreted as indigenous environmental theory, expressed as binary oppositions, metaphors and so on, amenable to

---

[5] The Wola are no different to many farmers the world over in this regard; many English farmers for instance would respond in a similar vein, knowing their farms intimately, having secured their livings off them for years.

structural interpretation. But post-modern criticism has revealed structuralism a clotheless emperor of a theory, and it is extremely difficult today to make such interpretations appear convincing, and not mere products of an author's imagination. It is no longer credible to try and reduce lived and practical knowledge to some obscure allegorical scheme.

This book attempts a more modest account of what I think I have learnt from the Wola about their knowledge of the environment. It is necessarily somewhat disjointed from our cultural perspective, not addressing indigenous knowledge of environmental processes within some integrated ecological-theory- equivalent paradigm. The account includes both verbalised information and descriptions of activities that have environmental ramifications, sometimes apparently unremarked by the actors. It attempts to cover articulated concepts and practical actions. The pragmatic foundation of some Wola knowledge presents problems. They follow practices which impact on the environment, without articulating a theory of why. The practices work, which is sufficient for them as farmers. But I wish to go further and try to explain why they work and have the effects observed. Here I turn to western natural science for a theory and concepts.

Perhaps I should address anticipated criticisms of ethnocentricism before proceeding any further, for some of those I have told about my work have levelled such at me, finding it odd that I should run indigenous and Western environmental perspectives side by side. The implication is *not* that we can translate another culture's conceptions about the environment into Western scientific discourse. In my discussion of the vegetation communities found in the Wola region for example, I give both local and scientific plant names. The purport is not that the Wola have a Linnean hierarchical system of plant classification related to evolutionary theory. It is to give a more comprehensive, and hence useful record. Someone ignorant of the Wola language reading this book can identify the plants, the botanical binomials relating them to our specialised knowledge system, whereas the Wola names alone would be just words with no specific reference to anything 'out there', except a 'tree' or whatever. This is accepted practice in the taxonomic discussions of ethnoscience. All that I am attempting is to extend the convention to contexts of environmental process and practice.

A straightforward account of Wola environmental lore, and their practices that intervene in nature's arrangements, cannot address some issues which I think pertinent in assessing their relationship with their natural surroundings. Indigenous knowledge and practices have implications which I cannot explore within the limited terms and idiosyncratic perspectives of Wola culture, so far as I have managed to apprehend these. The issue raised previously of how the Wola are able to cultivate some sites semi-permanently within the context of a nominally shifting cultivation strategy illustrates the point. I can describe Wola horticultural practices which allow this cultivation regime. I can report their comments on the soil and its behaviour under cultivation, about 'grease' levels and so on — to the extent that I comprehend them. But the question remains: what is it about the soils and crops of this region that allows such a cultivation regime to exist? If you ask the Wola this question, they are as likely to tell you 'well, it just does exist', which is sufficient to their minds and for their purposes. Encouraged to elaborate further they may comment that they are following the ways of their ancestors, implying that time and tradition have proved their practices effective. Thoughtful individuals might describe as relevant changes observed in some soil physical conditions under cultivation. But no thoroughgoing explanation will be forthcoming equivalent to that possible using our scientific theory.

The perspective of natural science commands respect in our culture, regardless of some, often ill-informed, criticism about it overly dominating discourse in several branches of intellectual enquiry. It has allowed *us* to develop an impressively documented and rigorously assessed understanding of the natural world. It is not the absolute truth, of course, not God's understanding of the world, but it is nevertheless a highly effective body of theory for mere human-beings to have devised, which has proved very successful in addressing the kinds of questions that occur to those educated in our cultural tradition — not to mention the central role it has played in the development of our sophisticated technology, which allows us to intervene dramatically in the environment. It derives from European society, in which I am inculcated as an Englishman. It is entirely foreign to the Wola, who have no ideas traditionally (so far as I am aware) equivalent to nutrient ions, gas exchange, physical functions, and so on. We can legitimately use it I maintain, to further our understanding of their environment, their place within it and the impact of their activities on their natural surroundings.

Running parallel with this book's scientific account is one into local practices and knowledge. These relate, I think, to the *same* natural environment 'out there', although they are expressed in quite different terms and reveal concerns for somewhat different issues. It is the contention of this study that both perspectives taken together help us to achieve a more rounded and better understanding of the environment, both natural and cultural. Jointly, they further our understanding of the environmental impact of human activities, relating to both the effect people think they have on nature as agents and what we make of their practices as observers. The objective is not to assess the veracity of local ideas against scientific ones, both are relative, but to enrich our overall understanding of environmental interactions within cultural context. The study of ethno-environmental issues as advocated here also affords us an opportunity to compare indigenous statements and explanations against measurable data, as opposed to dealing only with qualitative social 'facts', the subjective status or otherwise of which, as suggested by some current postmodern theorists, has contemporary social anthropology experiencing acute epistemological doubt; we can for instance, contrast the distinction made by the Wola between two seasons with the weather experienced at these times, we can compare their statements about the behaviour of soils under cultivation with observations of them, and so on.

The academic objectives of this book are twofold and relate to the foregoing intermingling of disciplinary perspectives. It attempts to demonstrate to social scientists, notably anthropologists, who are currently mired in a post-modern theoretical wasteland, that a scientific approach can have validity in the study of other societies, even though emphatically derived from our culture's intellectual tradition. And it tries to show agricultural scientists, notably soil scientists, that what local people have to say about their environments, their understanding and management of them, are worthy of serious consideration, no matter how 'primitive' and technologically unsophisticated they may apparently appear. These people know their natural surroundings intimately, their appreciation of environmental issues will be well informed, and they may have something to teach us if we listen, albeit expressed in an alien idiom, perhaps involving spirits, mythical events, and so on.

Furthermore, in 'development' contexts, where we think that we can offer technical assistance to other cultures based on our scientific understanding of farming and the environment, people are more likely to heed advice if presented sympathetically with regard for their knowledge and understanding, building on their awareness of the world. It has recently become popular to point out that indigenous people have their own effective 'science' and practices, and that to assist them we need first to understand their knowledge systems (Warren 1991; Richards 1985; Inglis 1993). This is undeniable, as development agencies have recently acknowledged. But it is necessary to go beyond asserting that local knowledge and ethnoscience are effective, to demonstrating this convincingly to those in other, notably applied disciplines (who will, in all probability, have considerable sympathy for the practical element in this knowledge). This book addresses this need, relating local knowledge and practices to our scientific understanding, to the language of our development technicians. It tries to show in their terms how effective other people are at managing the natural world, in the belief that such demonstrations may foster more respect for indigenous knowledge. Until technicians take this seriously, the debate over farmer participation in development projects will have limited impact (Chambers *et al.*, 1989).

The approach taken here might be dismissed as logical positivism (although I prefer that label to the opposite illogical negativism!). Maybe it is intellectually unsophisticated to think about objective facts, and to seek to document and explain patterns in them. But I am no earthworm dreaming that he is a human-being writing this book. Distortions and omissions there are in this attempted record, like any other; I have to accept these, and try and control their extent. I cannot overcome them, as the post-modern critique affirms. We either try to engage with the world 'out there' and record something about life in other societies in our inevitably imperfect accounts, or we sit in our bookworm burrows fretting self-consciously about the unavoidable shortcomings of our work. This account is unavoidably a representation, and a biased one — as an investigation into natural resources and land management, pre-eminently physical and practical matters, reduced to a discussion on paper. The explanation of observations and interpretation of local prescriptions are only *attempts* to make sense of the activities of this small population as they affect their natural environment, to engender some appreciation of their understanding of these and related issues.

The applied and lived character of the local environmental management regime underscores the representational status. The Wola, as heirs to an effective system of natural resource exploitation that has evolved over many generations of experimentation, follow it without apparent need of analytical discourse. In using the written word to convey something about this system, this book clearly attempts to explain and interpret their actions on another, somewhat alien level. Employing foreign words and concepts to boot it further distorts whatever it is that I have managed to comprehend about their views of their actions. Any translation of another culture is unavoidably distorting, this study is no different to any other anthropological account in this respect. It differs in its struggle to combine our culture's science and humanities perspectives in understanding and interpreting another culture. Its attempt to bring together the quantitative approach of environmental and agricultural science with the qualitative approach of ethnography and social anthropology distances this study from any fraudulent pretence about achieving an understanding of a foreign population as it understands itself (although I am not convinced that many who have undertaken anthropological fieldwork have ever thought that this is possible — Quigley 1993). Nonetheless the proposed adoption of a natural science perspective to further understanding of another culture's place in its environment may alarm some anthropologists.

There are some strange notions current about what science amounts to among non-scientists. It is not about absolute truth, nor does it claim omnipotent objectivity, nor does it feature only hypothetico-deductive and quantitative methods. A defining feature of science is its rigour. It employs precisely defined terms, and attempts to develop and test ideas against observed, preferably measurable evidence. This contrasts with the subjectively informed interpretations of many social sciences. One reason why social anthropology has proved so vulnerable to post-modern criticism is that it has allowed a proliferation of short-lived fashionable theories to dictate field research agendas, instead of engaging less ideologically, albeit less tidily, with ethnographic reality. Each generation of social scientists has been indulging in a form of deconstruction for decades, seeking new pardigms to sweep away previous ones, even overlooking or surpressing unrelated or unexpected observations which sit uneasily with currently ground theory-axes, instead of engaging in the rigorous accumulation and verification of ethnographic data to develop theoretical insights. All social anthropology needs, I have heard it said, to save it from the current confidence-sapping imbroglio, with structuralism's charade revealed and marxism fallen, is to identify the next new paradigm. I disagree. The lesson of contemporary unsettling criticism is surely that we need to review and consolidate our ethnographic approach and methodology. Ethnography demands great rigour.

According to post-modernists, anthropology is inescapably ethnocentric, and there is an unnerving tendency for the discipline to reflect and project western society's contemporary concerns onto other cultures. Recent examples include the demise of marxism as an acceptable paradigm with the fall of the Wall and all it presaged; the emphasis on gender and the social position of women in reaction to decades of male intellectual bias; the recasting of the 1930's nature: culture debate following recent medical test-tube interventions with fertility; and post-modernism's conundrums as confidence increasingly wanes and doubts multiply over the social and other consequences of our culture's technological achievements. The environment and its conservation are further topical issues; few Europeans being unaware of environmentalists' concerns about rain forest destruction and ensuing loss of species diversity, about predictions of climate change, and the possible environmental repercussions of industrial atmospheric pollution, with the so-called greenhouse effect, global warming, and acid rain, to mention but three topical issues.[6]

The weakness induced in social anthropology by allowing the fashionable winds of our society's contemporary affairs to blow it around and determine its intellectual interests, in its quest to appear relevant, becomes evident when it tries to find a place in this current environment debate, its contribution resembling in part informed 'green' journalism (Knight 1992; Milton 1993). And environmentalism's disquiet comes to appear like a metaphor for the discipline's current post-modern concerns, pluralistic confusion threatening catastrophe. These criticisms may appear blinkered, even hypocritical, this study apparently hitching itself to the environmental bandwagon. It is in some small measure as guilty as any other of undermining the subject's identity. Perhaps this is another post-modern puzzle, we cannot off-load our cultural baggage and we inevitably have some concerns in common across cultures. But the charge pertains largely to subject matter, the emphasis is different. The current environment debate concerns the destruction of natural surroundings and the squandering of the earth's resources, and their possible deleterious consequences. This book documents the reverse: a sustainable regime of natural resource expoitation, affirming the existence of stable management systems.

The interpretation put on the changes wrought in the tropical montane environment of highlanders by their horticultural and other activities contrasts with others, who think it results in land degradation, precipitating people into an ecological crisis (Wood 1982, 1991; Allen & Crittenden 1987). It seeks inductive space for the ethnographic and environmental evidence, avoiding premature deductive closure. The information collected and presented suggests that we should take the comments of local people about their environment at face value, that the yield potential of some sites for example, improves with use (Figure 2.4).[7] This squares with my own observations of

---

[6] While these global concerns may affect highland New Guinea weather patterns, they do not currently matter to people living there (unlike the inhabitants of some of the low-lying islands of the Pacific who fear catastrophic loss of land if sea levels rise as some forecast). It is noteworthy that the Wola report that some animals previously restricted to lower altitudes (e.g. pythons, megapodes and spotted cuscus) are found in the 1990s, for the first time to their knowledge, in their region at higher altitudes. These animals are doubtless sensitive to climatic changes with altitude, and their upwards movement suggests some warming of the region's climate. The Wola attribute the migration to the recent establishment of an oil well at lower altitudes adjacent to lake Kutubu, the flaming stack of which they say has frightened the animals that inhabit this region to flee elsewhere.

[7] The Hageners make claims similar to the Wola about soil conditions and sweet potato yields improving with time under cultivation (M. Strathern pers.comm.). The comments of the neighbouring Huli about land degradation, declining soil fertility and diminishing sweet potato yields stand in stark contrast (Wood 1984,1991).

many gardens under near-permanent cultivation for more than two decades without apparent degradation. The environmental change induced by human activity need not be synonymous with degradation. We are in danger of imposing on others our own contemporary worries about the damage inflicted on the environment by our culture, just as the marxists sought in vain until recently to find class-equivalents in all societies. It is intriguing that the Wola are not only able to cultivate some garden plots almost semi-permanently, but that they are also able to follow this practice without producing the degraded landscape reported for other tropical regions like Amazonia following extended forest clearance. The Wola, like farmers the world over, think that human use of the land can improve its natural endowment. The creed of gardeners, whether Western, Oriental or Islamic, is that by bringing nature under control they can direct her productivity to their benefit. This contrasts with the contemporary 'green' view that whenever humans interfere with the environment they diminish it.

The Wola: Place and People

The climate of the Wola region features relatively high levels of rainfall and, for the latitude, moderate temperatures — due to the ameliorating influence of altitude (the majority of the population lives between 1600 and 2000 metres). Variations in topography and altitude give rise to numerous micro-climates locally. The weather is generally equable, many days featuring sunny mornings and rainy afternoons. There are no notable seasons sufficient to

Plate 1.1 The rugged terrain of Wolaland: looking along the Was river valley.

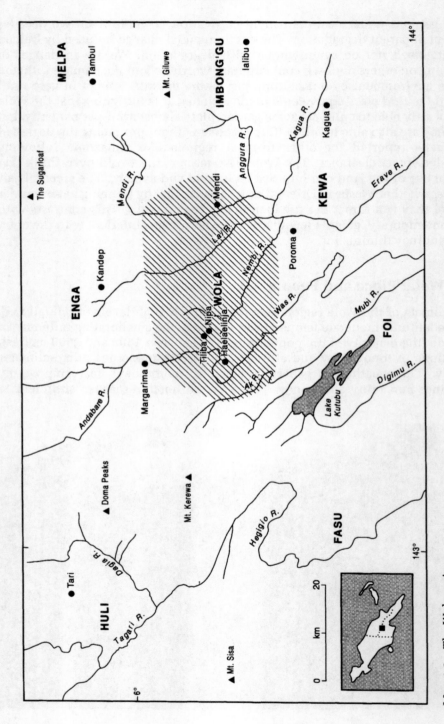

Map 1.2 The Wola region

influence crop cultivation. The same climatic conditions largely prevail throughout the year, although unpredictable perturbations can occur, such as overly dry or wet weather, which can adversely affect yields (see Section II).

The country is mountainous, rugged and precipitous, with turbulent rivers flowing along valley floors (Map 1.2). The Wola live along the valley sides, leaving the intervening watersheds largely unpopulated. In the valleys, where they have cultivated extensively, there are areas of dense cane grass interspersed with the grassy clearings of fallowed or recently abandoned gardens and the brown earth and dark green foliage of current ones. Lower montane rainforest occurs on the mountains and in the unpopulated parts of river valleys. The region's geology comprises sedimentary rocks largely, mainly

Plate 1.2   The fast flowing Was river at Kaerep with adjacent garden and house.

limestone, with igneous rocks of more recent volcanic origin on its margins. In the recent geological past the region was uplifted, and folded and faulted, and frequent earth tremors indicate that these earth movements continue today. The relatively recent occurrence of this folding accounts for the landscape's rugged and sharp relief. Contemporary geomorphological processes are changing the region rapidly, maintaining its youthful and raw topography; weathering proceeds apace, erosion is constant, and the occasional large-scale earth movement can dramatically change the local landscape (see Section III).

Previously I described the soils of the Wola region as relatively poor which, with the precipitous terrain and steep slopes, give somewhat unpromising agricultural land (Sillitoe 1983). This is not a characterisation that I should wish to maintain in the light of the detailed investigation of soils made

Plate 1.3 An exposure showing the region's deep blocky structured clay subsoil overlain by a friable dark topsoil, supporting *Miscanthus* cane.

here. Relative to some soils, such as certain Mollisols and Alfisols, the region's soils may appear somewhat unpromising, but compared to the soils that predominate over large areas of the tropics, notably the Oxisols and Ultisols, they are fertile. The youthful age of the soils, combined with several rejuvinating episodes of volcanic ash fall, results in fairly productive soils with appropriate management, which explains in part how the Wola are able to follow their semi-permanent horticultural regime (see Chapter 10).

Dotted across the landscape are the neat gardens of the Wola. They practice a sedentary variation of shifting cultivation, as noted, and subsist on a predominantly vegetable diet in which sweet potato is the staple. Their agricultural practices result in two broad classes of garden: those cleared and planted once with a wide variety of crops (the classic shifting regime), and

Plate 1.4 A profile of one of the major soils of Wolaland: a Tropept, called locally *hundbiy*.

those planted two or more times, sometimes over and over again for many years, with brief fallow spells, and supporting a narrower range of crops (the 'bone' garden regime). Gardens range in size from small plots adjacent to homesteads (av. 90 m^2), through taro gardens (av. 495 m^2), to large cultivations of mainly sweet potato (av. 1150 m^2).[8] When looking for a new garden site, men's choices are constrained by a number of factors. An important one is the land tenure system which restricts their choice to those areas where they have rights, traced through either consanguines or affines,[9] to lay claim to land. Other considerations concern the nature of the site; such as its slope and aspect, its distance from their homestead, the ease with which they can enclose it, and so on.

The Wola support pig herds of considerable size on the produce of their gardens, notably on sweet potato tubers. They hand these creatures around to one another, together with other items of wealth such as sea-shells, cosmetic oil, and today cash, in interminable series of ceremonial exchanges, which mark all important social events like marriages and deaths. These transactions are central to the ordering of their fiercely egalitarian, sometimes violent, society. The Wola live in a stateless society, where the importance placed on the *political* sovereignty and equality of individuals poses, at first sight, problems for the maintenance of social order. The exchange of wealth, elaborated into a complex institution, is a significant force in their accommodation, fostering an orderly social environment, while simultaneously allowing individuals comparative political freedom, unlike some other institutionalised mechanisms that maintain social order (Sillitoe 1979). Dihedral, it reconciles both individual co-operation and competition.

The Wola value exchange highly and accord high status to those who excel at it — as throughout the highlands of Papua New Guinea, where prominent men, the fabled big men of anthropological discourse, achieve their high social standing for their success in the management of wealth in the ceremonial exchange transactions which feature prominently in their lives (Godelier & Strathern 1991). Success in the exchange arena is paramount to achieving social renown; and they acknowledge the successful with the epithet *ol howma* (lit. man ceremonial-grass-clearing). It is even possible sometimes for these more able men, by virtue of their admired success, to exert a marginal degree of influence over decisions reached by those united for some purpose. But their success does not earn them the status of leader, as it does in some Melanesian societies; if they attempted to exert control, however subtle, over the actions of others this would offend against the ethos of equality that pervades their social life.

A key issue in any polity is who controls access to, and the use of, natural resources because, as marxism has affirmed, this is a potent source of political power. In an acephalous political order like that of the Wola it may be anticipated to be subject to subtle arrangements and tacit checks and balances to prevent any interest group extending undue claim to them. It is a theme that I have explored elsewhere and return to here, investigating the

---

[8] These average areas are calculated from a survey of over 500 gardens in the Was valley.
[9] Blood relatives, and relatives through marriage respectively.

Wola political economy further in the context of their environmental relations, notable access to land. The status of soil resources relates intriguingly to socio-political status. In this context, it is noteworthy that highly valued pigs, which feature prominently in exchange events, differ crucially from other inanimate valuables in that they have continually to be 'produced'. They have to be herded and fed daily with cultivated produce, largely sweet potato tubers, plus any other food waste. The Wola accentuate the ongoingness of pig production by not raising animals to maturity and then slaughtering them, but keeping adult animals alive for years.

While a man can in part sustain a good exchange record using ready-made shell and other inanimate wealth by astute 'financial' management of his assets and obligations, rather like a wealthy capitalist financier (Strathern 1969), and not concern himself with its production,[10] this is not possible when pigs enter the exchange equation. And they inevitably enter the equation because no man could meet his exchange commitments adequately, let alone achieve high social standing, by transacting in shells and such alone, he must handle pigs too as demanded in certain exchange contexts. Consequently, men of renown inevitably handle more pigs in exchanges than others. The implication is that they have access to, and exploit, more or better productive resources than other men (Modjeska 1982). Two ways in which they could possibly increase their porcine productive capacity are: 1 to have access to more labour, which implies more female relatives to cultivate, larger areas of land to supply the extra fodder required by their larger herds, or 2 to have access to superior productive resources, notably to more fertile areas of land.

Regarding the first option, it is common for men of esteem to have more wives than average, although not invariably, some successful men having access to the labour of other female relatives, such as widowed mothers and unmarried sisters, sufficient to meet their herding needs. In order to increase the size of their household's work force through marriage, it is noteworthy that men have to demonstrate beforehand a certain ability in exchange by mustering the bridewealth needed, for men intending polygynous unions receive markedly less assistance from their relatives in raising the wealth they require. They have in a sense to validate their access to increased female labour, which they may use to bolster their pig wealth productive capacity through cultivating larger areas of land, by demonstrating some ability in the exchange arena where they will transact the added wealth.

Scrutiny of the second option, that *ol howma* men of renown may use the influence that accrues to their social standing to gain access to more productive garden land and secure higher crop returns from their labour inputs, throws some interesting light on the constitution of Wola society. It is the option explored further in this study, confronting the issue of how in an egalitarian polity successful men might contrive to gain access to more fertile sites, if indeed they have the opportunity (see Chapter 11). This is an significant

---

[10] Many of these valuables originate outside the Wola region and are imported into it, hence men have nothing to do with their production (Sillitoe 1988).

consideration in a region where sweet potato, staple of humans and pigs alike, comprises by far and away the largest acreage under cultivation, and where people regularly feed something like 50% of the sweet potato tubers they produce to their pigs (Bourke 1985; Waddell 1972; Sillitoe 1983). Anyone with more fertile land will clearly have an advantage in the production of pig wealth. Land resources, it turns out, make an indirect contribution to the Wola political system, soil status impinges on social status, not to bolster it but to prevent its consolidation and efflorescence into something politically more hierarchical.

The Wola occupy small houses scattered along the valleysides of their moutain homeland, largely in areas of secondary regrowth. Their homesteads comprise variably composed family groups; ranging from a man and his wife or wives and their children, to three or four related men together with their nuclear families and other relatives (such as unmarried sisters, widowed parents, and so on). They call families *sem* and specify these nuclear social groups by the name of the male parent (e.g. Uwhay's *sem*), one of several manifestations of the public dominance of men. A marked separation of the sexes characterises their lives, men observing a number of conventions that keep them apart from women, one of which is the occupation of separate houses within the homestead.[11]

The larger, permanent social groups of Wola society come about through the aggregation of several variably related families (Ryan 1959; Sillitoe 1979; Lederman 1986). A few families that are genealogically related by male and female links, and reside and garden in the same area, constitute small named groups called *semg^{e}nk* (lit. family-small). Varying numbers of these small groups, which people think share a genealogical connection in the distant past (although they may not be able to remember its precise nature) and which occupy specific territories, comprise named communities called *semonda* (lit. family-large).[12] The members of these communities ought to observe a rule of exogamy because related, although inter-marriage occurs sometimes.

The Wola region comprises many territories identified with these small, kin-constituted communities (average population c.300 persons), subdivisions of which structure rights to, and tenure of, cultivable land. Access to arable land features prominently in the constitution of Wola society, these permanent and named groups resulting from the collection of several related families on territories where they have rights to claim garden sites. It bears on the sentimental attachment the Wola have for their land. Men can claim rights to land through a wide range of connections with a place (both kin and situational factors play a part), and the empirical constitution of groups is bilateral,[13] indi-

---

[11] These conventions are now in the 1990s breaking down, particularly among members of the younger generation, with couples increasingly sharing the same house.
[12] Sometimes between these small and large sized groups there are named mid-level *sem* groupings, which appear to be evidence of fission as *semg^{e}nk* groups grow beyond a certain size.
[13] Descent traced equally through women and men.

viduals' exercising their rights eclectically (regardless of men's spoken preference for residence where their father lived — where, in all probability, they may decide to remain, having lived there before marriage, knowing the region and its population, and jurally able to insist that their wives move).

The related members of these permanent territorial groups do not recognise any obligations to support them in certain situations as corporations representing their collective interests. They do not unite on this basis for political action. The Wola talk and behave in ways that nullify the assumptions of corporate theory that obligations to groups maintain social order — modified New Guinea versions of descent corporations[14] or otherwise. The action groups of their society are more akin to temporary coalitions which come together to pursue some end jointly (arranging pig kills or staging collective rituals) as individuals decide that it is in their, not the group's interests to participate in the mooted event. People do not speak about participating to support the group, nor any 'big man leader', although kin considerations and mutual obligations may well influence their decisions. While some of those resident on the same permanent *sem* group territory may comprise the nucleus of an action group, relatives and friends from elsewhere may join it. This pattern of recruitment reflects the apparent freedom extended to individuals politically to pursue their own ends within socially accepted bounds.[15] It negates the possibility of political authority vested in any leader, as reflected in people's relation with, and access to land resources.

---

[14] These are political groups common to uncentralised societies, notably in Africa, recruited according to descent, either through males (patrilineal/agnatic) or females (matrilineal/uterine) or both (bilateral/cognatic).

[15] The contrast between the interests of individual persons and social groups, by which I have tried to account for political action in Wola society, I continue to think useful, although today I might phrase parts of the argument somewhat differently (Sillitoe 1979). Others appear to think likewise, for although some have criticised the validity of the antinomy posed by the venerable conundrum of individual versus group interests, this opposition nonetheless appears to have featured prominently on the agenda of subsequent debates over the nature of Melanesian polities, even among those who remain wedded to the idea that obligations to corporate segmentary groups structure political relations in these stateless orders (albeit far removed from the African derived descent models that dominated up to the 1970s — Strathern 1988; Merlan & Rumsey 1991; Biersack 1995). I am not persuaded that a large gulf separates us, more differences in the use and understanding of the words we employ (a painful conclusion emphasising the need for terminological rigour in anthropology). Nevertheless it frequently occurs that what is obvious to an author may be obscure to readers — and here I can sympathise with the post-modernists' arguments about alternative constructions. I have never suggested that the Wola behave in some primeval sense as free-wheeling, unsocialised individuals out for their own selfish ends regardless of the interests of others, uninfluenced by opinion within their local neighbourhoods. They are inevitably conditioned and constrained by the shared expectations, values and so on of their socio-cultural environment — as demonstrated for example, by their observance of the ramifying obligations of their society's ceremonial exchange institutions — whether we choose to describe them as socialised to behave according to their society's norms, or pressured by mutual obligations, or more fashionably, as persons whose identities are created and continually moulded through interaction with other members of their society (subscribing to common beliefs, precepts and so on).

The supernatural or religious conceptions of the Wola centre on beliefs in the ability of their ancestors' spirits to cause sickness and death, in various forest spirit forces (see Chapter 8), and in others' powers of sorcery, among others. They believe that the ghosts of their ancestors have malevolent powers, visiting sickness, even death, on their descendants. These spirits lead a nomadic existence they think, and resided temporarily at any one of a number of defined places (such as in ancient stone artifacts, the skulls of ancestors, and some deep water pools). If someone fell seriously ill, a person with the required knowledge divined the current residence of the attacking ghost and the victim's relatives offered a pig there to placate the offending ancestor spirit and, as the Wola say, persuade it to stop 'eating' the sick individual. On other occasions many people combined to perform a large ritual for their collective benefit (see Chapter 4). Many today subscribe to Christianity, while continuing to fear that these traditional spirit forces inhabit their home.

CHAPTER 2

ROTATING LAND NOT CROPS: SHIFTING CULTIVATION AND SWEET POTATO

Shifting cultivation is a traditional strategy of crop cultivation and soil management widespread in the humid tropics, which maintains fertility by alternating brief periods of food production with extended periods under natural fallow. It features rotation of land, as opposed to crops, to ensure continued soil productivity, the regime subjecting parcels of land to oscillating phases of fertility depletion and restoration. Farmers are generally thought to crop temporary clearings for fewer years than they remain fallow (Sanchez 1976). Under shifting systems in the strict sense, people clear small areas of natural vegetation — which may vary from moist evergreen rainforest to grassy savanna to semi-deciduous woodland — and crop them for one or two years before leaving them fallow, to regenerate under natural regrowth for fifteen years or more (e.g. Siebert 1987; Juo & Kang 1989; Jordan 1989).

Shifting cultivators clear the land manually, by a combination of felling vegetation and burning the plant residues, this 'slash-to-ash' strategy releasing nutrients locked up in the natural cover and making them available to crops, reducing surface acidity and enhancing organic matter mineralisation too. Farmers exploit the short-lived boost in nutrient supply before it declines and natural vegetation recolonises the site, racing against time in their cultivation. The stumps and roots which they leave help maintain soil structure, promoting water infiltration and so on, and also facilitate rapid regrowth after the cropping interval (Kang & Juo 1986). Rapid regeneration following cultivation is reckoned important to reduce nutrient losses, reinstating the nutrient cycling equilibrium characteristic of mature plant communities (Jordan 1985, 1989; Vitousek & Sanford 1986). During the fallow period nutrients accumulate, to be made available to crops in the next round of cultivation. So long as land is plentiful, shifting cultivation is thought today to be a stable agricultural regime in tropical ecosystems.

The swidden regime of the Karimui of the New Guinea highlands illustrates the cycle. Forest clearance breaks the natural cycle, drastically reducing the supply of organic matter, which contains the bulk of plant available nutrients, notably N and P, some released in firing. Heavy rainfall and high temperatures facilitate the rapid breakdown of plant remains and leaching loss of plant nutrients, and these decrease under cultivation, phosphate is rapidly fixed in unavailable forms, and soil structure declines as aggregates breakdown, increasing erosion risks. "Thus the removal of the rainforest vegetation and the cultivation of the soil will lead to a decline in soil fertility. ... Few gardens are cultivated for more that two years and this suggests that the change in the amount of available nutrients is fairly rapid" (Wood 1979:6).

The Karimui cite declining soil fertility as the main reason for abandoning gardens. The sites gradually regain fertility under fallow, organic matter levels building up with the restoration of the natural nutrient cycle, until after 15 to 20 years under fallow they are ready for recultivation.

I. SHIFTING CULTIVATION

An extensive literature exists on shifting cultivation around the world (Robison & McKean 1992), a considerable part of which discusses its impact on soils.[1] In 1960, Nye and Greenland published their well-known review of the evidence that had accumulated to that date,[2] and subsequent research has largely built upon their findings, although questions are increasingly asked today about their worldwide applicability. It remains the accepted view of shifting cultivation, and the following account largely conforms with its characterisation of soils and cropping under this husbandry regime. While they note that the crops grown, husbandry methods practiced, and lengths of cropping and fallowing intervals vary greatly under different shifting cultivation regimes — with clearance of a range of vegetational successions, on a wide variety of soils and by people of markedly different cultural backgrounds — Nye and Greenland maintain that many of the processes behind the decline in soil fertility under cropping and its subsequent restoration under natural fallow are broadly similar.[3]

When shifting cultivators clear the natural vegetation and burn the debris, they deposit on the soil surface in the ash large quantities of nutrient ions from the standing vegetation and the litter layer (mainly calcium, magnesium, sodium and potassium, largely as carbonates, phosphates and silicates of these cations). They lose considerable amounts of the nitrogen (as ammonia, gaseous nitrogen and nitrogen oxides), sulphur (as sulphur dioxide), and carbon (as carbon dioxide) in the burned material to the atmosphere, but not all, because a deal of the larger woody fraction (tree trunks, branches, stumps etc.) is not consumed (Sanchez 1976:364). The intensity of the burn is important regarding the extent of losses, which may be considerable (according to Pivello & Couthino 1992, 95% C and N, 56% P, 44% K, 52% Ca and 42% Mg may be lost to the atmosphere in burning off savanna). A rise in soil pH occurs, corresponding with the substantial increase in exchangeable cations, which Nye and Greenland (1960) consider one of the most important effects of burning on acid forest soils (see also Juo & Kang

---

[1] Summaries of this work may be found, for Africa in FAO-SIDA (1974), and Newton (1960) besides Nye & Greenland (1960), and for Latin America in Sanchez (1973, 1976) and Watters (1971), and for India in Ramakrishnan (1984).
[2] Previous to this Kellogg & Pendleton (1948) had advanced an overview of the effects of shifting cultivation on the soil.
[3] They make a broad distinction between shifting systems in forests and those on savannas because although the general principles are similar, they differ in some particulars.

Table 2.1
The elemental composition of, and nutrients added in, burned Peruvian secondary forest. (*Source: Sanchez 1976, Table 10.6*)

Element	Composition	Total Additions (kg/ha)
N	1.72%	67.0
P	0.14%	6.0
K	0.97%	38.0
Ca	1.92%	75.0
Mg	0.41%	16.0
Fe	0.19%	7.6
Mn	0.19%	7.3
Na	180 ppm	0.7
Zn	137 ppm	0.7
Cu	79 ppm	0.3

1989). These changes may also help promote microflora populations different to those originally present.

The first rains wash the ash into the soil; the occurrence of pH changes to considerable depths suggests its rapid downward movement (Sanchez 1976). Some of the ash may be floated off the site with exceptionally heavy rain and a relatively impermeable soil, or blown away in a strong wind, and the nutrients lost (Toky & Ramakrishnan 1981; Ramakrishnan 1989 — Trenbath 1989 calculates that the loss of half of the ash off a site will reduce crop yields by a quarter). The magnitude of the increase in nutrients relates to the type of vegetation destroyed and its age, and the thoroughness of the burn. Table 2.1 gives an indication of the sizeable boost, listing the nutrients added with the burning of South American secondary forest. The amounts released from savanna vegetation are considerably less than from forest. Although mature wooded communities are larger in volume, the quantities released from them are proportionally less than from immature communities because of increased nutrient storage in less readily combustible ligneous material.

The burn destroys much of the litter layer, exposing the soil surface to sun and rain, but there is no loss of humified organic matter from the soil through incineration. The finely divided, partially decomposed litter not consumed in the fire is rapidly decomposed, and the rate of humus decomposition also rises with insolation and increased aeration due to surface tillage. On savanna, the thorough cultivation needed to clear coarse grass tussocks results in a more thorough aeration of the soil and accelerated decomposition (Nye & Greenland 1960). Another consequence of the burn is the heating of the immediate soil surface which, while its direct effects are confined to locations where vegetation is piled up and undergoes prolonged burning, may locally have profound consequences, in the top five centimetres or so, affecting soil colloid chemical and physical properties, nutrient ion availability, and microbiological populations (Zinke *et al.* 1978).

The Reasons for Site Abandonment

The large addition of nutrient cations in ash following the clearance and burning of a site, and the liberal increase in nitrate furnished by the mineralisation of humus (which is marked even in acid soils, where nitrification is rapid following the pH rise promoted by the burn), raises the overall availability of important plant nutrients to levels satisfactory for a good crop. The land also, following clearance, is free of standing weeds, is in excellent physical condition, and requires minimum tillage. But this ideal agronomic state is usually short-lived. "It is generally thought that fertility declines rapidly with successive seasons cropping" (Nye & Greenland 1960:73). They cite six reasons for these yield declines, which prompt farmers to abandon sites, Sanchez (1976:379) adding a seventh more recently:

1. **Multiplication of Pests and Diseases**: This is not a common reason for site abandonment. Pest and disease build-ups seldom curtail the cropping period, reflecting the effectiveness of rotation in their control. The practice of intercropping and the mixing together of cultivars with different disease tolerances, the scattered distribution of small plots, and long fallow intervals all serve to check their spread. Pest and disease problems only arise when people attempt continuous cropping or allow their crop:fallow ratios to decline too far.

2. **Weed Proliferation**: The difficulty farmers experience keeping land free of weeds as cropping proceeds — including grasses and other herbs, forbs, woody and herbaceous climbers, shoots from stumps and roots, and tree seedlings — is cited as a common reason for garden abandonment. "The explanation of native practice [of abandonment] is as often to be found in the problem of weed control as in the physical and chemical properties of the soil ... it is frequently difficult to decide whether falling yields are due to weeds or to declining fertility" (Nye and Greenland 1960:76). In S.E. Asia for example weeds are particularly difficult, due largely to the aggressiveness of *Imperata cylindrica*, and farmers here cite them as the reason for abandoning cultivations, as they do in humid savanna regions. It is thought that when the return a farmer anticipates on his labour in establishing a new cultivation exceeds that from weeding his current one, he will as a rule abandon it. But in many regions, notably where they crop semi-perennials, also in drier savanna regions, and the New Guinea highlands too, weeds are not a prominent reason for land abandonment.

Excessive weeding may even threaten the system (Sanchez 1976:378), which is well adapted to a temporary and minimal weed control regime, permitting good forest regrowth. After the burn the land is bare and at its cleanest. The thin cover of low-growing weeds common after a few months from planting is probably benign, helping with soil protection and nutrient conservation rather than hindering crop growth (Saxena & Ramakrishnan 1984; Lambert & Arnason 1986, 1989). When weeding, farmers, aware of the benefits, sometimes leave a proportion of weeds alone (Swamy & Ramakrishnan 1988 estimate 20% of weed biomass under the *jhum* regime of N.E. India, reducing sediment and labile nutrient losses by 20% compared with clear weeding). Sometimes they encourage woody invaders to a certain extent, as the start of all-important fallow regrowth (in the Bismarck Mountains of Papua New Guinea farmers protect them — Rappaport 1972).

The shade cast by overlapping crops is usually sufficient to make any necessary weeding relatively light. And aggressive, shade-tolerant, woody competitors are easily slashed down, their litter further protecting the soil, and returning organic matter and nutrients. The intercropping of plants of varying canopy heights, a common practice with shifting cultivators (Sanchez 1976:376) — questionably likened to apeing the structure of the rainforest (Beckerman 1983) — gives a competitive edge and further facilitates weed control. Another advantage of swidden cultivation practises is that minimum tillage does not encourage dormant weed seeds to germinate. And weeds such as grasses, which demand pulling up by their roots to eliminate them, are unlikely to gain a hold, if the preceding fallow was sufficient to ensure their suppression before cultivation. Farmers sometimes judge a site to be ready for cultivation again only when its vegetation cover indicates that cultivation weed species have been thoroughly suppressed.

3. **Deterioration in Soil Physical Condition**: A decline in soil physical conditions may contribute to site abandonment. It is usually a symptom of other problems too. Low nutrient status may for example exacerbate poor soil structure, impoverished yields affording inadequate cover and giving low litter and root residue returns. Physical deterioration may be particularly rapid and severe during the interval that the soil surface is exposed to sun and rain, following clearance and burning, resulting in an increase in soil bulk density (Juo & Lal 1977; Ewel *et al.* 1981; Clarke & Street 1967). Rain drops impact on the surface, shatter soil aggregates and seal pores with suspended silt, until the crop canopy, and any slashed weed litter, afford effective cover and arrest the process. (Pore size and continuity are critical to soil productivity, determining soil permeability and drainage characteristics, water storage capacity, and gaseous exchange in the root zone.) The rapid decay of the unhumified organic matter in the immediate surface, which helps keep pores open, worsens things further. Soil structural degradation is also closely linked to erodibility.

4. **Erosion of Topsoil**: The exposure of the soil with cultivation may lead to accelerated erosion, with loss of nutrients and humus from the topsoil, especially on the steep slopes characteristic of many forested tropical regions (Watters 1971; Lal 1974). The primary process of mantle transport under natural vegetation in humid regions is gradual soil creep. But when cultivation exposes the soil, it promotes more rapid sheet or rill erosion, even gullying if particularly serious. A rapid decline in fertility, sufficient to prompt garden abandonment, can only be attributed to erosion where it is catastrophic, otherwise it results in gradual depletion, with some losses during each cultivation cycle. These erosion losses can cumulatively be serious, contributing in the long term to the steady loss of nutrients, particularly as their largest concentrations occur in the top eight centimetres or so of the soil in many tropical forests (Mishra & Ramakrishnan 1983c; Ramakrishnan 1989).

Regardless of steep slopes and heavy rainfall, erosion is seldom serious under traditional shifting regimes, with small clearings and good crop cover (the intercropping of crops varying in height provides particularly good canopy protection). The porous, granular structured topsoil, little disturbed by the minimal tillage characteristic of shifting cultivation, is usually stable enough to withstand the shattering action of heavy rain until crops afford some protective cover. And the undisturbed natural vegetation that surrounds the dispersed small swiddens further off-sets erosion risks by offering run-off little opportunity to build-up to catastrophic flood proportions. Some lowering of the topsoil may even have a positive consequence, bringing fresh subsoil and released mineral nutrients within root range, particularly of fallow tree vegetation.

5. **Changes in Soil Fauna and Flora**: Marked changes occur in microbial populations following site clearance. A range of environmental factors, including soil aeration, moisture content, temperature, pH, and vegetation composition, determine micro-organism activity, which is responsible for the decomposition of soil humus and organic matter. A 'flush' of activity follows the partial soil sterilisation of the burn, increase in soil pH, and increase in bacteria species (Laudelot 1961; Ramakrishnan 1989). By exposing the surface directly to the sun, clearing also increases soil temperature, and contributes to the rise in the rate of biological activity (Cunningham 1963). An overall decline eventually occurs in soil microbiological populations with cultivation; the less acidic the soil the slower the fall. But it is unclear how far these changes in soil fauna and flora contribute to deteriorating soil conditions, and hence site abandonment, and to what extent fertility decline is responsible for them.

6. **Deterioration in Soil Nutrient Status**: The decline in plant nutrients features prominently in shifting cultivation accounts as a primary reason for plot abandonment. "There is no doubt about the decline in the level of nutrients during the cropping period ... levels in the soil will fall during cropping because of their removal in crops, and loss by erosion and leaching" (Nye & Greenland 1960:115). Many studies confirm this conclusion (e.g. Ewel *et al.* 1981; Krebs 1975; Siebert 1987; Ofori 1973; Mishra & Ramakrishnan 1983d; Jordan 1989).

The initial fall and subsequent rise in topsoil acidity with site clearance affects nutrient availabilities. Large decreases seem unlikely after a single, one to three year, cropping interval, but deficiencies may be anticipated if cropping continues longer, or intervening fallows are too short for loss replenishment. Changes in soil pH, indicating the percentage saturation of the exchange complex by nutrient cations (which gradually changes over time with leaching of bases), are usually slow, even when cropping proceeds for several years, implying that the fall in exchangeable nutrients is small and that their availabilities may hold relatively steady. High aluminium levels may exacerbate matters, depressing soil pH (aluminium can be a particular problem in highly weathered, acid soils; the ash from firing may briefly overcome any toxicity problems, but these soon reappear — Brinkman & Nascimento 1973).

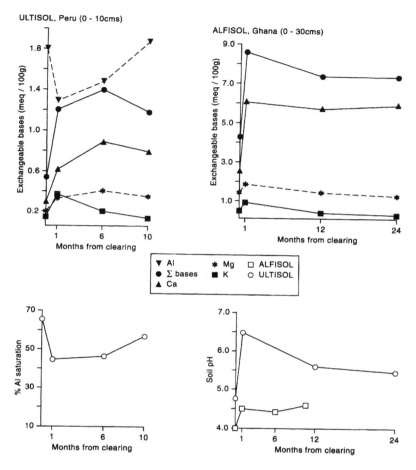

Figure 2.1 Changes in exchangeable bases and soil acidity following clearance of an Ultisol and an Alfisol (*source Sanchez 1976:367*).

The cations furnished by the ash resulting from the burn react rapidly with the soil and are adsorbed onto the exchange complex. Both leaching and plant-removals during cropping then reduce their quantities. The rapid loss of nutrient cations in the humid tropics is generally attributed to the intense rainfall (the brake factor is restricted nitrate anion availability to give required equilibrium solutions — the leaching rate depending on the product of soil solution ion concentration and percolation rate.) The clearance of a plot, dramatically reducing the amount of rainfall intercepted and previously transpired by vegetation, increases leaching rates, resulting it is thought in rapid losses of nutrient elements. Figure 2.1 illustrates the extent of these losses. These diminish as the crop establishes groundcover, although crop uptake now removes some nutrients.⁴ Release from non-exchangeable sources may compensate, at least in part, for these losses. Changes in exchange capacity may contribute too.

A reduction in effective cation exchange capacity (CEC) occurs under shifting cultivation with the decline in pH and organic matter content that result respectively from cation leaching and increased microbial activity — particularly on highly weathered soils, where pH dependent colloidal charge and negative organic matter radicals make a significant contribution to CEC (Brams 1971 reports a 30% reduction in CEC following a 50% reduction in soil organic matter five years after site clearance on West African Oxisols). These reductions in CEC may compensate somewhat for the fall in percentage cation saturation of the exchange complex, keeping up availabilities correspondingly (the proportion of cations on the exchange complex remaining constant when exchange capacity decreases). But the system is less well buffered, the decline in cation availabilities accelerating with leaching and crop-removals during the cropping interval.

During a brief cropping interval, changes in total soil phosphorus are usually small, and cannot always be detected. After a fallow, the availability of readily extracted phosphate increases, with its addition in the ash deposited following clearance, adsorped into the exchangeable pool (Sanchez 1976:372 cites additions of 7 to 25 kg.P/ha). This is particularly important regarding the fertility status of phosphate-deficient soils. During cropping, some phosphorus is also released by mineralisation of humus. The subsequent decline in soil phosphate status under cultivation occurs because of crop uptake, and a fall in availability due to diffusive equalisation of phosphate levels throughout the soil mass and fixation of phosphate into insoluble crystalline complexes. Leaching losses are negligible. Phosphate may sometimes be the limiting nutrient under slash-and-burn cultivation (Siebert 1987), its deficiency prompting site abandonment, although more commonly it is limited potassium availabilty that reduces yields, particularly in forested regions.⁵

The overall effect of swidden cultivation is to push soil organic matter contents down to new equilibrium levels, although humus contents of soils cultivated for several years under shifting regimes may still be relatively high (Popenoe 1957), albeit relentlessly forced downwards with increased cultivation intensity (Figure 2.2). The rate of humus decline, which relates to microbial population activity, approximates in the first years of cultivation to a straightforward exponential decrease,⁶ when calculated from changes in soil carbon, but in later years it levels off with increasing time under cultivation,

---

⁴ The removal of nutrients in harvested crops affects potassium in particular, the most vulnerable of cations to leaching losses and deficiencies; its depletion increases with the cultivation of carbohydrate crops like sweet potato.
⁵ Potassium is rarely deficient in savanna due to low nitrate levels and reduced cation leaching; less weathered soils may also compensate to some extent for declines by releasing more non-exchange potassium.
⁶ $dC/dt = -k_c C$ (where k_c = decomposition constant, and C = soil humus carbon content after t years cultivation)

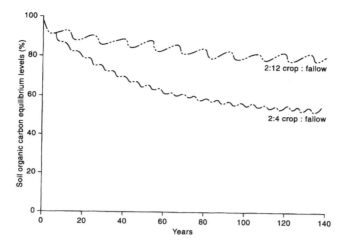

Figure 2.2 Percent changes in topsoil organic carbon content with different crop: fallow ratios (dotted line = 2 year lag allowed for fallow re-establishment: kc = 3%) (*source Nye and Greenland 1960:105*).

approaching some equilibrium point, dependent on soil type and crops grown (Nye & Greenland 1960:98–107). The depletion of organic carbon and nitrogen also depends on cropping intensity and the ratio of cultivation to fallow periods (Ramakrishnan 1989).

The availability of nitrogen depends on amounts stored in the soil and micro-organism mediated decomposition of organic matter (Mueller-Harvey *et al.* 1985; Tergas & Popenoe 1971). Large increases occur in nitrifier numbers with the clearing and burning of forest (associated with soil pH increase, and perhaps the removal of chemical inhibitors and competitive interference by litter-decomposing heterotrophs too), but they decline thereafter with subsequent cultivation (Saxena & Ramakrishnan 1986).[7] Somewhat reduced nitrate-N levels under cultivation could be advantageous, for while nitrogen supply must be adequate for crop growth, high nitrate levels are undesireable, leading to unwelcome leaching losses of both nitrogen and cation nutrients, particularly potassium (leached in proportion to their equilibrium soil solution concentrations — Ramakrishnan 1989; Toky & Ramakrishnan 1981; Mueller-Harvey *et al.* 1989).[8] The carbon:nitrogen ratio usually changes little under cultivation, so the rate factor for annual soil nitrogen conversion is similar to that for

---

[7] In contrast, under savanna, a steady improvement in nitrification occurs for one or two years after grass clearance, and nitrogen mineralisation rates change little for several years; but organic matter accumulation is less and may be insufficient to ensure adequate available amounts, particularly in older grasslands where there is some limit to nitrate formation. Nevertheless, it is mineralisation rate, not nitrification which ultimately influences nitrogen supply (Wild 1972).

[8] Mueller-Harvey *et al.* 1989 calculate that the equivalent of 1 t ha^{-1} calcium carbonate would be needed to replace the calcium leached with nitrate and sulphate over 2 years on a site in southern Nigeria; the bicarbonate ion, unlike the nitrate ion, is insignificant in holding cations in the topsoil solution.

humus decomposition. These changes in organic carbon, nitrogen and C:N ratios occur almost exclusively in the topsoil (Sanchez 1976:371).

Soil organic matter also exerts a large influence on the amounts of phosphorus and sulphur available to crops. There is little evidence that shortage of other elements, notably micronutrients, reduces the yields of subsistence crops.

7. **Social and Cultural Considerations**: Farmers may abandon their fields for social reasons, which have no apparent agronomic connection, such as a death in the family, disputes with nearby kin, and so on. The cultural response to an anticipated reduction in yield after the first crop may also prompt a careless attitude, with less attention paid to weeding and other management tasks. And seasonality in labour requirements might also reduce attention to husbandry, contributing to movements of site.

The Fallow Period

A combination of falling yields, increasing weeds, and so on, their relative importance varying from one region to another, dictates a shift of cultivation. The ash additions, and rapid organic matter mineralisation, that follow clearance, which promote a sharp increase in available nutrients to the first crops (frequently the most demanding — Sanchez 1976:376), is inevitably followed by a fall, and decline in subsequent crop yields, the rate varying with cropping system, soil properties and management, and fallow cover (declines occurring faster on savanna, where fewer nutrients are released). According to Latin American evidence, farmers abandon plots when they anticipate the yield of the next crop to fall below half their first crop, this decline occurring on average by the fourth to sixth consecutive planting, more rapidly with low soil pH (Sanchez 1976:374).

Natural vegetation regenerates quickly at the start of the fallow period, particularly in forest, regeneration having begun while the garden was still under crops, in part from the stumps and roots of the previous natural cover and partly from new seedlings.[9] The ever-increasing volume of vegetation stores many of the nutrients taken up, returning some to the soil in litter, timber fall, root decomposition and rain-wash (throughfall and stemflow). The litter returned decays rapidly, given high biotic activity in surface layers, adding nutrients to the soil (Aweto 1981; Toky & Ramakrishnan 1983a,b; Mishra & Ramakrishnan 1983a,b). And the vegetation cover reduces leaching losses, by cutting down rainfall percolation. The dense shallow roots help too by reabsorbing nutrients as they pass through the soil, immobilising them in plant tissue. This steady accumulation of nutrients can result over time in lush forest on otherwise infertile soils.

---

[9] Manner 1981 gives an account of the succession in the Bismarck Range of Papua New Guinea.

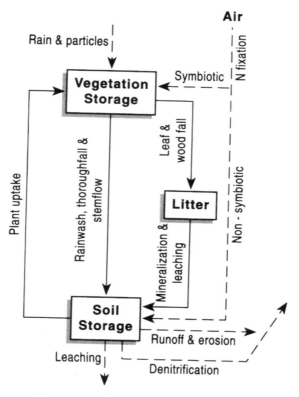

Figure 2.3 Diagram of forest — soil nutrient cycle (*after Sanchez 1976*).

A more-or-less closed nutrient cycle between soil and vegetation results eventually under mature forest, with nutrients pumped up from the subsoil root zone and stored in standing vegetation and topsoil (Edwards 1982). Several nutrient pathways connect the biomass and topsoil (Figure 2.3). Eventually the nutrients in the cycle level off, additions from vegetation to topsoil balancing plant uptake from soil.[10] The humus content of the topsoil builds up to an equilibrium characteristic of the virgin forest (Greenland & Nye 1959; Nye 1961), with the incorporation of leafy and woody litter remains deposited on the soil surface, following their rapid decomposition by a range of organisms (including termites, worms, bacteria and fungi — with considerable carbon respiration losses).

Beneficial physical developments occur too under fallow. The burgeoning vegetation cover reduces the loss of soil particles in erosive run-off by checking the kinetic energy of raindrops, and by promoting the development

---

[10] A position reached after eight years on an Ultisol under tropical forest in Zaire — Bartholomew et al. 1953.

of a porous granular soil structure (Newton 1960). The fine surface network of roots encourages soft porous crumbs, and the constant passage of macrofauna, such as worms and termites, creates numerous drainage channels, thus maintaining good soil surface structure, which permits rapid water infiltration and is resistant to erosion. And the dense shade reduces soil temperatures, moderating biological activity.

The Relativity of Shifting Cultivation

The widely held assumption throughout the colonial era and beyond that shifting cultivation cannot sustain soil fertility indefinitely, because it inevitably degrades the environment whatever the fallow period, is no longer credible (FAO 1957; FAO-SIDA 1974; Watters 1971). Several studies have shown that, with an adequate fallow interval, the system can adequately replace those nutrients removed in crops, lost through leaching, and taken by erosion (see Robison & McKean 1992), and it is now generally accepted that the one time widespread condemnation of shifting cultivation, as destructive of natural vegetation and soil fertility, was largely misplaced.[11] Nevertheless, the idea that shifting cultivation is somehow inefficient in its use of land resources remains widely current, as does the image of a particularly vulnerable agricultural system, delicately balanced environmentally, easily destabilised by population increases and attempts to intensify it (Charles 1976; Arnason et al. 1982).

It is widely accepted that short cropping periods and adequate fallow intervals are important to the stability of shifting systems, permitting rapid regrowth of natural vegetation and restoration of site fertility, otherwise savanna and coarse grasses will take over (Kellman 1969; Uhl 1987; Trenbath 1989). If farmers maintain a succession of crops on a swidden, yields decrease, even where they have cleared virgin forest. The supply of the major nutrients — nitrogen, phosphorus and potassium — declines with reductions in the length of the fallow interval, with resulting establishment of savanna. The amounts of potassium and phosphate added in the ash decline, as do humus levels, and nitrogen and phosphate mineralised from it. The stumps of felled trees lose their coppicing ability and the surface of the soil may be physically degraded. But few people crop their land to this extent, the problem of weeds rather than dramatically declining fertility more often obliging them to move.

The restoration of humus and nitrogen levels following a period of cultivation depends on the productivity of the natural succession, but so long as this is able to re-establish itself adequately, levels will return to virgin proportions, albeit over a long period perhaps (assuming no other essential mineral

---

[11] The FAO advocated until recently for example, the elimination of shifting cultivation as a wasteful land use strategy, causing soil deterioration.

nutrients are restricting growth — Greenland & Nye 1959). Regarding other nutrients, those held both in the exchangeable pool and in non-exchangeable forms feature in replenishment, the total nutrient reserve extending to the effective rooting depth of the fallow vegetation. During cropping the exchangeable pool may undergo rapid depletion, particularly of potassium and phosphorus, but slow release throughout the fallow from non-exchangeable reserves will replenish it (Aweto 1981; Toky & Ramakrishnan 1983b). Over the long period of a fallow nutrient elements tend to equilibrate in a characteristic ratio again, between topsoil and subsoil. The subsoil makes good a proportion of losses, successive cultivations depleting it (uptake from the subsoil may account for twenty percent of the total in the natural cycle — Sanchez 1976:358). The seriousness of subsoil depletions varies — on highly weathered soils (Oxisols and Ultisols) characteristic of ancient land surfaces (e.g. parts of Africa and Latin America) it is a considerable problem to the continued long-term maintenance of soil fertility; potassium reserves being the most vulnerable to exhaustion (Greenland & Okigbo 1983; Juo & Kang 1989). Nonetheless people may return many nutrients as waste, albeit erratically, to the land around their settlements, and may recultivate these refuse-enriched sites, following adequate fallows, in sustainable, albeit delicately balanced, cropping systems.

A striking feature of shifting cultivation is its universality (Robison & McKean 1992), but it is mistaken to think that this implies uniformity (Brookfield & Brown 1963). Many studies imply that what happens to the natural environment under cultivation is uniformly predictable, emphasizing effects on soil fertility, notably chemical properties. The above account has become the conventional wisdom found in overview texts, recently attracting the attentions of systems modellers predicting exponential fertility and yield declines under different intensities of cultivation (Trenbath 1989). It is the regime reported in many studies of traditional agriculture in highland New Guinea (Reynders 1961; Rappaport 1972; Clarke 1971; Wood 1979). But there have been signs for some time, which are growing in number, that this coventional view, while applicable in some parts of the world, is considerably manipulated in others.

Land management strategies in some parts of the New Guinea highlands evidence such manipulation, demanding some reinterpretation to account for them. The suggestion that different interpretations of soil behaviour in shifting contexts may be in order in some places is not new. In their overview of soils under shifting culitivation, Nye and Greenland (1960:23) noted several years ago that fallows were considerably less important for maintaining soil fertility in Alfisols and Vertisols than other soils. They concluded that the inherent qualities of soils determine to a considerable extent the intensity of cropping feasible without noticeable deterioration, giving some examples of short cultivation-fallow cycles (such as 3 years cropping followed by only 3 years fallow over many decades in some savanna areas — although fertility declined to low levels at this intensity). In S.E. Nigeria, Vine (1954) reported a 4 year cycle ($1^1/_2$ years cropping followed by by $2^1/_2$ years of fallow) on soil inherently low in mineral reserves. And in Guatemala Popenoe (1957) reported a 5 year cycle (1 year under cultivation followed by 4 years fallow) on soils derived from limestone and metamorphosed material enriched by some volcanic ash deposits (similar to those reported in this study), where a rapid restoration occurred in soil physical and chemical properties under secondary regrowth, and he concluded that there are "certain anomalies in our concepts of shifting cultivation in relation to soil properties ... one must exercise care in extrapolating data from one tropical area to another" (1957:76).

In his later overview of soils under swidden cultivation, Sanchez (1976) too concludes that the magnitude and speed of changes when any soil is cleared and cropped vary with soil properties and ash inputs. He reports that in some places yields have not declined 50% after 8 to 15 years under cultivation, and that in others they have even increased under cropping because of reduced pest problems or weather improvements. He notes that while there is unequivocal evidence that fertility depletion causes yield declines on low-base-status soils (Oxisols and Ultisols), there is little such correlation on high-base-status soils (Alfisols, Andepts and Mollisols), where changes in soil properties measured before and after cropping are too small to argue that they are the primary cause of falls in yields (see also Juo & Kang 1989). In his opinion, the major reason why farmers abandon fields on high-base-status soils is the need to control weeds.

The Papua New Guinea soils reported here fall into the high-base-status class, but weed control is not a significant problem in the highlands because the dense creeping foliage of the staple sweet potato soon smothers any weed competitors (Siebert 1987 reports the same in the Philippines). These soils are also capable of supporting extended periods of cropping, equal to any of those reported elsewhere in the context of shifting regimes. In the Tari Basin of the Southern Highlands, Wood (1984, 1991) reports gardens under continuous cultivation for 150 years, and pH, calcium, magnesium and phosphorus levels increasing on some sites after ten years under cultivation. And at Aiyura in the Eastern Highlands, Kimber (1974) reports that sweet potato yields held up well in a long-term soil exhaustion experiment, particularly with brief grassy fallows between cultivations. The more recent research on swidden cultivation on young, often volcanic-ash-rejuvinated soils in S.E. Asia supports these findings. Some studies of soil fertility at the time of site abandonment show topsoil pH, available phosphorus and exchangeable calcium and potassium values higher than before clearance, even nitrate levels too on occasion (Nakano 1978; Sabhasri 1978; Zinke *et al.* 1978; Driessen *et al.* 1976). Research on fertile soils in Central America also indicates that periods of cultivation do not result in dramatic declines in soil nutrient status (Popenoe 1957; Lambert & Arnason 1989). And in West Africa, good crop yields have been obtained on Alfisols, without any bush fallow intervals, for over 10 years with appropriate soil management practices (Juo & Kang 1989; Jaiyebo & Moore 1964)).

The evidence suggests that fertility is not everywhere the problem presupposed: "nutrients may actually be less important for shifting cultivators than has previously been believed" (Nakano & Syahbuddin 1989). The ability of secondary vegetation to regrow at fast, almost exponential rates on many abandoned sites is further evidence of continued high post-cultivation soil fertility in some places (Nakano 1978; Sabhasri 1978); although it is arguable that trees and shrubs are able to tap to some extent a supply of nutrients beyond the reach of shallower rooting crop plants, recovering leached nutrients back into the natural cycle (Sanchez 1976). And while secondary vegetation is thought to be highly efficient at utilising nutrients under the conditions of reduced soil organic matter that pertain after a cropping period, it may not always out perform crops. In Guatemala for example, neither young secondary successions nor mixed vegetation stands were more efficient than crops in nutrient uptake and dry matter production on soils derived from limestone and serpentine parent materials (Tergas and Popenoe 1971). Native vegetation may grow more effectively in the long-term because of its superior pest and disease resistance and more efficient use of space, rather than any adaptive tolerance to low levels of some soil nutrients. The rapid regrowth observed during the initial fallow period may also reflect the initial selection of comparatively fertile sites by shifting farmers (Dove 1985), besides the relative abundance of nutrients in the soils of abandoned sites in some regions (Nakano & Syahbuddin 1989).

The Highlands 'Shifting' Variant

On some sites the Wola follow the classic shifting strategy, abandoning them after one crop, but others, their 'bone' gardens, they cultivate repeatedly for decades. They practice semi-permanent cultivation within the context of a shifting regime, farming perhaps one-fifth of their gardens repeatedly five or more times:[12]

Table 2.2
Semi-permanent cultivation within the context of a shifting regime

	No. Times Garden Cultivated					
	1	2	3	4	>5	>10
No. Gardens (%)	46	21	10	5	11	7
Area Gardens (%)	35	27	13	7	10	8
n = 348 gardens; 386807 m^2						

[12] These statistics only give an approximate indication of the extent of Wola continuous cultivation, some of the gardens currently cultivated only once or twice becoming long-term cultivations.

Once a man has chosen a new garden site, his first task is to clear it. This is men's work. The vegetation men have to clear will be either secondary regrowth, of trees and *Miscanthus* cane grass, or primary montane forest, with its associated tangle of understory vegetation. Today they clear this with machetes, using axes on the larger material (up to the 1960s they used polished stone axes), and it takes them about 250 man-hours of work to clear a hectare. When they have cleared the required area, men have to enclose it. Often they situate their gardens in localities where they can use natural features (such as rivers, limestone bluffs and steep-sided gullies) for this purpose. Otherwise they have to erect a fence to keep pigs out, which they make from stakes split from hardwood trees (they favour beech in particular), sharpened and driven into the ground, with a sapling runner lashed along the top edge to give added strength. They use timber from trees felled clearing the site, where suitable. It takes them about 342 man-hours to enclose a hectare in this fashion.

Following its enclosure, they prepare the site for planting. This involves firstly the pollarding of any trees left standing and the pulling up of as many roots as possible. Men usually arrange joint work parties to execute this tedious and hard work quickly, which involves the use of heavy pointed sticks to lever out the roots of trees and cane clumps. It takes some 749 man-hours to clear a hectare ready for burning off. Men and women work together to burn the cut vegetation when it has dried sufficiently in the sun, men usually handling the larger pieces of wood and women burning the smaller

Plate 2.1 Burning small pieces of vegetation remaining on a garden site before starting to plant it.

plant refuse. This firing is important, as stressed, in releasing nutrients and returning them to the soil for crop uptake. It is time-consuming work, taking some 1014 work-hours to burn the vegetation spread over a hectare.

Once the burning is completed, the garden is ready for planting. This is largely women's work, they are responsible for planting the staple sweet potato which predominates in most gardens. Other crops include taro, bananas, beans, cucurbits, and various greens and shoots (Sillitoe 1983). It takes about 678 work-hours to plant a hectare. This brings the overall labour-input total for establishing a newly cleared garden to some 3033 work-hours per hectare. The cultivation will resemble a classic swidden, having a wide variety of crops, planted in partly tilled soil, intercropped amongst a jumble of partially fired tree stumps, roots and logs.

The duration of the mixed cropping period depends on yields, but rarely exceeds two or three crops. Later cultivations involve more soil tillage and a reduced range of crops, dominated by sweet potato grown in large circular mounds, composted with the weeds and grasses that colonise sites during brief fallows. The total time spent preparing and planting such an established fallow garden for a second or subsequent time is around 2545 work-hours per hectare. This is largely women's work. They spend some 371 hours uprooting vegetation, and the other 2174 hours on the time-consuming tasks of breaking up the soil, heaping it up into rows of mounds and planting these (largely with sweet potato vines). Women largely tend gardens (except those of taro) and harvest the crops.

A remarkable feature of this farming system is that inputs for all gardens are internally derived, whatever their age. Nothing is added to sites from

Plate 2.2 Heaping up dried fallow grasses to compost a sweet potato mound.

Plate 2.3 Breaking up soil to build a mound.

outside during their productive lives, other than planting material brought in to stock them. When cleared initially of natural vegetation, all cut material of suitable size is burned; other than the larger material used in enclosing the plot (for fence stakes, log barricades or whatever). Nutrient elements locked up in the vegetation are returned quickly to the soil, as described, in a form in which they may be rapidly further broken down for plant uptake, except for that fraction lost as gas to the atmosphere. When a plot is recultivated and planted, all weedy regrowth and any remaining crop residues are composted into the earth mounds which characterise cultivation of the region's staple sweet potato crop. Similarly, if a site is left fallow for a longer period, until grassy and herbaceous vegetation cover it densely, the coarse grasses and associated herbs are uprooted and incorporated into mounds during recultivation, either green or burnt, as compost or ash (Floyd et al. 1988).

The foregoing review of the literature on tropical farming systems would lead us to expect some decline in the productivity of sites over time, due to nutrient losses, weed proliferation, soil depletion through erosion, and so on. If no manure or other amendments are added to gardens from outside, we should anticipate a steady decline in soil fertility. All farming systems follow practices to maintain soil productivity. In Europe today we rely heavily on industrial fertilisers and biocides, whereas our feudal ancestors employed the in-field/out-field system and our yeomen ancestors various crop rotations, like the well known Norfolk-four-course. Shifting cultivation is the equivalent found throughout much of the tropics. While a deal of our understanding of the

Plate 2.4 Heaping up turned soil into a mound for sweet potato propagation.

system comes from research in the lowland tropics, it occurs not only here but in a wide range of environments including tropical montane regions, even parts of cool temperate Europe until recently (Raumolin 1987). All swidden farmers follow the same broad practices to maintain land productivity. In referring to shifting cultivation, I am not comparing the Wola to some inappropriate lowland tropical model, when they live in a climate which, although tropical in its lack of seasonality, is more akin, at their altitudes, to moist warm temperate.

The question is how are the Wola able to continue cultivating those sites from which they do not shift for decades, even generations, when they move on from others after only one crop. We may take the decrease in the variety of crops cultivated on a site after it has been planted once or twice, to reflect some change in soil nutrient status. When first cleared and cultivated, gardens support a wide range of crops (Sillitoe 1983), including various tubers (sweet potato and taro among others), cucurbits (such as pumpkins and gourds), beans (like the common bean and hyacinth bean), many green-leafy 'spinach' crops (including amaranths and crucifers), plus a range of longer term crops (like bananas, sugarcane and pandans [*Pandanus julianettii*]), together with various recently introduced plants (including brassicas, onions, and tomatoes among others). After one or two cropping cycles, a markedly narrower range of crops is grown, with many gardens passing under a virtual monocrop of sweet potato, perhaps with some longer term crops here and there, and the occasional patch of pumpkin, edible pitpit (*Setaria palmifolia*) and acanth greens (*Rungia klossi*).

II. SWEET POTATO

The sweet potato occupies a central place in this farming system (Yen 1974). The region's staple crop, comprising something like 75% of all food consumed by weight (Waddell 1972; Sillitoe 1983), it makes up by far the largest area under crops. The agricultural regime, notably the semi-permanent garden plot, depends on this crop's ability to continue yielding adequately, regardless of changes in soil fertility status with time under cultivation. A general discussion about what happens to soil properties and so on under shifting cultivation, divorced from a consideration of the nutritional and other demands of the particular crops cultivated, would be partial (akin to worrying about the health of ageing persons without regard to the demands life makes on them). The suitability, or otherwise, of crops to soil, and ability to accomodate to any changes in it under cultivation, are critical issues. The changes that occur in the soils of the Wola region under cultivation do not necessarily reduce sweet potato yields. Contrary to expectations, farmers maintain that the soil on some sites improves with use, becoming better for sweet potato cultivation with time. Data on sweet potato yields from gardens cultivated for differing periods of time supports their claims, tuber yields not declining dramatically over time (Figure 2.4).[13] What is it about sweet potato physiology that fits the crop to this 'stationary shifting' regime?

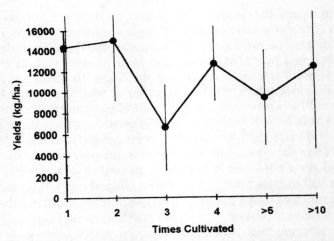

Yield: overall mean=12,248; pooled σ=7673; C.V.=62.7; F=0.84 with 5, 31 d.f.:N.S.

Figure 2.4 The yields of sweet potato from gardens cultivated semi-continuously for different periods of time.

---

[13] On this graph, and ones of similar design throughout this study, the circles represent sample mean values joined by a zig-zag line, and the vertical lines plot the sample standard deviations. The length of the standard deviation lines on this graph indicate a considerable range of values about the means (some newly cleared garden yields being well below established garden yields, and vice versa). See Chapter 13, note 11 for details of the data used in compiling this graph.

The sweet potato (*Ipomoea batatas*), cultivated for its edible tubers, is a creeping plant with trailing and twining vines (Figure 2.5). Its leaves, borne on long stalks, vary greatly in size, shape and colour, depending on cultivar. There are many cultivars; the Wola name at least sixty-four. The plant flowers freely, producing bell-shaped, pale-purple corollas from the base of leaf stalks. Insects cross-pollinate the plant, which produces small black seeds in a dehiscent capsule. But these are rarely evident in the Wola region due to photoperiodic influences on flowering, and are not used locally to propagate the crop; the Wola practice vegetative propagation, using stem cuttings, largely planting them in earth mounds. The plant produces tubers by the thickening of certain roots, largely in the top twenty-five centimetres or so of the soil. Long days, which do not occur in the Wola region, reduce the formation of tuberous roots, and delay flowering too. A marked difference between night and daytime temperatures also encourages tuber formation. The tubers contain a white, sticky latex, as do the plant's stems. They vary considerably in shape, size, colour and organoleptic qualities. They are of relatively high nutritive value, rich in starch and contain some sugar, protein and fat; pigmented tubers are rich in vitamin A, and contain some B and C (Bradbury and Holloway 1988).

The sweet potato is a versatile and vigorous crop, amenable to cultivation in many different environments, under a range of edaphic and climatic conditions and diverse horticultural regimes (Hahn 1977). A montane tropical American crop in origin, it is well suited to the soils of the highlands of Papua

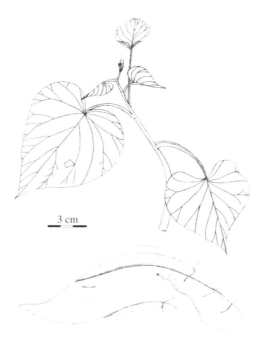

Figure 2.5 Sweet potato *(Ipomoea batatas)*.

New Guinea. Its nutritional demands allow it to continue thriving regardless of the changes that occur in soil conditions and fertility status under cultivation, which makes it a key component in permitting the Wola agricultural regime to exist. But there are costs. In contrast to its edaphic adaptability, the sweet potato cannot tolerate wide fluctuations in climatic conditions, such as can occur in the highlands, either too wet, too dry or too cold. Losses due to diseases, like fungal black rot (*Ceratocystis fimbriata*), are generally slight due to the cultivation of highly resistant cultivars. Viral diseases probably cause the largest reductions in yields, after those due to extreme climatic perturbations, although carelessness can cause trouble with insect pests, notably by spreading foliage and tuber damaging larvae.

Sweet Potato Physiology

Sweet potato productivity correlates with dry matter accumulation, the tuberous root comprising a high proportion of total plant dry weight. (In a series of measurements on tub grown glasshouse plants, the mean tuber yield comprised 45% of the total mean plant dry weight.[14]) Any increase in total dry matter production should consequently increase tuber yield, although the rate of increase may vary depending on environmental conditions which influence partitioning between aerial and subterranean parts of the plant. It is necessary that the environmental factors are in balance, with their positive and negative effects in appropriate correspondence to maximise tuber yield. Several environmental factors influence dry matter production, interrelated in a complex feedback relationship, notably supply of mineral nutrients, radiation, temperature and moisture (Figure 2.6).

The higher yielding a sweet potato cultivar, the greater the proportion of its dry weight that is diverted into tubers, the percentage partitioned into the vine correlating negatively with tuber yield (Enyi 1977). Variations between plants in tuber yield occur because of differences in both the number of tubers to a plant and the size of individual tubers, that is high yields are associated with both more and larger tubers. Regarding the relative importance of these two factors in determining yield there is apparently some variation between cultivars, with yield in some cultivars correlating more with mean tuber weight (Enyi 1977), and in others it depending more on total number of tubers per plant (Bourke 1984). This is perhaps to be expected with large differences reported between cultivars for such parameters as tuber yield, number of tubers per plant, mean tuber weight, leaf area, and total dry weight per plant.[15] The rate and the duration of tuber bulking are also important determinants of yield, and earliness of tuber initiation too, higher yielding cultivars having earlier initiation.[16]

---

[14] Sample of four plants grown to maturity at Wye, range of tuber yields 37.7% to 51.6% of total plant dry weight.
[15] Photoperiod sensitivity is not reported to influence tuber yield, only flowering and breeding (sweet potato is a short-day plant); it is not considered further here.
[16] Wilson & Lowe 1973 consider these two stages to be particularly significant and propose basing on them a distinction between two phases in sweet potato tuberisation: 1) tuber initiation, 2) tuber bulking.

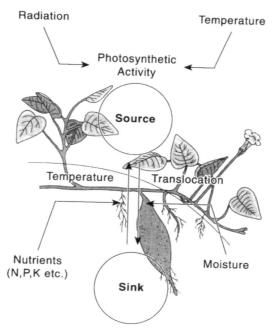

Figure 2.6 Environmental factors affecting sweet potato dry matter production (*after* Hahn 1977a:239).

The growth of the sweet potato plant may be divided into three stages (after Enyi 1977): 1) vegetative phase, when most assimilates are used to produce stem and leaves; 2) mid-vegetative phase, when stems and tubers compete for assimilates (the proportion of dry matter diverted into each varying with organ growth rates and differing between cultivars); 3) tuberisation phase, when most of the dry matter produced is diverted to the tubers. The plant follows a sigmoid growth curve over time, with number of leaves and mean leaf area increasing until mid-season and then declining, crop growth rate increasing to a maximum and then falling off, and net assimilation rate (NAR) decreasing with time (Bourke 1984). The leafy top growth reaches a maximum some two to three weeks before maximum storage root development. The formation of storage roots need not necessarily retard vine growth; when rapid increases in root storage weight occur, vegetative top weights may remain constant or decline (Austin *et al.* 1970). The maintenance or otherwise of sufficient leaf area is significant for the undiminished supply of photosynthates to the expanding tuber.

Leaf area development depends on rate of vine growth, to support green leaves, and leaf expansion, balanced against rate of leaf senescence. The better yielding cultivars have significantly higher leaf area indices (LAI) than lower yielding ones (Bourke 1984). There is also a relationship between leaf area duration and tuber yield. The consensus of experimental observations is that LAI starts to decline some two months after the plant is established. And it has been suggested that the lower the vine:tuber ratio at harvest, the higher tuber yield (Enyi 1977). But an appropriate balance needs to be maintained between leafy photosynthetic machinery and root storage capacity, and considerable disagreement exists over the optimum point.

The sweet potato root is specialised as a starch-storage organ.[17] Its principal tissues are periderm, cambium, vascular tissue and storage parenchyma. Development of the modified storage root is due anatomically to the successive activity of secondary cambia. The accumulation of photosynthates in the root, largely carbohydrates, signal tuber formation. The vascular cambium of potential tuberous roots has the capacity to generate storage parenchymatous cells instead of xylem cells, which together with the activity of anomalous meristematic tissues associated with isolated vascular bundles in the central pith, contribute, as sources of cells for starch accumulation, to increasing the size of the tuber (Wilson & Lowe 1973). Frustration of tuber initiation occurs where primary xylem elements develop which, during secondary thickening, lead to central stele lignification.[18] Regarding tuber development, four types of roots may be distinguished (Togari 1950): 1) young roots, 2) fibrous roots, 3) hard roots, and 4) tuberous roots. All roots start off young, and become either fibrous, hard or tuberous depending on primary cambium activity and degree of stele cell lignification, which in turn are influenced by environmental conditions. The desired tuberous roots form when primary cambium activity is high and stele cell lignification low, which is encouraged when potassium supply is high and temperature relatively low (Figure 2.7).

Regarding source-sink relationships in sweet potato, it is thought that sink capacity is the primary limiting factor with respect to yield, the rate of source photosynthesis depending on demand for photosynthates by the tuber sink (Hozyo 1970; Wilson 1982). Sweet potato has the highest solar energy fixing efficiency of any food crop, with a tremendous capacity to produce dry matter for a long period of time (Hahn 1977), but in order to maintain high photosynthetic activity, it is necessary that photosynthates are translocated from the leaf into the tuberous root sink. Sink capacity affects the translocation of photosynthates. The consequent reduction in leaf carbohydrate content stimulates further photosynthetic activity, and dry matter accumulation. But this widely held proposition that sink activity is more significant in determining sweet potato yield than photosynthetic capacity of foliage has recently been called into question. It appears that the relationships between root yield and source-sink partitioning are more complex than this attractively straightforward model suggests, rates of translocation and assimilate sharing between different organs vary between cultivars and even change during the course of the growing season. The main differences between plants are thought to relate

---

[17] Regarding storage sink sites in sweet potato, tuberisation is not limited to the root system but can occur at one of three alternative locations: 1) adventitious roots growing from planted stem cutting nodes, 2) replanted mother tubers, and 3) stems of plants.
[18] Wilson (1982) distinguishes between two types of lignified root, which he calls 'string' and 'pencil'.

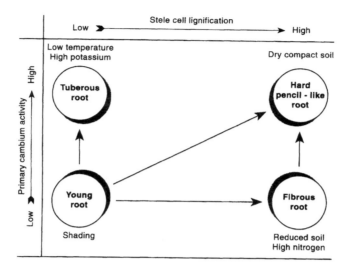

Figure 2.7 Development of sweet potato roots in relation to cambium activity and stele cell lignification (*after Togari 1950 and Hahn 1977*).

to the differential effects of root sink strength on partitioning and yield. Cultivars with relatively stronger root sink effects on partitioning have reduced vine growth later in the season, which may in turn reduce yield because of diminished photosynthetic capacity. Plants showing strong root sink effects consequently give lower yields, reductions in canopy area negating storage root demands on assimilate translocation, reducing final root yield (Bouwkamp 1983; Bouwkamp & Hassam 1988).

It is evident that tuber yield depends on a fine balance between both rate of dry matter production and the distribution of dry matter within the plant. One suggestion is that distribution of assimilate within the plant may be the primary determinant of yield where cultivars have adequate LAI, but that area of foliage has a greater effect on yield where they have a low LAI, tuber yield here depending more on total dry weight production than on redistribution of photosynthate within the plant (Bourke 1984). It is possible that the factors limiting sweet potato yield may differ between cultivars, with assimilate production important for some and capacity for tuber growth with others. High sweet potato tuber yields depend on both high total plant dry matter production and diversion of a large proportion of dry matter produced into tubers, the relative importance of these two factors varying between cultivars, and with environmental conditions.

The sweet potato, featuring prominently in the highland agricultural regime, is more sensitive to some environmental variations than others, which markedly influence the complex feedback on tuber production. It is remarkably tolerant of different soils, so long as not too wet, dry or cold. It will produce acceptable yields even on land of low fertility and relatively low pH. And it is particularly well adapted to the low input subsistence farming system of highland New Guinea, some plots cultivated long-term with no external amendments, on soils having inherent fertility problems in the supply of some essential plant nutrients (see Chapter 12). It is not unique. The evidence from slash-and-burn regimes elsewhere of high soil fertility levels after site abandonment, of crops performing comparably to secondary regrowth, of extended periods of cultivation with limited fallow periods and no external inputs, and so on, all indicate the widespread existence of natural resource management strategies that do not always demand frequent rotation of land to sustain productivity. The relative rotation of land by the Wola, as an aspect of their ecological relations, features as one theme in this soils-focussed ethno-environmental enquiry. The following section deals with the climate, its contribution to soils and agricultural practices, and subsequent sections work through the land and biotic factors, to the soils and their behaviour under cultivation.

SECTION II

THE CLIMATE FACTOR

Sabkabyinten obuw ten mboliy Jackie nonbiy hat bort burubtuw kison. Chay diy hakuwl diy nguwp nguwp bayor g^embom iyb mael layor ten ngon. Shiyort haek diy obuw yuw kalow berayor. Waek ebenjip ebayor tomb obokor serep iyl haek nobtuw hondba buw marakor. Iysh iyl ora sayor ngo tomb yuw waek pokor Sabkabyinten yin bort. Ngo tomb ngo bort nae harakor. ... Horwar Saliyn ngo obuw ten mboliy menday obuw hokay hombuwn buwm hayn kisor. Hokay day eben kisor, ngo tomb iyshpondamahenday hondaesmiy.

Sabkabyinten, she is a white-skinned woman like Jackie [the author's wife] who they say lives up in the sky. She makes the rain and thunder, by splashing with her digging stick in a large pool of water up there. She looks after the cassowaries and other birds too. During the *ebenjip* season they come to eat various fruits [which ripen then]. When the fruits are finished they go back up there to Sabkabyinten. They are not to be found in our region then ... Horwar Saliyn, now she is another white-skinned woman, and she made sweet potato and everything disappear. When such a sweet potato hungry time came, everyone looked to the *iyshpondamahenday* ritual.

Maenget Kem

CHAPTER 3

ETHNOMETEOROLOGY: THE CLIMATE

The climate sets the conditions for the reactions that occur in any environment, by determining temperatures, dilutions, hydrolysis conditions, and so on for interactions that take place in nature. And it is responsible for some of the physical forces that feature in moulding the natural world too, by setting preconditions that decide the course of weathering and subsequent soil forming processes. But the Wola are not particularly interested in, even fully aware of, these long-term processes. The interest they evince in the climate is more parochial, issues of the local weather forecast type preoccupying them. They share these with all humankind; namely, what is the weather going to be like at the place where I live, and why? They show a keen interest in the weather, similar to those who live close to the land the world over. It is a common opening gambit in conversation, individuals frequently commenting *chay ebay* "the rain is coming", or *baen bayor* "it's set fine", or whatever. But they are interested in the weather not merely as a subject for idle chat. It is central to their farming activities, the weather having profound, and sometimes dire, effects on crop cultivation and food supplies. Their staple sweet potato crop being particularly susceptible to certain climatic fluctuations.

The impression gained by visitors to the New Guinea highlands is of an equable climate year round, no noticeable seasonal variation in the usual, though not invariable, daily pattern of warm sunshine in the mornings followed by rainfall in the afternoons. According to one climatologist the "most conspicuous feature is the lack of seasonal contrast" in the weather (Fitzpatrick 1965:56). Nevertheless, according to the Wola, their region not only experiences two seasons, which they call *ebenjip* and *bulhenjip*, but also the weather can sometimes prove so inclement that food shortages occur, even famine in extreme circumstances.

The climate not only sets conditions that influence soil development, it also sets parameters on what human-beings can achieve in any environment, while not determining what they do within its constraints. This chapter presents an overview of the climate of the Wola region, within the context of an ethno-meteorological account. This is a neglected but noteworthy approach to climatology (McAlpine *et al.* 1983:3–4). It offers other cultures' perspectives on issues concerning the weather. We need to know something about a region's climate because annual and seasonal rainfall patterns, daily and yearly temperature ranges, and wind, sunshine, humidity and so on, all set limits on the activities of those living there. How the human population perceives of these limits and explains the weather regime responsible for them will have some bearing on its behaviour, notably how it seeks to manage them.

People contrive to use technologies of varying sophistication and environmental menace to manipulate conditions to gain some control over climatic fluctuations; build reservoirs and irrigation channels to regularise water supplies where rainfall is seasonal, manipulate plant species to shade crops from intense direct sunlight, manage the soil by terracing, contour ploughing and such like to reduce erosion risks, and so on. The digging-stick-technology of the Wola employs earth mounds. These have, among other things, significant microclimatic effects regarding soil wetness, temperature and so on (see Chapter 13; Waddell 1972). But large perturbations in the weather are beyond their technical competence to manage. On these occasions individuals sometimes resort to allusive gestures, like squashing sappy plants to encourage rain. And in times of severe climatic fluctuation communities sometimes resorted to ritual measures to appease a white-skinned female spirit believed responsible.

High annual rainfall and seasonally uniform tropical temperatures characterise the climate of the S.W. Pacific region, dictating the weather over the Southern Highlands of Papua New Guinea, though modified by local controls. The seasonal shift north and south of the thermal equator, together with the associated positional changes in the southerly subtropical anticyclone centres and northerly monsoonal wind system, are the principal seasonal controls on the Oceania region. The seasonal variation in rainfall is small nearly everywhere, best described according to one authority "as a change from 'fairly wet' to 'very wet' " (McAlpine *et al*. 1983:3). High humidity and cloudiness are predictably associated with the combination of high rainfall and temperature. There is a seasonal alternation in the direction of prevailing winds, from N.W. monsoons to S.E. trades, they are sometimes strong but rarely violent. The climate of the Wola region itself is of the 'Lower Montane Humid' type (according to McAlpine *et al*. 1983:160). It is characterised by high rainfall, the absence of soil moisture droughts, cool temperatures due to the moderating effect of altitude, and the relative absence of any seasonality.

The Daily Weather Pattern

The weather usually follows one of three fairly predictable broad daily patterns in Wolaland. In the commonest, the mornings normally start bright and clear, with some low cloud settled in the bottoms of the valleys, clearing by about 8.00 am, the temperature then rising quickly as the sun, generally unobscured by any cloud rises to its zenith. In the early part of the afternoon clouds generally build up and the sky becomes overcast with a consequent fall in the temperature. It usually rains in the latter part of the day, if not earlier. This pattern occurs on about 46% of the days in a year (Table 3.1). On the other days it might be either sunny all or most of the day (23% per annum), or overcast all or most of the day, probably with rain (30% per annum). The nights may be clear and cold, through to heavily overcast and wet.

Table 3.1
The daily gross weather pattern each month (percent days/month)
[Haelaelinja observations]

	Jan	Feb	Mar	Apr	May	Jun	Jul	Aug	Sep	Oct	Nov	Dec
Sunny all/most of day	15%	10%	56%	28%	15%	18%	19%	45%	17%	10%	21%	27%
Sunny morning, overcast afternoon	55%	41%	25%	52%	53%	43%	47%	25%	58%	55%	49%	48%
Overcast morning, sunny afternoon		4%		1%			3%		2%	2%		3%
Overcast all/most of day	30%	45%	19%	19%	32%	39%	31%	30%	23%	33%	30%	22%
No. Records	93	82	51	94	66	93	78	57	43	60	61	69

The Wola give names to the sunny and overcast extremes of weather. Dry days throughout which the sun shines are *baen biy* (lit. fine make), whereas wet days that are overcast, perhaps with low cloud-mist and damp atmosphere, are *chay kung* (lit. rain ridge-top [i.e. cloudy and wet like on mountain summits]). These terms have a wide range of reference, applying to both several consecutive days of either weather, or parts of the same day. A sunny morning is *baen biy* and an overcast afternoon may be *chay kung*. This common sunny morning and overcast afternoon daily pattern has no name as such, although the Wola say that it more characterises the *ebenjip* than the *bulhenjip* season, though this is not evident from Table 3.1 (50% of the days in both seasons featuring this daily weather pattern). The monthly distribution of different daily weather patterns shows little regularity overall. Scrutiny of the different weather sequences each month reveals no obvious pattern. There is no monthly nor seasonal correlation. Yet the Wola maintain that they can distinguish two seasons from monthly weather patterns; whatever they perceive is subtle (compare Clarke 1971:39–50 on Maring region seasonality).

The Seasons

The Wola divide the year into two seasons of equal length. The *ebenjip* season extends from October to March, and *bulhenjip* from April to September.[1] The differences between them are not marked they say, as the data presented so far suggest, nor is the weather pattern entirely predictable according to season.

---

[1] The Wola did not traditionally join both seasons together to give a twelve month year. This concept is an introduced Western one; today the Wola refer to a year as a *mol* (from the Pidgin use of Christmas, pronounced *Krismol* and abbreviated to *mol*.). See Meggitt 1958 for comparative ethnography on the Enga calendar.

Similar weather conditions largely prevail throughout the year. Regardless, the weather pattern varies sufficiently according to the Wola, for them to distinguish between the seasons. The weather during *ebenjip* is less variable they say, and more often follows the predominant pattern of sunny mornings and overcast, sometimes rainy, afternoons. The weather during *bulhenjip* features more dry and sunny days, or wet and overcast ones, occurring in consecutive sequences, such that alternate drier and wetter periods characterise this season.

The weather alone is not the sole marker of the seasons. The meteorological data presented here suggest that the Wola would be hard put consistently to distinguish between them by just changes in the weather. The position of the sun in the sky, as elsewhere in the world, is also an important, and an invariable, seasonal indicator. The sun rises and follows a different path across the sky in the two seasons. In *ebenjip* the sun rises towards the southeast, called the *nat aenaen* (lit. sun S.E.-direction), and moves across the sky to a culmination point more-or-less directly overhead, whereas in *bulhenjip* it rises in a more north-westerly direction, called *nat aumuwnaen* (lit. sun N.W.-direction), and reaches a lower culmination point in the sky, to the north.[2] The changeover between the seasons occurs around the equinoxes, when the sun rises due east and reaches a mid-point culmination (Meggitt 1958:76). According to the Wola the changeover is not abrupt; if several people are asked between March and April or September and October to name the season they are likely to disagree or hesitate over pronouncing which it is. The gradual changeover extends over four weeks or so, as the sun changes its position in the sky.

In addition to watching the sun in the sky, the Wola watch for changes on earth too. Some of the trees and animals in the forest undergo changes, nature regulating some of her rhythms according to the seasons. The few trees in the Wola region that shed their leaves synchronously — like the *hog* cola nut (*Sterculia* sp.), the *sabok* (*Meliosma pinnata*), the *huwgiyt* red cedar (*Toona sureni*), the *injil* umbrella tree (*Schefflera* cf. *hirsuta*), the spiny *mak* tree (*Harmsiopanax ingens*), and the *pak* and *piyp* water gums (*Syzygium* spp.) — all do so during the *bulhenjip* season, possibly as sensitive photoperiodic responses. During this season's *baen* sunny period the leaves on these trees dry out the Wola say, become brittle and fall to the ground. This is taken to herald the start soon of a rainy spell, when the trees begin to sprout new leaves. This time at the back-end of *bulhenjip* is sometimes referred to as clenched-paws-time, after the resemblance between the tight new buds of these trees and the clenched paws of marsupials.[3] People say that the rain at this time hinders insects eating the tender new leaf shoots. When the leaves are fully grown, soft and flexible, the weather improves as the *ebenjip* season

---

[2] The seasons distinguished by the Wola thus equate with the Southern Hemisphere's summer and winter. Living near the equator, and traditionally without an accurate measure of time, the small variation in day length that occurs between the seasons passes unremarked.

[3] Another analogy is sometimes made between buds and a leper's stunted or digitless hand.

begins, with more sunshine and less rain, even perhaps some *baen* fine spells.

During *ebenjip*, on the other hand, several species of tree fruit. People say that the regular rainfall at this time causes many fruits to swell up (although not all, many species fruiting in both seasons, with some individuals producing fruits in one season and some in the other). The plants that fruit in *ebenjip*, indicating this season's presence, include: the cultivated pandans *aenk* (*Pandanus julianetti*) and *wabel* (*Pandanus conoideus*), and the wild pandans *pundin* (*Pandanus archboldianus*) and *paym* (*Pandanus antaresensis*); several broad-leaved trees including *munk, munkiyrit* and *wok* gamboges (*Garcinia* spp.), *shongom* and *shwimb* bead ashes (*Elaeocarpus leucanthus* and *E. dolidrostylus*), *piyp* water gum (*Syzygium* sp.), *dorok* (*Helicia oreadum*), *natnat* (*Aceratium tomentosum*), and *serep* (*Platea excelsa*); plus some shrubs and grasses like *hezambul* bramble (*Rubus moluccanus*) and *henj* sword grass (*Miscanthus floridulus*).

These fruiting plants in turn attract migratory birds to the region during *ebenjip* to feed. These birds include the *shiyort* dwarf cassowary (*Casuarius bennetti*), the *wola'uw* purple ground dove (*Gallicolumba jobiensis*), the *ibiyabuwk* cinnamon ground dove (*Gallicolumba rufigula*), and Goldie's lorikeet (*Psitteuteles goldiei*). The return of these birds marks the start of the *ebenjip* season, the Wola say, and their disappearance the onset of *bulhenjip*. They think that these birds migrate high up into the sky in *bulkhenjip*, into the care of a white-skinned female sky-being called Sabkabyinten (Sillitoe 1981), who has some control over the weather too, as mentioned in Maenget Kem's opening comments to this section. The *ebenjip* season is also the time when most birds build nests, lay eggs and hatch chicks. And the *huwmbuwga* black butcherbird (*Cracticus quoyi*) only sings during *ebenjip*, it is not heard in the *bulhenjip* season. The marsupial population of the forest also exhibits physiological and behavioural changes according to the season. During *bulhenjip* these animals have a thicker layer of fat on their bodies, and females carry suckling young in their pouches. When *ebenjip* arrives the young are large enough to ride on their mothers' backs, and when eaten the animals' flesh and bones are "stronger", tougher and more sinewy.

The variation between the seasons is quite marked when we consider all the factors, meteorological, solar and biological, that exhibit some change. Nevertheless, how discernible are the seasons in the annual patterns of rainfall, cloud cover, temperature, and so on, given their relative absence in the broad daily weather summaries? Furthermore, what is the weather experienced across the Wola region, and how does it impinge on people's agricultural activities, and how do they contrive to manage these constraints? The following meteorological and ethnographic data address these questions.

Rainfall

Rainfall is high overall in Wolaland, as anyone who has visited will vividly recall, walking through either torrential downpours or fine misty-drizzle,

Plate 3.1 A deciduous *hok* tree (*Sterculia*) that has shed its leaves in the *bulhenjip* season.

which infiltrate clothing and soil, running over the body as the contours of bare slopes. It is expectable, Papua New Guinea being one of the wettest regions on earth (Chang 1968). Annual averages across the Wola region run from 2644 mm to 3378 mm. Rainfall fluctuates from month to month, following no strikingly noticeable annual pattern; the same randomness characterises daily fluctuations (Figure 3.1). We can discern a drier period in the middle of the year between June and August, followed by a slight peak in rainfall in September–October, a pattern which corresponds with the *baen* and *chaykund* sequences described by the Wola for *bulhenjip*. The rainfall in other months is consistently high generally, corresponding with the more uniform *ebenjip* pattern.

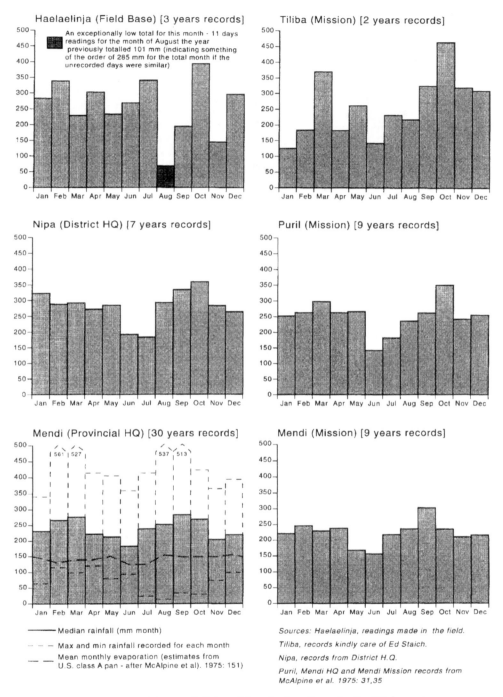

Figure 3.1 Mean monthly rainfall (mm) for six locations in the Wola region.

Plate 3.2 Heavy rain turns *howma* ceremonial clearing into a muddy expanse.

It is noteworthy that the seasonal rainfall pattern distinguished by meteorologists for the S.W. Pacific region corresponds to the Wola seasons. Although seasonality is low in the Southern Highlands, maximum rain falls in the "north-west" season between January and April, equivalent to *ebenjip*, and the minimum in the "south east" season between May and August, equivalent to *bulhenjip* (Fitzpatrick *et al.* 1966; McAlpine *et al.* 1983:65). And monthly rainfall is more variable in the drier season throughout the central highlands. Level of rainfall depends on local physiography too, valley and basin centres receiving less rain than surrounding hills (Barry 1978).

The overall impression, modest seasonality notwithstanding, is one of a region where rainfall varies largely unpredictably from one period and place to another. The Mendi Mission annual average for instance is considerably lower than that recorded at the Provincial headquarters, only about one kilometre away. Although not directly comparable (covering different time periods), this serves to illustrate the unpredictable and variable rainfall pattern of the region. It may be clear at one place, while raining heavily nearby. These variations between different places are not surprising, given the annual, monthly and daily fluctuations that occur at the same place. They are due in large measure to local orographic effects. The weather folklore of the Wola reflects this regional variability. They say, for example, that when someone handles poison with malicious intent, there will be a large rainstorm over that place. Or if someone is thoughtless enough to yodel in the forest in the Olaem region, they will cause heavy rain and earth tremors in that part of the Was river valley.

In view of the variability in rainfall, to talk in terms of mean annual rainfall and report that it averages around 3000 mm across the Wola region is to give only a gross indication of precipitation. Monthly rainfall probability, together with monthly evaporation rates, are agriculturally more significant (Figure 3.1). These indicate that while conditions of high rainfall are the norm throughout the region, dry *baen* periods occasionally occur, especially in *bulhenjip*, when rainfall may remain substantially below mean values for some weeks (usually due to a regional inflow of dry air from the high pressure zone to the south of Papua New Guinea). They are sometimes associated with frosts, and even without these, crops may suffer moisture stress with the partial depletion of soil moisture reserves. But this is uncommon. The region's normal hydrological situation is one of water excess not shortage (Keig, Fleming & McAlpine 1979). The ratio of median monthly rainfall to mean monthly evaporation varies from 1.3 to 3.1, and rainfall exceeds evaporation in nearly all months in 75% of years. The annual evaporation at Mendi totals 1702 mm, with relatively little variation throughout the year. Under these conditions, soil moisture is usually around field capacity throughout the year, giving a perudic soil moisture regime according to the U.S.D.A. (1975) scheme.

Rain is not simply rain to the Wola. Living in a fairly wet region, and regularly experiencing rainfall, they distinguish between several different kinds of rain. Rain in general is *chay*. Drizzle and fine rain is *chay pokotpokot*, whereas a storm of heavy rain is *chay onda*. A hailstorm is *taendok hobor ebay*, which can damage the large broad leaves of some plants (e.g. crop plants like bananas and sugar cane), tearing them into strips; it is a rare occurrence. A thunderstorm is *hatsab bay* or *hakuwl kay*. Persistent fine rain that comes with low cloud shrouding everywhere in mist, is *chaykund*; the same as a period of rainy weather with no sunshine. If there is a rainshower and the sun continues to shine through it, this is *chaysuw*. The Wola believe this to be a threatening combination, saying that if someone is visiting elsewhere and arrives in a *chaysuw* shower he will soon die. A related potentially dangerous phenomenon that occurs in sunny-rainy conditions is the rainbow or *ningiylbap*. When small, children are told not to look at a rainbow or else their parents will fall ill, even die.

The reasons for these superstitions are unclear. They might be related in some way to the white-skinned sky woman Sabkabyinten.[4] According to the Wola it is she who causes the rain to fall. She makes it rain by splashing in a celestial pool of water with a stick. When she splashes gently, fine rain falls, and when she beats on the pool hard there is heavy rain. The sound of her hitting the pool hard is the clapping of *hakuwl* thunder. The cause of accompanying *buwbiyaeb* lightning is not agreed; it might be the glistening of the raised wet stick. Thunderstorms, people say, occur only in the *ebenjip*

---

[4] See Vicedom & Tischner 1943–48; Strauss & Tischner 1962; and Strathern 1970; on female sky-being spirits in the Western Highlands. The Hageners attribute thunderstorms to sky-beings (Strauss & Tischner 1962).

season; a claim not borne out by daily weather records, although they occur more often in *ebenjip*. The reason for this seasonality some surmise is that their hunting of Sabkabyiten's migratory bird charges in *ebenjip*, notably the flightless cassowary, provokes her to beat the celestial pool hard in anger.

The forecasting of rainy spells is of interest to the Wola because they interfere with their subsistence activities. They are unable to work so often in their gardens (cannot burn off cleared vegetation, fear puddling the soil and damaging its structure, and so on), and their crops, notably the sweet potato, grow less well they say in continuously rainy periods. When rain is approaching, people maintain that the *aluwmb* bird Reichenow's honeyeater (*Melidictes rufocrissalis*) cries out loudly. When it rains in the night and the following morning is overcast, they forecast a spell of wet weather. Further indications that the rain has settled in for a *chaykund* period are the loud croaking of *duwduw* (*Xenobatrachus* sp.) and *kinjabuwl* (*Litoria iris*) frogs, together with the increased chatter of *kuwntok* skinks (*Scincella elegantoides*). A sign of rain during *ebenjip* is the *paeraeb* megapode (brown-collared brush turkey, *Talegalla jobiensis*), scratching together vegetation to build up its mound prior to laying eggs.

The intensity of rainfall is another important consideration regarding human activities, notably in relation to its effects on soil conditions. In a region where steep slopes are common and soil is periodically exposed when newly planted under cultivation, there is some risk of topsoil erosion. But the evidence indicates that the chances of receiving high rainfall in a 24-hour period, of sufficient intensity to cause serious erosion are low (see Chapter 6). The monthly distribution of rainfall of different intensities is broadly even, as with the mean annual rainfall statistics, although intense rain is perhaps less likely in the middle of the year, which corresponds with the *baen* spells of the *bulhenjip* season.

While a certain amount of rain is necessary for the successful cultivation of crops, too much, either spread out over many *chaykund* wet days or concentrated into high intensity downpours, is undesirable, even deleterious to agricultural activity. The Wola follow a curious practice to halt too much rain and promote finer, sunnier conditions. They take a piece of old netting from a disused bag or man's apron and hang it from a fence stake or tree branch, setting fire to it along one edge and leaving it to smoulder. The practice is referred to as *nuw* (or *haenaep*) *piriy haeray* (lit. bag (or apron) worn-out burn). The dirt and grease impregnated piece of netting does not blaze quickly with a flame but burns slowly with much smelly smoke. There is no associated spell, or other action. Why a piece of smouldering string netting should end a rainy period is unclear. One man suggested that the malodorous smoke "fights" the clouds and drives them off, in the same way they believe that fighting a pain with pain cancels it out and aids recovery (for example, rubbing an aching stomach with painful stinging nettles). Others thought that the smoke rising up from the earth is a confusing signal, for this usually occurs during fine weather spells when they burn off newly cleared gardens, and this somehow causes the rain clouds to disperse and give way to the clear skies associated with burning off; the inference, although no-one ever suggested it, is that Sabkabyinten is fooled by the rising smoke and moves off elsewhere on her clouds. On the other hand, some people do not think that burning old netting is efficacious at all, several individuals pointing out that it does not necessarily result in fine weather. It is ambiguous, like many symbolic gestures. Perhaps it is an act of defiance, "making" more cloud, or an impotent demonstration, expressing a helpless wish for the weather to change, although what the smouldering material symbolises is unclear.

Cloud Cover

The Wola region is fairly cloudy throughout the year, which is expectable with its high rainfall. The cloud cover shows relatively little fluctuation from month to month (Table 3.2). There is an indication that mornings during

Table 3.2
Cloud cover data from two locations:– mean daily total cloud cover, and percentage occurrence of cloud types each month

	Jan	Feb	Mar	Apr	May	Jun	Jul	Aug	Sep	Oct	Nov	Dec
HAELAELINJA (Field Base) 09.00 Observations												
Mean daily cloud (oktas)	6	6	5	5	5	5	5	4	4	5	5	5
Range: highest (oktas)	8	8	8	8	8	8	8	8	8	8	8	8
lowest (oktas)	2	1	1	1	1	0	1	0	0	0	1	1
Cloud type (% days):												
cirrus	25	23	46	39	35	48	51	65	49	41	53	42
cirro-stratus	51	27	23	32	35	26	25	0	8	17	27	34
cirro-cumulus	11	15	9	24	26	16	8	7	4	10	9	8
alto-stratus	4	6	9	3	6	8	0	0	0	2	12	3
stratus	43	49	41	23	24	37	27	28	25	25	14	33
strato-cumulus	8	16	4	11	10	16	10	0	8	22	9	12
alto-cumulus	2	6	0	1	0	0	0	0	0	3	0	3
cumulus	51	61	57	67	50	51	55	49	51	70	68	69
cumulo-nimbus	1	2	0	0	0	0	0	0	0	3	0	2
nimbo-stratus	2	1	0	0	0	0	0	0	0	0	0	0
Visibility (% days):												
misty (100 to 400 m)	0	6	4	1	0	1	0	0	4	3	0	3
mist in valleys	14	18	13	9	15	14	16	23	6	0	3	7
mist on mountain tops	12	12	7	11	12	18	18	0	8	19	10	9
No. Records (days)	93	83	56	90	68	90	49	43	49	59	59	67
MENDI (Provincial H.Q.) 09.00 Observations:												
Mean daily cloud (oktas)	6	6	6	5	5	5	5	5	5	5	5	6
15.00 Observations:												
Mean daily cloud (oktas)	7	7	7	7	7	7	7	7	7	7	7	7
No. Records (yrs)	13	13	13	13	13	13	13	13	13	13	13	13

Sources: Haelaelinja, observations made in the field; Mendi H.Q. records from McAlpine *et al.* 1975:159

Plate 3.3 Wood smoke seeps through the thatch of a men's house *enza* to drift into the sky and become cloud.

Plate 3.4 Cloud fills valleys in the early morning, "feeding" on the night's fire smoke.

ebenjip might be marginally more cloudy than in *bulhenjip*, which accords with fine *baen* spells occurring more frequently then. But in all months some days are completely overcast (8 oktas), and the Mendi data suggest that by 15.00 hours the sky is more-or-less fully clouded over almost every day of the year. These observations concur with those for the Melanesian region as a whole (if we allow for the orographic influences of the central highlands, the local effects of which, superimposed on the broad-scale regional circulation, result in persistently heavy cloud cover (McAlpine *et al.* 1983:120–23)). Satellite imagery shows the S.W. Pacific covered by cloud for most of the year, with maximum cloudiness at the beginning of the year.

The Wola distinguish between different types of cloud. Low cumulus cloud and mist they call *muwlol* or *nonow*, and high stratus and cirrus cloud they refer to as *sowhat*. White round cumulus cloud in an otherwise blue sky on a fine day they call *soltukay*. A completely overcast sky of grey, perhaps rain laden clouds is *saykay* or *suwsaway*, which translates as "ground mumu" (the land covered by cloud like an earth oven); these terms also refer to chilly weather. Heavy black storm clouds from which rain is anticipated to fall soon are *muwlol chay maenmaen bay*, which refers to "clouds readying themselves to rain". When cloud comes down to earth and forms mist on mountain tops and hillsides, or fills valleys in the early morning, it is *muwlol mungum bay*, cloud that "hides and covers" everything up.

The monthly morning occurrence of different cloud types shows little pattern. The cirrus and stratus clouds of finer weather occur perhaps a little more often in the middle of the year, corresponding with the *baen* fine spells of the *bulhenjip* season. And the presence of nimbo-stratus storm clouds in the morning, which accompany thunderstorms, are restricted to the *ebenjip* months, which accords with Wola observations that thunder is more likely in this season. Mist too is frequent throughout the year. The cool mountain air, high atmospheric moisture and nocturnal heat radiation all contribute to early morning mist forming in the valleys. And rainy spells can be accompanied by a lowering of the cloud ceiling to envelop peaks and slopes and reduce visibility at any time of day.

According to the Wola, cloud is the accumulation of smoke from their earthly fires; suggestive perhaps of our association of smog with the burning of fossil fuels (without attendant worries over the possibly deleterious effects of excessive atmospheric pollution). In the early mornings, when the cloud is in the valleys, they say that it has come down to earth to "feed" itself on the smoke of the night's fires, to build itself up before returning aloft for the day. While we may not agree with this theory of cloud formation, the gaseous elements put into the atmosphere by their fires are not to be dismissed, playing a part in the cycling of nutrients through the Highland ecosystem (see Chapter 9). The Wola do not associate cloud formation with other meteorological processes, having no notion of the ascent and cooling of masses of damp air, expanding as pressure falls with their rise through convection, turbulence, frontal activity or topography. While they comment on the build up of cloud before rainfall, and associate its accumulation with rain, they do not seem to think that this has anything to do with the sky woman Sabkabyinten. And no one is clear where cloud goes on fine days when the sky is clear, although some surmise that the wind blows their smoke accumulation to and fro, away elsewhere and then back again.

Wind

The seasonal movement of the intertropical convergence zone (ITCZ) dictates the regional pattern of air circulation over New Guinea, such that north-westerly winds characterise December to March and bring more rain than the south-east trades which dominate from May to October, giving rise to generally drier conditions (McAlpine *et al.* 1983:15–29). Local data on wind direction and force only weakly reflect these air movements (Figure 3.2), topography modifing the effect of regional air flows on the local climate. The ranges and valleys of the mountainous Wola region markedly influence air movements, the orientation of the folded terrain having a conspicuous effect on air flow, generating and directing winds along the grain of the country (Map 3.1). The wind roses clearly reflect the influence of topography, with the prevailing winds at Haelaelinja aligned along the curving Was river valley, or coming over through a gap from the neighbouring Nembi valley.

In the night, 'mountain winds' develop due to the funnelling of cool air down the valleys (the night cooled dense air of higher altitudes flowing towards lower elevations under gravity). In the day, reverse 'valley winds' supercede them (the solar heated air at lower elevations becoming unstable during the day and rising, flowing up slope). The warm up-valley afternoon airstreams commonly bring precipitation with them, accounting in part for the lack of seasonality in rainfall. Other terrain induced moderations of air flow include associated local katabatic-anabatic wind reversals up and down high mountain slopes (such as those of Mount Giluwe, east of Mendi), and to some extent föhn winds produced when large-scale atmospheric flows force air up across mountain ranges in short periods of time, the air cooling on its ascent of the windward side, causing some precipitation, and warming as it descends the other, giving a dry wind (see McAlpine *et al.* 1983:44–47).

The Wola have no apparent explanation for wind; where it comes from, is they say, just one of those mysteries with which life abounds. They refer to a light wind or breeze as *popo*, and a strong wind as *posaebsuw* or *popo onda*. Gale force winds are uncommon, and winds rarely, if ever, reach destructive speeds. People maintain that the wind may blow from any direction the year round, no direction predominating in either season. This accords with the constant effect of topography on air movement. They also say that strong winds only occur in *ebenjip*. Both claims agree with data collected: stronger winds are more likely between October and March, and it is difficult to see any seasonal patterning to direction, although westerlies (NW through to SW), which contravene the topographic control of wind direction, are perhaps

Ethnometeorology: The Climate

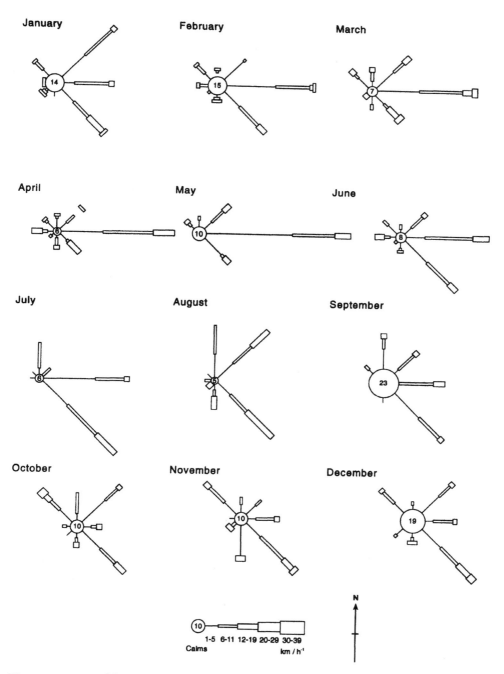

Figure 3.2 Monthly wind roses (direction and speed) for Haelaelinja (*wind force estimated using Beaufort scale*).

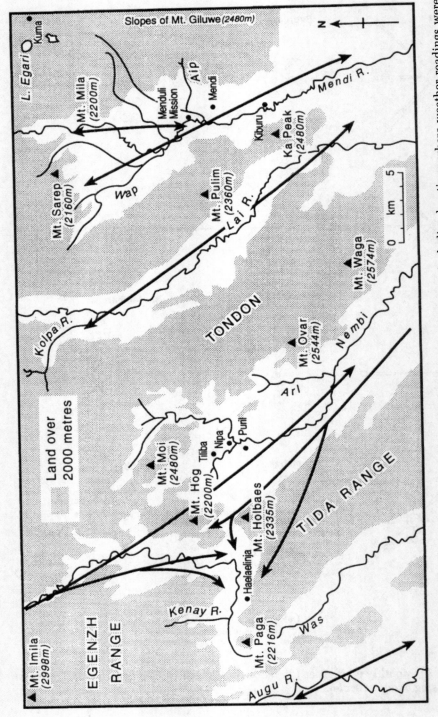

Map 3.1 The effect of topography on wind direction across the Wola regions including locations where weather readings were made

more likely in *ebenjip*, reflecting the influence of ITCZ movements at this time of the year on air circulation in the upper atmosphere.

A noteworthy feature of the annual pattern of equatorial, trade and monsoon winds over the S.W. Pacific region, including circulation in the upper atmosphere, is again that the two "seasons" distinguished by climatologists correspond closely to the seasons discerned by the Wola (Fitzpatrick 1965:56; McAlpine *et al.* 1983:23–29). The "south-east season" of S.E. trades runs from May to October, correlating with *bulhenjip*, and the "north-west season" of equatorial westerlies runs from December to March, corresponding to *ebenjip*, with April and November as transition "doldrum" months. The variation in pressure due to atmospheric circulation is small throughout the year.

Temperature and Humidity

Over the year, mean daily temperature is around 18°C, with the mean maxima between 20° and 25° and mean minima between 20° and 15°.[5] The diurnal variation is of the order of 10°C, and the difference between the hottest and coldest months about 2°C. This mild temperature regime shows relatively little seasonality (Figure 3.3), although there is a slight trend for marginally lower temperatures in the middle of the year during the *bulhenjip* season, equating with the increased incidence of both chilly *chaykund* overcast days at this time of year and warmer *baen* fine days (when temperatures fall to lower levels during the associated clear nights). Regarding temperature variation, altitude is far more significant than season. Higher locations have noticeably lower temperatures than lower ones.[6] The mountainous and irregular terrain of the Wola region results in a range of temperature regimes, with its considerable altitudinal variation (people cultivating gardens regularly between 1600 m. and 2400 m. — the upper limit for sweet potato). But the averages quoted are a fair indication of the temperature range over the greater part of the populated area.

Besides the temperature of the air, the temperature of the soil is important regarding agricultural activity, notably crop growth. Information on soil temperature is unfortunately limited. The World Bank funded Agricultural Field Trials, Studies, Extension and Monitoring Unit (AFTSEMU) stationed in the Southern Highlands recorded temperatures for eight months at 2.5 cm. depth in a peaty soil (Typic Tropohemist) at their experimental farm at Kuma on the Kombie swamp north of Mendi (Radcliffe 1985a:8–9). They found the median soil temperature to be around 19°C–20°C with a diurnal variation of 3.5°C and a monthly variation of 1.5°C–2°C between the hottest and coldest

---

[5] The readings for Haelaelinja and Kiburu were not made in Stephenson screens and need to be interpreted cautiously.

[6] According to McAlpine *et al.*'s 1983:95 analysis of temperature against altitude, annual maximum temperatures decrease at a rate of 0.672°C per 100 m. increase in altitude and annual minimum temperatures decrease by 0.535°C over the same altitudinal range.

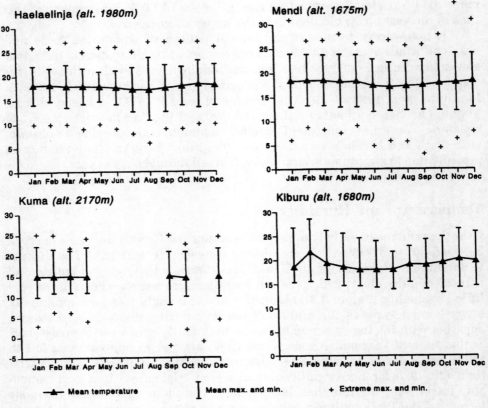

Figure 3.3 Mean, maximum and minimum monthly temperatures (°C).

months. This pattern predictably compares with air temperatures, while exhibiting less variation. The maximum values (range 20.2°C to 22.2°C) are similar to maximum air temperatures, while the minimum ones (range 16.7°C to 18.2°C) are considerably higher than minimum air temperatures. If we assume that surface soil temperatures parallel air temperatures in a similar manner across the region, with altitude the dominating control factor, and also presuppose that air-soil temperature relationships of a peat reflect those of other soil types occurring in the region, we can characterise the soil temperature regime as isothemic according to U.S.D.A. (1975) criteria.

Humidity is likewise related to both air temperature and altitude, which affect solar evaporation and plant transpiration, the pathways by which water enters the atmosphere. The island of New Guinea has a humid climate with its equatorial location, surrounded by oceans; dew point temperatures elevated by warm tropical air, it enjoys high atmospheric humidity levels throughout the year. Vapour pressure decreases with altitude. Seasonal variation in mean monthly relative humidity is slight with no discernible pattern (Table 3.3). Nevertheless, it is perhaps possible to detect an inverse relation-

Table 3.3
Humidity data for two locations: mean monthly wet and dry bulb temperatures (°C at 09.00 hrs, 15.00 hrs.), and relative humidity (%)

	Jan	Feb	Mar	Apr	May	Jun	Jul	Aug	Sep	Oct	Nov	Dec	Mean
HAELAELINJA													
09.00 Reading:													
Dry bulb (°C)	17.9	17.7	18.4	18.5	18.1	17.4	16.9	18.8	18.6	18.9	19.0	18.4	18.2
Wet bulb (°C)	15.9	16.0	16.2	16.4	16.2	15.6	15.2	16.3	15.9	15.9	16.3	16.3	16.0
Relative Humidity %	82	84	78	81	83	83	83	77	76	74	76	80	80
No. Records	93	83	56	90	68	90	49	43	48	59	59	67	
MENDI (Prov H.Q.)													
09.00 Reading:													
Dry bulb (°C)	18.2	17.9	18.1	18.2	18.3	17.1	16.6	16.8	17.7	18.4	18.8	18.4	17.9
Wet bulb (°C)	15.8	15.8	15.9	15.9	15.9	14.8	14.7	14.9	15.3	15.4	15.7	15.8	15.5
Relative Humidity %	78	82	80	80	78	79	83	83	78	73	74	76	79
15.00 Reading:													
Dry bulb (°C)	21.5	20.9	21.2	20.8	21.4	20.4	19.6	19.4	19.9	20.2	20.8	21.2	20.6
Wet bulb (°C)	17.9	17.9	17.9	17.9	18.4	17.6	17.1	17.1	16.9	16.9	17.3	17.7	17.6
Relative Humidity %	71	75	73	76	75	76	79	80	79	73	72	72	75
No. Years	13	13	13	13	13	13	13	13	13	13	13	13	

Sources: Haelaelinja, readings made in the field; Mendi H.Q. records from McAlpine *et al.* 1975: 138, 144.

ship between humidity and air temperature, as noted elsewhere in Papua New Guinea (McAlpine *et al.* 1983:106–107). The annual cycle of relative humidity tends to show a peak during April to August, corresponding to *bulhenjip* when air temperatures are at their lowest, and a trough between September and March, equating to *ebenjip* when monthly temperatures reach their highest levels. But the variation is small, beyond detection by the human senses.

Daily variation in humidity is more noticeable, although still small — diurnal variations being greater in the highlands with their drier atmospheric conditions, than elsewhere. Relative humidity reaches a daily peak at dawn, when temperatures are lowest and air close to saturation. It decreases steadily as temperatures increase through the morning, reaching its lowest level in the early afternoon when air temperatures are at a maximum. It rises again through the evening and night as temperatures fall. The amplitude of the daily humidity cycle increases with altitude; physiographic position (valley floor, slope or mountain ridge etc.) is also significant in determining its magnitude (McAlpine *et al.* 1983:108–11).

The frequent occurrence of mists and the generally heavy cloud cover over the central highlands result in relatively low amounts of sunshine for their equatorial position. Yet sunshine shows a predictable inverse seasonal relationship with rainfall, lower hours of sunshine occurring through *ebenjip* or the north-west season, when rainfall is higher. The Wola refer to the sun shining as *nat day* (lit. sun burns). They maintain that they have no myth relating to the sun's origin, why it moves in the sky and so on. They distinguish between different kinds of sunshine. The soft warm sun of the early morning they call *nat turiybiy* (lit. sun pleasurable), and the intenser hot sun of midday which can be painful, they call *nat taendabiy* (lit. sun hurts). They judge the time of day by both the strength of the sun and its position in the sky. On days when the sun becomes particularly intense they say that it will be short-lived, this is the kind of sunshine that precedes heavy rain, and is called *chay nat* (lit. rain sun). On the other hand, if the sun remains pleasantly warm throughout the morning it heralds a fine day, even a spell of clear weather and is called *baen nat* (lit. fine-weather sun). Another term for an interval of warm, sunny weather is *turuwpiy tomb* (lit. warm time); the antonym for cooler, overcast weather is, *huwpiy tomb* (lit. cold time) or *kuwmb tomb* (lit. shadow time).

In addition to their somewhat hit-or-miss forecast of sunshine or rain according to the feel of the sun and the season of the year, the Wola look out for other intimations of fine weather. A favourite is the condition of the flowers of the ubiquitous *gaimb* sword grass (*Miscanthus floridulus*), which abounds in their region. When the tassles are emerging the weather is variable, sometimes wet, but when they dry out and their fluffy windborne seeds are released and blown about, people say that the weather turns fine with a *baen* period and plenty of sunshine. While they acknowledge that cloudiness determines sunniness, no one has ever suggested that they could perhaps induce fine weather by burning fewer fires and reducing their cloud-forming smoke emissions, although sometimes individuals try to achieve the opposite and bring rain. Another sign of a period of fine rainless weather is heat haze filling the valleys, which the Wola insist is wispy wood smoke refusing to rise up into the sky. The thin mist-like cloud obscures distant vision. It is misplaced cloud-smoke, especially common in the mornings, and heralds hot sunny days, until it drifts off to its proper place in the sky.

The solar radiation received at any location is affected by topography (slope, aspect etc.) and cloud cover, east-facing slopes receiving more radiation than west-facing slopes because clear conditions occur more often in the mornings (Chang 1968; Clarke 1971); this may account in part for differences in soil temperatures (Smith 1975), and it is significant regarding the siting of Wola gardens. Seasonal variations in solar radiation are due to differences in cloud cover, for there is little variation in extraterrestial radiation over Papua

Table 3.4
Solar radiation data: mean daily total radiation (MJ/m^2) (after AFTSEMU data)

Jan*	Feb*	Mar*	Apr*	May	Jun	Jul	Aug	Sep	Oct	Nov	Dec*
18.7	18.9	16.6	15.0	14.6	12.5	NR	NR	15.3	14.9	NR	17.7

*Average of two year's readings, all other months are one year's readings only (NR = no readings)

new Guinea as a whole in the course of a year (Kalma 1972). Again data are limited for the Southern Highlands, but AFTSEMU made readings of solar radiation for fourteen months at Kuma (Table 3.4), which we may take as representative of the region (Radcliffe 1985a:12). The trend is for total daily radiation to reach a maximum in January to March and a minimum in June. In the middle of the year, during *bulhenjip*, day long drizzly rain and overcast conditions are more likely, reducing radiation, whereas for the remainder of the year the weather pattern is one of clear days with high intensity radiation, or more often clear mornings followed by a build up of cloud in the afternoons, probably with rain.

A prolonged period of dry *baen* weather is as undesirable as an overly long spell of wet *chaykund* weather, both having a deleterious effect on crop growth and agricultural activity. In dry conditions plants may suffer water stress and yields fall, run-away bush fires are a considerable danger when burning off and frosts are a threat. The Wola have an unusual practice to promote rain, analogous to the burning of some old netting to promote fine weather. In dry times, a person may take some *woluwmsaeren* plants (*Elatostema* sp.), which grow in damp locales and have a copious wet slimy sap, and place them on a stone and squash them by beating them. There is no other action nor associated spell. Why the practice should promote rain is unclear, and some people I spoke to pooh-poohed the idea. The rain seems to be encouraged by a process of association, the squashed plant covering the stone with slime so that it looks dripping wet as if there has been rain. The practice is more an allusory expression of desired events than thought an efficacious act. A more effective practice is thought to be the controlled burn off of vegetation cut down and drying out on newly cleared garden sites. The smoke bellows into the sky to form new clouds from which rain might fall in the near future. And the ashes which result from the burn are among the driest of things imaginable, and spread on the soil they give it a desiccated aspect, as if there has been an extended sunny spell thoroughly drying up the earth. When Sabkabyinten looks down, people say that she sees how inordinately dry it is, and this may prompt her to beat her celestial pool and send rain.

Frost

The greatest danger from a fine *baen* period is not so much drought seriously affecting crops, because rainless periods are usually too short, but frost (Brown & Powell 1974; Brookfield & Allen 1989). If there is no cloud build-up and a clear night follows a sunny day, temperatures may fall to uncomfortably low levels, even on occasions low enough to cause radiation frost (at 4–5°C), when the atmospheric "window" is open to the transmission of long-wave radiation. The Wola associate frost, which they refer to as *pibiyhond tingilay* (lit. dew-frost covers), with fine *baen* spells. It comes or "hits" they say, during cold nights that follow sunny, cloudless days. Conversely, they think a cold night indicates a fine day, and forecast such. Frosts occur during the *bulhenjip* season. When frost strikes hard it kills off above ground vegetation, leaves curling and dropping off, foliage scorched and dying back, the drying and shrivelling up the Wola call *gaip say* (lit. dry become). Crop damage varies from slight to catastrophic depending on the length and severity of the cold period and the siting of gardens. The consequences can be dire.

Again altitude is a significant consideration, as the Wola acknowledge. The risk of frost increases with height above sea-level and land above 2000 m. is particularly susceptible. Topography and vegetation cover also influence the likelihood of frost, gardens on steep slopes or near forest are less susceptible than those at the foot of slopes or in grassland (Waddell 1972:133–134). The Wola are well aware of this; those living in the Haelaelinja area, for example, point out that frost rarely affects them seriously because they are obliged to cultivate on steep land adjacent to extensive forest. The vulnerability of flatter or grassy or higher places to frost has implications regarding the location of gardens. The gentler the slope, the better the soil on the whole, so long as drainage is not impeded to produce a gley; the risk of serious erosion and topsoil loss is also less (see Chapter 6). But people have to balance these benefits against the increased risk of frost damage. The canny Wola farmer tends to establish gardens at several different topographical locales, exploiting a range of environments and insuring against loss at any one; from frost, for example, on a highly fertile level site.[7]

Nevertheless some communities occupy territories with a higher proportion of flatter or higher, frost vulnerable land than others (for example the place Paym occupies a small basin-like feature in the Was valley which is a known frost pocket — Map 5.2). They are more susceptible to frost losses. If they are engaged in hostilities with a nearby community that suffers less damage because it is at a lower elevation or farms steeper land, their enemies celebrate their misfortune. They may decorate themselves and dance and sing, infuriating and taunting their foes. To celebrate others' misfortune in this manner is called *liywakay*.[8] The phrases of one such frost celebrating refrain go as follows:

hael	*waembow*	*yor*[9]	*layor*	*pibiyhond*
pandan variety[10]	pandan variety	leaves	hit	frost

---

[7] Waddell 1972:152–59; 1975 argues convincingly that the agricultural practice of mounding soil for sweet potato cultivation also reduces the frost hazard.
[8] The Wola also perform *liywakay* when an enemy dies from arrow wounds or sorcery, to celebrate the death; in the event of frost they are performing it to celebrate misfortune and anticipated future hardship.
[9] This is an alteration of the word for leaves *shor*, to rhyme.
[10] These pandan varieties are of the cultivated *Pandanus julianettii*.

gaimb	hael	ay	yor	pibiyhond
sword grass[11]	pandan variety	rhyming word	leaves	frost

The occurrence of frosts relates to the aforementioned regional pressure system with May to October the principal risk period when inflows of dry air into the Highlands lead to large losses of radiated heat on clear nights (Brown and Powell 1974). Conditions sufficient to induce severe and prolonged frosts, with associated drought conditions, enough to scorch crop foliage seriously and lead to food shortages, occur about once every decade (McAlpine et al. 1983:98–101). Nonetheless severe frosts sufficient to destroy the staple sweet potato crop are only likely to occur at higher elevations, below 2000 m. less devastating mild frosts are more probable. Only a small part of the region occupied by the Wola is at high risk, notably parts of the upper Mendi valley and the high watershed between the Lai and Was/Nembi valleys.[12] But their position is not so dire as it appears, some of the people cultivating at these high altitudes having other gardens lower down, to which they can retreat.

The Weather and Cultivation

While frosts may seriously disrupt crop production, substantial ones are mercifully uncommon and yields rarely badly affected. Marked variations in the climate, seasonal or otherwise, of which frost is a dramatic instance, are infrequent. The same weather prevails largely throughout the year, any seasonal variations exert little influence over subsistence activities, notably crop cultivation. The planting and harvesting of crops follows a constant and regular daily pattern, unless there is an adverse perturbation. The Wola maintain however that crops planted during *ebenjip* tend to grow better and faster, especially beans, cucurbits, sugar cane and greens, and men intending to establish new gardens may aim to clear and fence them during *bulhenjip* ready for planting the next *ebenjip*, but this is only an expressed preference not a rule, and men can be found clearing gardens at any time of the year, just as women can be found heaping and planting sweet potato mounds (compare Meggitt 1958; Crittenden 1982).

The weather, whatever the season, can disturb the everyday agricultural rhythm. An extended wet period during *ebenjip* or *bulhenjip* can inhibit the burning off of newly cleared areas, and hinder the planting of crops in all gardens. When the weather clears up, a burst of activity might follow as a backlog of plots are prepared for cultivation and much planting occurs. This increased activity may occur as one season changes to the other, reflecting

---

[11] *Miscanthus floridulus*.
[12] See Radcliffe 1985a:18–19 for some data on frost occurrence at one of these locations, Kuma in the upper Mendi valley.

their sometimes differing weather patterns, and have a seasonal appearance. But strictly speaking there is scant seasonal patterning to cultivation. Variations in the usual pattern of weather can also markedly influence the growth and yield of crops. The sweet potato is particularly susceptible to changes in the weather, especially extreme fluctuations away from the conditions it favours, either too wet, too dry or too cold.

Prolonged periods of either rainy *chaykund* or sunny *baen* weather can adversely affect crop cultivation and yields, whatever the season (Roberts 1982). During overly wet weather not only are people prevented from attending to normal cultivation activities, but also some crops, notably sweet potato, grow less well. The inclement weather physically inhibits both men and women; burning off damp vegetation is obviously not feasible nor is breaking up sticky wet soil. People fear puddling the soil at field capacity by clearing, fencing or planting it, markedly reducing its cultivation potential. Indeed they are chary of walking across gardens at all. The yields of crops that dislike wet conditions decline as the soil becomes increasingly saturated, even anaerobic, although the relatively steep, well drained slopes farmed and use of mounds reduces the chances that extreme soil conditions will develop (Waddell 1972). The slowdown in growth is not immediately apparent, its affects on yields becoming evident some weeks later when it is noticeable that tubers for instance, have not set nor expanded as usual. Similarly, the interruption to everyday agricultural work does not show up until sometime later in the cultivation cycle, when yields fall due to decreased planting some months previously. The effects of inclement weather consequently can be felt as food shortages sometime after the event, giving rise to what seem inexplicable hungry times.

The occurrence of apparently climatically unaccountable food shortages has attracted some attention. According to Bourke (1988, 1989), there are parallels between sweet potato tuber fluctuations and the infamous 'hog-price cycle'. He argues that periods of sweet potato shortage prompt higher than normal planting rates, and that the resulting increases in tuber supplies, when they arrive some months later, then lead to a fall in planting rates below average, giving rise to tuber shortages sometime later which have no direct climatic cause.[13] This proposal seems to underestimate the importance of fluctuations in pig herd size in influencing planting rates (Rappaport 1968), and also the shrewdness of local farmers, particularly women, in assessing planting rates relative to their households' projected needs.

Although high rainfall is usual throughout the year in the Southern Highlands, periods of water shortage sometimes occur too, and adversely affect crop yields. When extended dry conditions prevail, crops suffer water stress, and yields decline as plants reduce transpiration to conserve water.

---

[13] Just as pig farmers in an unregulated market expand and contract their pork production, the ensuing meat gluts and scarcities ensuring the perpetual up-and-down motion of prices that prompt them to vary production.

The Wola explain that the sun dries the soil out and bakes it hard so that roots, particularly of tuberiferous crops, cannot penetrate it easily and grow. While gardening activities are less disrupted than in extended wet periods, burning off can result in fires raging out of control and destroying houses and gardens with subsequent claims for damages, and women's planting activities may be slowed down because the hardened soil is more difficult to break up and work. Sometimes rainfall remains below mean values for some weeks, giving rise to "drought" conditions; during the extended dry *baen* interval that occurred between July and November 1982 the rainfall at Mendi was only 27% of the median values and all monthly totals were below 100 mm (Radcliffe 1986:12). These infrequent "drought" conditions usually occur because of a regional influx of dry air from the high pressure zone to the south of Papua New Guinea. They may be accompanied by frosts, seriously reducing yields in a short space of time, whereas dry or wet spells alone result in more minor, sometimes delayed food shortages.

It is possible to make a crude estimate of the number of consecutive rainless days crops might endure without suffering moisture stress and a check on growth and yield, by taking the readily available water content of soils (defined as water held at tensions between 0.1 bar and 1.0 bar) to depths of 30 and 100 centimetres and comparing them with average daily evapotranspiration water losses (estimated from daily evaporation readings using a conversion factor — after Radcliffe 1986:90–92). Shallow crops, rooting to 30 centimetres, including those with roots confined to mounds, would deplete readily available water reserves on soils derived largely from volcanic ash after about 10 days and on soils derived largely from sedimentary material within about 8 days. Deeper rooting crops to 100 centimetres would not be affected for sometime longer, between 15 to 25 days on both soils. Rainless periods of eight to ten consecutive days are rare occurrences in the Wola region; these calculations indicate that crops are only likely to suffer serious moisture stress in the periodic regional droughts that occur perhaps once a decade or so.

Periods of either extreme wet or dry weather, and induced food shortages, are unpredictable in occurrence, they do not necessarily happen at certain times of the year and not others. The severe "droughts", with accompanying devastating frosts, seem to occur about *bulhenjip*, around once every ten years (McAlpine *et al.* 1983:100; Allen *et al.* 1989), records existing for droughts in 1941, 1950, 1962, 1972 and 1982.[14] The Wola themselves associate famine conditions with the *bulhenjip* season when overcast and damp *suwsaway* misty and *chaykund* drizzly weather are more common. They say that conditions remain wetter at this time and crops do not grow so well; they point to those trees which shed their dead and dry leaves as a sign of no growth. They also maintain that insect pests like *ed* caterpillars (*Pieris* sp.) are

---

[14] No frost has occurred in the western Wola region since 1972, and the frost in that year was of limited occurrence, being restricted to notorious frost hollows.

more devastating under these conditions. The unpredictability of adverse weather and ensuing food shortages is caught in Wola folklore which associates them with the random occurrence of earth tremors. According to one of their sayings, when an earth tremor occurs in the afternoon there will be plenty of sweet potato and no hunger, but if one occurs in the morning a shortage of tubers will occur and a hungry time. They also tell children not to touch the grass called *dayngeltay* (*Agrostis avenacea*) or else they will suffer a famine period; the name of the grass meaning 'hungry-time-origin'.

CHAPTER 4

COPING WITH CLIMATIC VARIATIONS: THE SPECTRE OF FAMINE

Hungry times in the New Guinea highlands vary from brief periods of hardship following prolonged wet or dry spells when food, notably sweet potato, is in short supply, to thankfully infrequent times of great suffering and distress when people may die of starvation, after extended *baen* "droughts" probably featuring severe frosts. The Wola refer to any time of food shortage and hunger as *day ebay* (lit. hunger comes) or *dayngel* (lit. hungry-time). The last time people died of starvation was in the late 1920s or early 1930s, when there was an extremely bad "drought" and frosts according to those who can remember it, which reduced them to collecting any wild famine foods they could find like *bawiy* yams, *horon* pueraria, and *mokombez* cane grass infloresences (Table 4.1).

It is probable that adverse climatic perturbations have characterised the highlands of New Guinea for millenia. But their impact on human populations may have varied, particularly with prehistoric changes in agricultural regimes, notably that postulated with the arrival of sweet potato (Golson & Gardner 1990; Bayliss-Smith & Golson 1992). While the nature of famine and malnutrition may have changed post-sweet potato, and subsequent conjectured population growth have exacerbated them (Crittenden 1982), speculation on these long-term changes bears only tangently on the contemporary ethnographic theme of this book. Except in one regard. It is arguable from an eco-archaeological viewpoint that periodic major famines could have contributed to the sustainability of the current day agricultural system by ruthlessly culling human and pig numbers, keeping them in check such that a population explosion of a magnitude sufficient to destabilise the system was unlikely ever to occur. While demographic data are unavailable, anecdotal evidence from older people and oral history accounts suggest that starvation to death was certainly a feature of major "droughts" and frosts.

I. HUNGRY TIMES

There are degrees of privation, from food shortages to outright famines. Hunger is a relative matter for the Wola, as is the weather's ups-and-downs. There is not having enough to eat of their staple tuber, and there is not having enough to eat of any food whatsoever.[1] The response of crops to weather

---

[1] Failure to appreciate this distinction resulted in an over-response to the 1972 "drought" by the then Australian colonial authorities, which supplied tinned fish and rice to communities where the threat of famine, as opposed to sweet potato hunger, was not serious; people received so much that they even fed some of the rice to their hungry pigs (see Allen & Brookfield 1989).

Table 4.1
List of plant foods resorted to in times of famine

Wola Name	Botanical Identification	Parts Eaten	Wola Name	Botanical Identification	Parts Eaten
IYSH (trees):			**EBEL (bananas):**		
Haiyow	*Ficus pungens*	new leaf shoots	Kat	*Musa sp.*	pseudostem heart
Haiyowma	*Ficus porphyrocaete*	new leaves			
Iyshgemb	unidentified	fruit	**YA (vines/climbers):**		
Kaeriyl	*Lithocarpus rufovillosus*	fruit	Bawiy	*Dioscorea sp.*	tuberous roots
Mbuwp	*Artocarpus vriesianus*	drupe kernal	Dinbuwm	*Mucuna schlecteri*	fruit
Pay	*Castanopsis acuminatissima*	fruit	Hezaembuwl	*Rubus moluccanus*	fruit
Poiz	*Ficus wassa*	young leaves & fruits	Homat	*Melothria belensis*	fruit & roots
Shongom	*Elaeocarpus leucanthus*	fruit	Horon	*Pueraria lobata*	tuberous roots
Shuwat	*Ficus dammaropis*	new leaves & fruits	Huwlhaeruwk	*Mucuna tomentosa*	fruit
Shwimbset	*Elaeocarpus ptilanthus*	fruit	Tat	*Trichosanthes pulleana*	fruit
			Tombel	*Stenomeris dioscoriifolia* (?)	leaves
HENK (tree ferns):			Waenuwkuwnguwp	*Aristolochia cf. engleriana*	fruit
Bobaiyow	*Diplazium archboldii*	)			
Hongok	*Cyathea magna*	)	**GAIMB (cane grass):**		
Kabiyp	*Dicksonia grandis*	)	Mokombes	*Saccharum robustum*	immature inflorescence.
Kilakila	*Cyathea aff.nottofagorum*	)			
Lorwalorwa	*Cyathea aff. macgillavrayi*	) tender	**DEN (grasses/herbs):**		
Meshmesh	*Cyclosorus aff.archboldii*	) fronds	Deraennomoniyl	*Rubus ferdinandi*	fruit
Shumbuwhond	*Polystich sp.* (?)	)	Dikiytagot	*Setaria palmifolia*	stem heart
Taendbiyaib	*Cyathea hunsteiniana*	)	Loliy	*Physalis peruviana*	fruit
Tombogaim	*Diplazium latilobum*	)	Mbolinomoniyl	*Rubus niveus*	fruit
Wem	*Diplazium dilatatum*	)	Hombiyhaem	*Commelina diffusa*	leaves
Wolhenk	*Cyathea pilulifera*	)	Honmunk	*Inpatiens sp.*	fruit
Yaegarom	*Kypolipis sp,*	)	Mamuwn	*Rubus rosifolius*	fruit
			Mondkaend	Uriticaceae	leaves
GOIZ (palms):			Oluwng	*Cardamine sp.*	leaves
Goiz	*Gulubia sp.*	)	Omok	*Dicliptera papuana*	leaves
May	*Heterospathe aff. muelleriana*	) terminal	Suwtaguwt	*Solanum americanum*	leaves
Mbet	*Areca aff. macrocalyx*	) heart	Taziy	*Oenanthe javanica*	leaves
Zin	*Heterospathe elegans*	)	Torwatorwa	*Desmodium repandum*	leaves & fruit

conditions that are out of the ordinary varies. On the one hand, longer than average spells of sunny or rainy weather may adversely affect sweet potato yields more than most other crops, while on the other, extreme though rare conditions of very low night temperatures and frosts may devastate nearly all crops. All plants are touched to some extent in frosts, and occasionally devastated, whereas in times of excessive rain or sun only some crops are seriously affected. During the 1982 "drought" sweet potato was seriously affected, although continuing to yield after a fashion, but several other crops were not too badly affected, if at all; cabbages thrived in the conditions, presumably due to vernalisation with the low night temperatures, and were in surplus supply. Although these other crops continued to yield fairly well, the Wola insisted that they were hungry. Sweet potato is of over-riding importance in their diet (Sillitoe 1983), and any fall in its yield spells hunger, even if certain other crops are still available.

The Wola response to hunger is further unanticipated in some respects, with regard to pigs. While they show the stoic resignation of people familiar with periodic empty bellies, called *tombow nae hae* (lit. stomach not stands), they do not cull their pigs for meat. When asked if they would kill pigs for food during a "drought", men looked surprised and asked why they should slaughter animals for "nothing", by which they meant not in fulfilment of exchange obligations. Sweet potato is the staple of humans and pigs alike, and while men may not slaughter pigs for food when it is in short supply and they are in direct competition for what there is, they do stop feeding them, with the same results under extreme conditions, animals dying of starvation. During the 1982 "drought" several pigs died for lack of fodder and their owners butchered them and shared out the pork in formal or informal exchange events or, a recent practice, sold it in *bisnis* (Pidgin 'business') transactions for cash, so long as the animal showed no signs of disease or "bad blood" (if an animal's blood is thick and turgid when butchered it has "no blood" and is thought to have died of disease). In the catastrophic "drought" of the 1930s people say the pigs died first, men losing their entire herds, and then humans started to die for want of food.

Hunger is culturally as well as biologically defined. It is both a psychological and physiological condition. It starts for the Wola with a shortage of sweet potato, whatever happens to the yields of other crops. And sweet potato seems particularly susceptible to climatic perturbations. Sometimes adverse weather causes both the vines and tubers of sweet potato to grow poorly, or not at all. Other times sweet potato vines grow prolifically with much foliage above ground at the expense of the tubers which remain small and stringy. According to the Wola, sweet potato tubers are small when vegetative growth is prolific because the dense foliage shades the soil and prevents it warming up, encouraging tuber growth; the soil remains too wet and "strong" (i.e. aggregated) for sweet potato tubers, which prefer drier and more friable soil conditions.

Climate and Sweet Potato

Periods of hunger among the Wola, whether serious and life-threatening or less hazardous and discomfiting, depend critically on the vulnerability of their staple to certain climatic variations. There are three aspects of the weather which particularly affect sweet potato growth and yield: 1. sunshine, 2. temperature, and 3. rainfall. In general, sweet potato grows best under high light intensity, high temperature, and fairly high rainfall conditions, provided drainage is good.

The sweet potato is a sun-loving crop, growing best where light intensity is relatively high. When exposed to increasing light intensities, sweet potato leaves, exhibiting typical kinetic responses, reach saturation photosynthesis at about 140 J m^{-2} s^{-1}. The solar radiation received daily in the Southern Highlands is twice or more this level in all months (Table 3.4). And the sweet potato is particularly suited to the sunny morning/overcast afternoon weather pattern that is common to the region, having its highest rates of photosynthesis between 9am and 1pm, decreasing gradually thereafter up to 5pm by about five-sixths. But it would be incorrect to suppose that because sunlight is at or above optimal levels for a goodly part of most days in the Wola region (except for particularly heavily overcast and misty days), that sweet potato plants are fixing the maximum atmospheric carbon of which they are physiologically capable. The upper leaves in the canopy may be light saturated but not lower ones, mutual shading reducing the intensity of light reaching them. The plant's creeping habit and leaf arrangement results in poor light penetration, which is good for smothering out competing weeds, but there comes a point when more leaves, increasing leaf area index, result in a decrease in net assimilation rates. This relates to the balance between foliage and tuber growth, which recurs as a central theme in maximising sweet potato yield (see Chapter 2). Light also affects tuberous root development directly. The low light intensity of shaded conditions delays tuber differentiation and development, decreasing both cambial activity and stele cell lignification. But this has to be set against the promotion of tuber formation over vine growth that occurs under low intensity light. The part photoperiod plays, if any, in sweet potato energy transactions is unclear, some report that long days stimulate storage root formation, others short days (Hahn 1977; Kays 1985).

While the relationship between solar radiation and root size is positive, yield increasing with higher light intensities, it is complicated in the field by the interaction of other environmental factors, notably temperature and rainfall (Sajjapongse & Wu 1989). It is fluctuations in temperature and rainfall that have a more direct impact on sweet potato yield variations at different times in the Wola region than levels of sunlight which are invariably sufficient over a crop's life to ensure an adequate yield.[2] The growth and yield of

---

[2] In trials in the Western Pacific, only on the Solomon Islands has any relationship been reported between solar radiation and crop yield — Gollifer 1980.

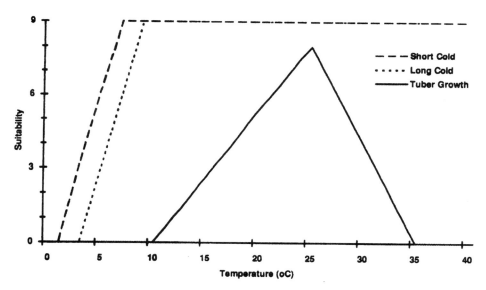

[Suitability classes range from 0 (immediate death), through 2 (negligible yield), 5 (moderate yield), 7 (high yield), to 9 (maximum yield).]

Figure 4.1 Responses of sweet potato to different temperatures (after Hackett 1985).

sweet potato is markedly temperature dependent (Figure 4.1), temperature influencing not only root size and number but also, beyond a certain range, plant survival. The crop yields optimally at temperatures around 25°C, cool weather restricts growth and below 10°C the plant is damaged, growth severely retarded and tuberisation halted; likewise in excessively hot weather. The crop clearly requires a frost-free growing period. It is probable that low, though not fatal, daytime temperatures reduce photosynthetic activity and hence translocation rates to roots, whereas high ones favour vine growth and leaf expansion, again reducing photosynthates available to roots (Sajjapongse & Wu 1989). A warm soil also promotes dry matter production. The temperature regime of the Wola region is on the whole well suited to sweet potato production, with the exception of the occasional incidence of damagingly low night time temperatures, and infrequent frosts. The maximum average daily temperatures are perfect for the crop. And the warm days and cool nights that characterise this montane environment are ideal too, thermoperiodism influencing the allocation of assimilates within the plant. Warm days and cool nights facilitate storage root growth, the low temperatures increasing cambial activity and lessening stele cell lignification, whereas continuously warm temperatures night and day stimulate vine growth at the expense of tubers.

It is rainfall above all else that most often noticeably affects sweet potato yields in the Wola region. Both excessively dry and wet conditions can hamper growth and yield. Low rainfall represses vine growth, reducing plant photosynthetic area and so decreasing the supply of photosynthates to roots for tuber growth. High rainfall stimulates vigorous vine growth to continue during tuberisation at the expense of storage roots, reducing the proportion of dry matter diverted to subterranean parts (Enyi 1977; Kays 1985). Overly wet conditions also promote tuber rot. High rainfall is frequently the problem in New Guinea. A series of trials conducted in the Western Pacific indicate the scope of the negative relationship between excessive rainfall and tuber yield, all showing that high rainfall conditions at any time during the cropping period reduce yields, these decreasing linearly as the number of weeks the soil is fully saturated increases (Gollifer 1980; King 1985; Enyi 1977; Hide et al. 1984; Bourke 1988).[3] The crop ideally requires about 50 cm of rainfall during its growing period. The rainfall in the Wola region is likely to exceed this by three or more times which, while less than perfect, may not be a serious hinderance where drainage is good, which it often is on the steep slopes people cultivate. Adequate drainage is necessary because sweet potato cannot withstand waterlogging for any time (Figure 4.2). It retards tuber initiation and growth, reducing yields. This is one reason why mounding markedly improves yields, by promoting good soil drainage.

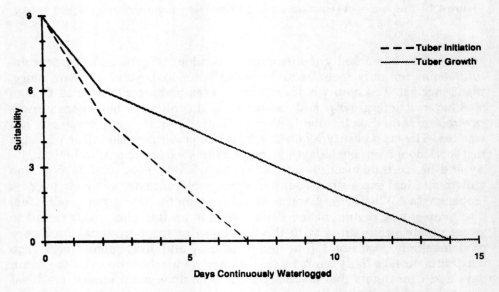

Figure 4.2   Effects of waterlogging on sweet potato (after Hackett 1985).

---

[3] The trials were conducted on the Solomon Islands (Gollifer 1980); the Papuan coast (King 1985); Waigani in Port Moresby (Enyi 1977); Simbu Province in the highlands (Hide et al. 1984); and Aiyura experimental station in the Eastern Highlands (Bourke 1988).

Overly wet conditions also severely diminish soil oxygen concentrations, repressing tuber formation. Respiration by tuberous roots is rapid under normal conditions. Poor soil aeration reduces it considerably, cutting oxygen levels within roots and checking cell division and expansion, inhibiting both existing tuber enlargement and new tuber initiation. While a root zone environment of <2.5% oxygen totally represses tuber induction and development, it substantially increases shoot growth and top growth proliferation, again to the detriment of roots regarding the proportion of photosynthates diverted to them. Furthermore, excessive soil moisture and reduced aeration adversely affect fleshy root formation because oxygen deficiency early in growth increases stele cell lignification and suppresses primary cambrium activity, resulting in fibrous as opposed to tuberous root development (Hahn 1977; Kays 1985 — see Chapter 2).

The sweet potato is better able to tolerate periods of drought than waterlogging, although if water shortage occurs six weeks or so after planting, with storage root initiation started, it reduces yields considerably. The results of field experiments demonstrate the extent to which the crop can withstand drought, yield responding to applied water only when soil moisture levels are well below field capacity; for example, no significant yield differences were found irrigating at moisture levels higher than 20% of field capacity at any stage of growth, nor were any yield increases recorded keeping soil moisture up to 50% as opposed to 25% (Hammett et al. 1982). Severe moisture stress depresses both tuber size and number, tuber production increasing with irrigation when available water is very low, particularly both later in the cropping period, when shortages most depress tuber swelling, and during early growth and establishment of young plants, when tubers are initiated (Sajjapongse & Wu 1989). Water stress reduces photosynthetic activity due to the closure of stomata. It may also result in the accumulation of what photosynthates are produced in vines and leaves at the expense of tubers due to reduced translocation rates with water in short supply.

Following a meticulous comparison of data on soil moisture conditions at times of reported food shortages in the highlands region, Bourke (1988, 1989) concludes that drought[4] alone is rarely responsible for shortfalls in sweet potato supply, as many people have previously assumed, unless it is extraordinarily prolonged, or a protracted wet period precedes it, or for some reason a large fluctuation in planting rate occurs. In contrast, extended periods of wet weather and water surplus,[5] or short-term occurrences of extreme water excess,[6] often accompanied by local flooding, are frequently associated with food shortages, when it is common to hear people

---

[4] Drought defined as a continuous period when soil moisture <20% of field capacity for >10 weeks.
[5] Surplus defined as a continuous period when rainfall exceeds evapotranspiration, with soil fully saturated, for >20 weeks.
[6] Excess defined as a week in which calculated water surplus >170 mm.

complaining that edible tubers have failed to develop. The depression in yield that occurs when a fairly mild drought follows a prolonged wet period is something to which he draws particular attention.[7] The period soon after planting, when tubers are initiated, and that towards harvest time, when they rapidly swell, are widely recognised as critical to yield (Wilson 1982). A long wet spell during the former and drought during the latter apparently depress yields to such an extent that shortages occur. According to Bourke, an early wet spell reduces the number of tubers initiated, and hence yields, and also renders the plant more vulnerable to any subsequent drought because it induces shallow rooting, concentrating roots in that region of the soil that is the first to dry out. But his connection of this weather sequence with reduced sweet potato yields disagrees with the findings and recommendations of other workers, who maintain that it is advantageous to have a wet period coincide with the early phase of plant growth (when substantial rainfall facilitates the rapid growth of vine and foliage to give maximum photosynthetic capacity quickly, and cooler conditions enhance root formation), followed by a drier spell during the enlargement of storage roots (to hold down top growth and divert a higher proportion of assimilates to tubers, the warmer weather also stimulating root enlargement — Kays 1985; Sajjapongse & Wu 1989; Hahn 1977).

Whatever the conflict of interpretations over sweet potato's response to variations in water supply, the evidence is that rainfall has the largest influence on sweet potato yields in the highlands of Papua New Guinea, and extremes in it are more often responsible for shortages than any other climatic factor; some commentators have perhaps overplayed the danger of frosts in this regard across the highlands region, except for those relatively few settled high altitude locations.

II. HARD TIME RITUALS

When crops fail and famine threatens, the Wola resorted to rituals such as *iyshpondamahenday* and *ngoltombae* to appease the spirit of a white-skinned woman called Horwar Saliyn,[8] whom they believed somehow responsible for adverse weather ruining gardens. The performance of the rituals and the offerings made "pleased her", persuading her to relent and allow conditions to

---

[7] He cites two trials as further evidence of the significance of this weather pattern, in support of his comparative data on highlands-wide food shortages, one at Aiyura and the other in Simbu Province, which show that the highest tuber yields occur when sweet potato is planted into soil which has relatively low moisture content following a dry spell for the tuber initiation phase, followed by more saturated conditions, with a wet spell towards the end of the cropping cycle during the tuber bulking phase.

[8] Horwar is a place in the Was valley where some people say she resided, and Saliyn was her personal name.

improve. She was not related to the white-skinned woman Sabkabyinten who resides in the sky and sends the rain; no one has ever associated them together.⁹ Why two apparently unrelated fair-skinned women should be thought responsible for controlling aspects of the weather is unclear to those I have questioned, and exactly what Horwar Saliyn does to influence the weather and crop yields for the worse is not something that any of them could explain.¹⁰ She just does, people assert.

The apparent inability, even disinterest, of the Wola in citing any relationship between the two white-skinned female spirits has long perplexed me. Although the two characters are uncannily similar, it is perhaps wrong to expect any relationship between them, none may have ever existed. This is not improbable. It was over one Christmas that I was struck by a parallel in our own culture to the apparent lack of any connection between Sabkabyinten and Horwar Saliyn, in the absence of any obvious relationship between Jesus Christ and Santa Claus. Perhaps human beliefs and ideas not only fail to comprise connected wholes, but are also necessarily contradictory to accomodate our paradoxical and conflicting psyches. These two female spirits could be figures of independent origin, who over generations have somehow come to be associated with the same issue, namely the weather, regardless of being the same sex and strange skin colour.¹¹

The Wola have a myth about Horwar Saliyn, but it gives no clues about her control of the weather, causing crop failure and hunger. Her deeds seem entirely unrelated to these matters. According to people living in different parts of the Was valley she used to live in the forest, either at Horwar or Ngoltimaenda, and killed men until stopped by an individual called Ezaendael who thrust a hot stone up her vagina. One version of the myth¹² that tells of her exploits and demise is as follows:

> Once there were twenty men at a place called Kaerep. One of them decided to go and find a tree with bark suitable for girdle making, to collect a strip off it.¹³ He told the others to wait for him and set off alone. They sat waiting until the evening but he never returned. He was lost. A second man decided he too would look for girdle bark and seek for the first, and

---

⁹ The Enga, who think sky-beings control the weather, also believe that they are responsible for frosts, drought and too much rain.

¹⁰ When I suggested that their white skins perhaps related to the white of clouds and the whiteness of frost, this was denied.

¹¹ It is plausible that the sex of these spirit beings relates in some deep way to Wola men's fears of women, which centre on their reproductive role (Sillitoe 1979b).

¹² See Mawe 1982:42–50 for a Mendi version of a myth about a white woman, quite different to the one given here. See also Vicedom & Tischner 1943/48:25–31 and Strauss & Tischner 1962:436–37 for myths about a female sky-being spirit in the Western Highlands (in contrast to the myth given here, the Mbowamb say the woman was a virgin, her genitals closed off).

¹³ Wola men traditionally wear a bark girdle around their waists (Sillitoe 1988).

set off alone. He also failed to return. One after the other nineteen of the men set off alone and all failed to come back.

The last man remained behind and waited and waited for someone to return. Finally, he set off too like the others, looking for girdle bark, to find them. He followed the tracks of his brothers, walking fast without any stops. He walked up and up until he came to the top of a sheer rockface. When he peered over the edge of the cliff, the wind carried an awful stench up to him. It was from the rotting corpses of his brothers.

The man decided to hunt for some marsupials in the surrounding forest and cook them in an earth oven at the cliff top. Before he had arrived up there and looked over the cliff and smelt his putrifying brothers down below, before looking down on their bodies of decomposing flesh with exposed bones glittering in the sun, he had picked up a large basalt stone, which he now included with those he heated for the oven. While the stones were heating up for the oven, he took a length of cane internode he had cut for a water container to a nearby pool to fill it up.[14]

When the man returned, he looked up and saw a white-skinned woman sitting a little way off. He went over to her. The stones were now very hot, the fire glowing red like the neck feathers of the *oljomb* fairy lory,[15] scorchingly hot they had fired ready to be tossed into the earth oven.

"I'll steal her",[16] he said to himself. "Lie on the ground, I intend to climb up you like my brothers," he told her.

"I lay like this before, for the others" she said, as she reclined on the ground and parted her legs.

But when she opened her legs, wah! She made like this (the speaker used his arms to depict an enormous man-killing vagina). When she did that the man took the large hot basalt stone he had collected and hit her between the legs with it, like with a heavy wood-splitting-club, and poured water over it from the internode container. The steam hissed around her and she was in great pain.

The woman staggered off in agony, collapsing to the ground every so often and writhing around in torment. Wherever she fell, she created a pool of water. She forced herself to walk on, falling down repeatedly in agony at different places, like Ngol-Timaenda, Iyb-Huwliyp and Huwmiyba. She travelled very quickly over a large distance, on and on and on, creating many water pools where she stopped on her way, until she reached Shoba where she entered a house, and that's all.[17]

---

[14] It is usual to pour some water over the hot stones placed in an oven to generate more steam for cooking its contents.
[15] *Charmosyna papou*.
[16] The Wola refer to sexual intercourse out of wedlock as "stealing a woman".
[17] This version of the myth was told to me by Wenja Neleb.

The *Iyshpondamahenday* Ritual

The procedure and scale of the *iyshpondamahenday* ritual varied, depending on the severity of the food crisis.[18] If the "drought" was bad, maybe with frosts and serious crop failure, many communities in the lower Was valley and beyond would co-ordinate their ritual efforts, with representatives going first to Horwar, where they thought Saliyn lived, to take part in a large ritual, and then returning to the locations near their communities where she rested and left behind pools of water, to stage further local rites. The *iyshpondamahenday* was the only large-scale ritual likely to he held during the *bulhenjip* season, when extended *baen* fine weather spells with frosts are probable; it was normal practice to stage all other large communal rituals in *ebenjip*.

Each community that took part in the *iyshpondamahenday* cult had a pair of ritual representatives called *moray nais* (lit. *moray* boys). There was usually a considerable age difference between them, the younger representative possibly not having acted in the role before, for the large-scale *iyshpondamahenday* ritual was an infrequent event occurring only at times of dire crisis (perhaps once a decade, if the occurrence of serious frosts this century is any guide). If inexperienced, the younger man served as an apprentice to the older *moray nais*. When one of them died, the community nominated an adolescent male to take on the role when the next ceremony was in the offing, after informal discussions and a consensus had emerged as to whom everyone thought adequate to the task.

The role was not an onerous one hedged around by many restrictions on behaviour, although for the rest of their lives *moray nais* should not share tobacco pipes nor water gourds with others or else the unfortunate sharer would die with a swollen belly. The funeral of a *moray nais* also differs from other people's in that relatives only mourn for a day or so, over the corpse lashed to a horizontal pole, before depositing the body on a platform built inside a pothole. They wrap the corpse up in a bundle of tough *goiz* palm (*Gulubia* sp.) leaf spathes. The reason for the attenuated period of mourning and unusual corpse disposal is that if any of a *moray nais* man's putrefying bodily fluids fall on the ground a "drought" and famine would ensure. This, men say, is because he has had dealings with Horwar Saliyn. Also, if his sightless eyes look upon the night sky and become stars, as the Wola believe happens with the eyes of the dead, this would cause a famine too. The association here is between stars and the occurrence of "droughts" and frosts on clear and cloudless nights when they are visible.

When famine threatened and a large-scale ritual performance was mounted, the *moray nais* congregated at Horwar. At the last ceremony there

---

[18] See Mawe 1982:30–37 for an account of the white woman's ritual called *upia*, performed in the Mendi valley during times of hunger. See also Vicedom & Tischner 1943–48; Strauss & Tischner 1962 and Strathern 1970 on sky-woman spirit cults in the Western Highlands, where they were staged to ensure fertility and general well-being, not end a food crisis; these rituals differed considerably from that of the Wola with large dances and many pigs killed.

were fourteen of them.[19] The last *iyshpondamahenday* at Horwar occurred in the 1962 "drought" to judge from the comments of participants. A long house called the *iyshpondamahenday aenda* (lit. *iyshpondamahenday* house) was built there. It was along the lines of a conventional house only larger, with pitched roof of pandan leaf thatch, the ridge about three metres above the ground, and walls of beech wood stakes, around two metres high. It was ten metres or so long, with a crawl through doorway at either end in the gable wall. During the ritual, men decorated these gable ends with many pearl shells. Inside the house lengths of wood pegged to the ground divided up the floor area into rectangular sleeping areas along both walls, each occupied by a *moray nais*, and down the centre was a walkway extending between the two doors, along which there were fireplaces at intervals. In addition to the long house, two ordinary bush houses were built, one adjacent to it and occupied by the experienced *moray nais* and the other some way off for the apprentices.

When the *moray nais* took up residence they withdrew from everyday life, indeed other men and women should not see them, and if they heard somebody inadvertently approaching they would shout a warning or hide. They were preparing to intercede with the white woman's spirit on behalf of all and their close association with her makes them dangerous to others, who would, some people said, fall ill and die if they saw them, as one man expressed it in Pidgin, they are the spirit's *mankimasta* or 'domestic servants', readying things for her. The *moray nais* depended on others to bring them food, besides hunting for game for themselves in the forest. Relatives would bring vegetables and leave them at an appointed place on the path leading up to the long house, shouting out to the *moray nais* to come and collect them. They should only eat taro, sugar cane and ginger, besides marsupials and birds they caught themselves. The absence of sweet potato from their diet signalled famine men said, although they had probably already considerably reduced their consumption of tubers due to their chronic ritual-prompting shortage.

At an appointed time, the experienced *moray nais* congregated in their house adjacent to the long house and the apprentices in their's away from it, and a girl led a red-haired pig up to the experienced men's house, handing the pig's tether over through the door, without looking at the men inside. While one *moray nais* took hold of the tether, another painted a red line vertically down the girl's forehead, from hairline to the bridge of her nose. Only a pig with red bristles would suffice men said, its colouring parallelling that of Horwar Saliyn's pale skin, it was the only kind of pig acceptable to her.[20]

---

[19] My information on the ritual, which I have not observed, comes from conversations with men who have taken part in it, including several *moray nais*, and visits to several of the ritual sites, including Horwar.

[20] The Wola refer to the colouring of both pale-skinned humans, who are light brown, and red-bristled pigs, which are ginger-brown, as *hundbiy*; both are uncommmon. Horwar Saliyn was a *ten hundbiy* (lit. woman light-brown), and a red-haired pig is a *showmay hundbiy* (lit. pig light-brown).

The experienced *moray nais* tell everyone else that they kill and burn the entire pig as an offering to Horwar Saliyn's spirit, whereas in reality one of them told me that they cook and eat it (the secret was apparently well kept because an apprentice *moray nais* earnestly related to me a decade after the last ritual how the others had burnt the whole pig). No ceremony attended the pig's slaughter, although the blood spilled and the smell of singeing bristles was said to please the white woman's spirit (as blood and singeing bristles do other spirits, such as those of the ancestors and forest demons, when pigs are killed as ritual offerings to appease them — see Chapter 8).

The next event is a dance performed by the *moray nais*. It is the climax of the Horwar site *iyshpondamahenday* ritual and occurred a month or so after preparations had started there. It is the first time that others have seen the *moray nais* since they took up residence at the long house. People from surrounding communities far and wide anticipated the event and congregated at Horwar to see the dance, when they hear that it is about to occur. The evening before the dance all the *moray nais* collected together in the long house, where two of them stand back-to-back in the middle of the central walkway and paced to the door they face and back again. They recited the following ritual incantation twice, once as they walked away from one another and once as they come towards one another:

Bunba[21]	*diy*	*enjom*	*ma*	*siysiysa.*			
Wallabies	and	those	caused	to be present.			
Eb kuwmb	*diy*	*enjom*	*ma*	*siysiysa.*			
Salt bundles	and	those	caused	to be present.			
Ib	*taenjel*		*puw*	*diy*	*enjom*	*ma*	*siysiysa.*
Ritual word[22]	cowrie necklaces		long	and	those	caused	to be present.
Honayok	*diy*	*enjom*	*ma*	*siysiya.*			
Cosmetic oil[23]	and	that	caused	to be present.			

A free translation of the incantation runs:

And those wallabies they presented there
And those salt bundles they presented there
And those long cowrie shell necklaces they presented there.
And that cosmetic oil they presented there.

According to Moray,[24] the man who told it to me, the incantation recalls that long ago their ancestors amassed a great deal of wealth at Horwar and

---

[21] *Bunba* is special incantation term for the *maepun* class of wallabies (*Thylogale bruijni* and *Dorcopsis vanheurni*).
[22] Incantation word with no meaning, put in for rhyming rhythm.
[23] *Honayok* is special incantation word for *wombok* (*Campnosperma brevipetiolata* oily sap).
[24] The Wola commonly call men who fulfil ritual offices after their ritual title, like Moray after *moray nais* and Kem after *shor kem* (another communal ritual role).

presented it to the white woman's spirit. The incantation reminds her of these generous gifts, which symbolised the wish of the living to have an agreeable relationship with her, just as they structure sociable coexistence among themselves through institutionalised exchanges of wealth (Ryan 1961; Sillitoe 1979; Lederman 1986). The incantation attempts to persuade her to act properly by the canons of normal moral behaviour, and not capriciously and malevolently causing hardship and hunger.

The following morning, after their last night together in the long house, the experienced *moray nais* donned their best finery, with majestic bird of paradise plume headdresses set in wigs, faces brightly painted and torsos daubed with shining cosmetic oil. Before leaving Horwar, they staged a dance. They stomped in a large circle, chanting dance songs, clapping together two strips of pandan leaf. They held the rectangular strips of tough leaf together by their ends, sandwich-like one on top of the other, and by pushing their hands together so that they bulged apart and then sharply pulling their hands outwards, caused them to smack together. At about midday the apprentice *moray nais*, who had been hiding in the long house, burst forth to dance behind their mentors. They wear only white cockatoo feathers in their hair; throughout their stay at Horwar they dress restrainedly, not sporting any necklaces or armbands. At their appearance the spectators hurriedly leave, not looking behind to see what those who correspond with Horwar Saliyn's spirit on their behalf are doing, to do so would be to court death from her.

The *moray nais* adopted an ambulatory life next, visiting some of the sites at which Saliyn rested, leaving behind a pool of water. There are many

Plate 4.1 One of the up-welling subterranean pools associated with the injured female spirit Saliyn's flight, at Iybwimb.

Plate 4.2 A man decorated in full finery like the *moray nais* at the close of the large *iyshpondamahenday* sequence.

of these locations, six in the lower Was valley in the Horwar region at Iybhuwliyp, Wal, Huwmiyba, Poiyiyba, Makazil and Obanoba, and four on the adjacent Nembi valley watershed at Pongal, Hont, Paypay and Wabolkolkobay, plus another seven in the middle Was valley in the Ngoltimaenda region at Iybwimb, Wet, Bombet, Iybtiy, Kiyshol, Iybnaep and Kombiyab. In the latter region some of the pools connect to nearby watercourses by subterranean passages, sometimes reappearing at intervals as surface streams before flowing underground again, and the local people attribute this intriguing drainage pattern to it originating from Saliyn's scalding vaginal secretions dissolving the rock. Some people also associate the gas and oil seeps found across their region with her passage (see Chapter 5), these being more prominent during fine *baen* periods when food shortages occur, but no one staged any rites at these places.

It is noteworthy that the *iyshpondamahenday* ritual involved several local communities, each sending *moray nais* representatives and hosting part of the ritual cycle. The demands of this region-wide co-operation would be likely to reduce the possibility of intercommunal violence breaking out, deterring people from attacking others elsewhere for food. In a stateless society, with no authoritative offices to enforce order, this could be a significant bulwark against chaos. The Wola deny that they fight for food, or over land resources to produce it (except where two or more individuals disagree over tenure rights to a specific garden plot and come to blows with fatal results, sparking off revenge action). But under severe privation some people might be tempted to resort to violence to secure food. A large co-operative ritual concentrated what energies they have into negotiating with supernatural forces to end the unfavorable conditions, not using what hunger-sapped strength they have to make matters worse by falling anarchically on one another to grab what little food there is.

The communities adjacent to the water-pool sites to be visited built leafy enclosures there. They were rectangular in plan, fenced with fronds on three sides and left open on the fourth. In the screen opposite the open side they made a crawl through door-like opening, leading to a conventional style bush house built the other side. Any saplings and branches can be used to build the enclosure fence. They are left with their leaves attached and driven in close together to produce an impenetrable hedge through which people cannot see. On the inside of the enclosure men lined the screen with *shiyp* meliad and *bat* tree leaves (*Chisocheton ceramicum* and *Bubbia* spp.), poking them behind horizontal spars lashed there, their pale undersides facing inwards to give a light coloured wall.[25] On the day of the rite, when the *moray nais* have arrived and occupied the house at the rear of the enclosure, men hang pearl-shell crescents on the inside of the leafy screen. The large show of wealth, which features prominently in the ordering of Wola social relations, is again intended to urge the white woman's spirit into behaving morally and properly, to correct her life-threatening persecutions.

People from the adjacent communities gather inside the enclosure after the *moray nais* have arrived, on the morning of the rite. A length of vine was laid from the small opening in the screen across the enclosure and out of the open side, and those present, both men and women, stood on either side and held it up. Their outstretched arms formed a tunnel leading up to the doorway in the rear wall. A girl walked through it with a pig, again red bristled, and as she handed over the tether to a *moray nais* through the door one of them again painted a line in red ochre down her forehead with his finger. When she turned, everyone in the enclosure fled, men grabbing their valuable pearlshells as they left; again no one looked back for fear of death.

The *moray nais* slaughter the pig, cook and eat it. The blood and smell of singeing bristle are again said to appease Horwar Saliyn's spirit, coming from an offering of a valuable pig, made in good faith to her. The men who supply the animals derive considerable social esteem from having the capacity to donate pigs, being able to arrange and finance their exchange obligations such that they can afford to forgo wealth without defaulting on their exchange commitments to relatives. The donation of a red-haired pig for the ritual not only boosts social standing, it is also an opportunity to dispose of an animal which

---

[25] Although the pale leaf screen is temptingly reminiscent of the pale skin of Horwar Saliyn, no one made this association; furthermore the Wola use screens of similar construction in other contexts which have nothing to do with the white woman's spirit (see Chapter 8 on forest demon rituals; also Sillitoe 1987). See also Strathern 1970:581 who refers to similar leaf arches among the Melpa.

might otherwise die of hunger if famine conditions persist, so the cost is lower than under normal circumstances.

The *moray nais* stayed in the house for the number of nights they believe the white woman rested in the vicinity nursing her injury, usually one or two nights. The *moray nais* also find a supply of their restricted foods in the house, placed there before their arrival. After they have visited some of the places associated with Horwar Saliyn's agonised flight, the *moray nais* resumed normal life. No longer negotiating supernaturally with the white woman's spirit they were not dangerous and could return to everyday social interaction. It was believed that their intercession would assuage Saliyn's spirit and persuade her to relent, ending the "drought", if this had not already occurred.

People I spoke to said that even if conditions changed for the better during the *iyshpondamahenday* ritual they would not abort proceedings part way through once started because this would show disrespect to the woman's spirit and anger her, revisiting hard times upon them again. It is likely that weather conditions will change before the ritual is over, or even started, but the food shortage be set to persist for some time to come. It is due not to current weather but an inclement earlier spell. In this event, the intention of the ritual is less to affect the weather, than express hardship, and a communal attempt to ward off further disaster.

The *Ngoltombae* Ritual

There is no connection readily apparent between the weather over the Southern Highlands region of New Guinea and the *iyshpondamahenday* ritual, or its sister ritual the *ngoltombae*, also performed to appease the white woman's spirit. The people with whom I have discussed them have made no direct connection between the rain, clouds, sunshine, frost or whatever and the white woman's spirit and associated rites. They assert that her spirit is responsible for crop failure and hunger through the agency of inclement conditions, but how she manipulates the rain, frost and so on is unknown; nevertheless whatever she does, the ritual "pleases" her and "persuades" her to relent.[26]

It is pertinent that rituals and associated incantations are readily borrowed, sometimes bought, from others in this part of the world.[27] It is quite feasible that the *iyshpondamahenday* and *ngoltombae* rituals originated elsewhere, not only evolving as they have moved between places but also through time at any one location, people constantly modifying and adapting them in

---

[26] See Waddell 1975 for an account of the practical responses of the Enga to climatic perturbations, which largely involve migration to less affected regions.

[27] In a Sepik-centric interpretation of the movement of rituals in the Highlands, Harrison (1993) even suggests that it is the intangible equivalent of wealth exchange, an arena in which men vie for power.

Plate 4.3 At Ngoltimaenda Pelsuwol Njep points to the two ritual stones Hab (left) and Homok (right).

the absence of any fixed liturgy. This accretionary development renders expectations of complete ritual exegesis inappropriate. Perhaps it is misplaced to look for a coherent system of supernatural beliefs and ideas in such a cultural environment (Lewis 1980; Brunton 1980), which the excesses of structuralism have led us to expect. The ritual is a genuine mystery, adding to its efficaciousness in Wola eyes. The *ngoltombae* ritual illustrates the scope that Wola society affords for rituals about the same theme to vary. Communities further up the Was valley from those that stage the *iyshpondamahenday* performed the *ngoltombae* ritual. It too was intended to appease the spirit of Horwar Saliyn in times of hardship, but although staged for the same reasons by neighbouring communities in the Was valley it was quite different in form to the *iyshpondamahenday* ritual.

The *ngoltombae* ritual was held on a steep-sided ridge top at a place called Ngoltimaenda, adjacent to the locale of Tuwp. One side, which falls

Coping with Climatic Variations: The Spectre of Famine 97

Plate 4.4 Pelsuwol Kwimb holds a hammerstone used to crack the leg bones of marsupials against Hab and Homok.

away near vertically, is where local residents say Ezaendael, the hero in the myth, found the nineteen corpses of his 'brothers'. And adjacent to the ritual site is a flat area like an abandoned homestead site, which is where they say Horwar Saliyn had a house. There are two large stones on the site called Henja Hab and Henja Homok. According to the custodians of the place the former, Hab was the stone thrust into Saliyn's vagina, which has subsequently grown greatly in size. There is also a large circular earth oven pit on the ridge crest made by the writhing Horwar Saliyn with the hot stone inside her. Adjacent to it there was a small bush material house, used in the event of rain during the ritual.

The men living near Ngoltimaenda at Tuwp told me that when some of them decided that it was time to stage a ritual there, they would spend several days hunting for marsupials in the forest. They then congregated at the ridge top site with their catch. Only men were present. Women and children were expected to remain quietly at home; it was important that no one made a noise in the neighbourhood.[28] The marsupials' blood and singed fur are again said to 'please' the white woman's spirit, and men rubbed the two stones with charred fur and blood. They also broke the animal's bones on the stones, married men cracked the bones of those they had caught on the stone called Homok and the unmarried men theirs on Hab. One man told me that they killed any live animals by dashing their heads on the stones too. When they took pigs to the site, they smeared blood from the bleeding snouts of the clubbed animals over the Hab stone. If they took only a container of pig's blood to the site, as they sometimes did when visiting a number of ritual locations to make offerings at them all when someone was seriously ill, they similarly smeared this stone.

When administering pigs' blood or singed marsupial fur and blood to the stones, men chanted an incantation like the following:

Hokay	*njiybiy,*	*ma*	*njiybiy,*	*sab*	*diy*	*njiybiy.*		
Sweet potato	give,	taro	give,	marsupials	and	give.		
Showmay diy	*njiybiy,*	*ten*	*nonknais*	*kab diy*	*ma*	*ebay*	*sha*	*bay.*
Pigs and	give	women	children	two and	cause	good	become	do.
Suw	*diy*	*ma*	*ebay*	*sha*	*bay,*	*showmay*	*diy*	*njibiy.*
Land	and	cause	good	become	do,	pigs	and	give.

A free transalation of the incantation runs:

Give us sweet potato, give us taro, and give us marsupials.
Give us pigs too, and keep the women and children well.
Keep the land well too, and give us pigs.

The import of the incantation is fairly obvious, relating directly to the dire circumstances that have prompted the ritual's performance. The reciter is asking for adequate food supplies, citing sweet potato, taro, pigs and marsupials as particularly esteemed foods. He is also asking that women and children should remain in good health, with sufficient food to eat. And that the land, by which he intends the climate of the place too, should also remain clement, in good health and fertile, to supply the food everyone needs.

Besides the incantation very little is said during the ritual. During its performance, the married and unmarried men sat in separate groups. They

---

[28] When I visited the site, my guides became very angry when some of the younger men who accompanied us started to act boisterously, and furious when some women tried to follow us there.

did not talk until they had cooked and eaten the entrails of the marsupials, which they cooked in a separate small earth oven from the rest of the animals, the meat of which they cooked in the large spirit-made earth oven hole. Only men were allowed to eat this meat. While the entrails cooked they sat silently not talking to one another. They communicated by miming and pointing only, with the occasional whisper, until they had consumed them.

The *ngoltombae* and *iyshpondamahenday* rituals stand in stark contrast to one another, in duration, arrangements and numbers of people involved. The men who staged the *ngoltombae* maintain that they did not take part in the *iyshpondamahenday*, and vice versa, although they were aware when the other was staged. Those who put on the *ngoltombae* even assert that their performance of the ritual, at the place where all their troubles began with Horwar Saliyn and the hot stone episode, prompted those living lower down the Was valley to start the *iyshpondamahenday* ritual. Those lower down deny that there was any such association, that they needed any signal; and that anyway Saliyn lived near them at Horwar, not higher up at Ngoltimaenda. The scope for ritual improvisation and rival interpretations is considerable in this part of the world, contradictory assertions should be anticipated. They are another aspect of the region's celebrated social flexiblity.

Ritual, Climate and Soil

The room for variation in the performance and interpretation of rituals helps in part to account for the absence of any overt connection between Horwar Saliyn and the weather, or apparent association between her and Sabkabyinten, the other white female spirit believed to control aspects of the weather. Nevertheless, it is perhaps beside the point to look for a direct connection between the weather and the white woman's spirit cult. The parallels are on other levels, between her unforseeable and capricious behaviour and the unmanageable and unpredictable character of the weather. In the myth, the man Ezaendael is presented with little choice, he either allows the woman to kill him like his brothers or he incapacitates her. And in disabling her, he mortally injures her in her man-killing organ, which is apt revenge to the Wola, who place a high value on avenging relatives' deaths. But in justifiably incapacitating the white woman the man creates a wayward and spiteful spirit, over which the living have scant control. It is similar to the weather, with which the spirit is associated, being responsible for extreme conditions which are unpredictable and essentially uncontrollable too. People have no choice with the weather, like the man in the myth, they have to make do and cope.

Regardless of our advanced technological capabilities, we are more or less in the same position as these New Guinea highlanders, the weather remaining largely beyond our control. We may attempt engineering and agricultural innovations to palliate the effects of drought in Ethiopia or cyclones in Bangladesh, but the weather responsible exceeds our prediction and management. In their myth and ritual the Wola are perhaps expressing concerns common to humankind. The weather is something over which we all have little control, whatever we do has scant effect on it. We are akin to the man in the white woman myth, who hitting her with the hot stone may have saved his life but could not stop her persecuting the living, whom she no longer assaults directly with her killer-vagina but indirectly, by causing them hunger and famine, occasionally affecting the weather freakishly and disrupting food supplies.

These dire-time-rites are typical in many regards of Wola rituals generally which are part of a supernatural order that overridingly concerns itself with misfortune, treating or averting it. The occurrence of local rites involving small numbers of people at the white woman's ritual sites at other times, underlines this pervasive concern to control a typically capricious spirit world. On occasion, a few men from a community might decide to stage a small-scale local rite at the water-pool-resting-site nearest to them, in addition to taking part in the infrequent larger-scale rituals at times of widespread famine. They were prompted to do so when there was a food shortage, notably of sweet potato, at times of less dire hunger than follow dramatic climatic perturbations like frosts and "drought", such as come after an excessively wet spell that has interrupted normal gardening activity. Sometimes men were prompted to stage an informal rite by some other event; for example if someone was ill and relatives, casting around for the cause, were killing pigs at various sites to appease whatever spirit forces (ancestors and so on) were possibly responsible, they might decide to visit the nearest white woman spirit site, taking a pig to slaughter or a side of pork and container of blood there to offer to Saliyn's spirit; or if a man out hunting caught a marsupial adjacent to one of the spirit woman's resting places he would cook it there, offering the spilled blood and smell of singeing fur to her.

No matter how fertile your soil, under inclement climatic conditions farming is difficult. But the white woman's cult is more than a mere expression of fatalism about misfortune generally and the weather in particular. The Wola think that it is effective too, judging from their comments on it. They talk as if they believe it really had an effect on the weather, by persuading the white woman's spirit to relent. The import of the ritual is to induce her to act morally, and they believe that they can pressure her to act properly like anyone else. References to wealth exchange, which is central to the regulation of Wola social life, figure prominently here, with pigs slaughtered and pearlshells displayed.

These rituals are the Wola equivalent of our technological innovations to alleviate the worst effects of severe weather. We might consider them ineffective, with our scientific approach to coping with these problems, nevertheless the ritual could serve as an effective coping mechanism given their limited technological capabilities. While many contemporary social scientists consider functionalist arguments grossly inadequate, there is something persuasive, even comforting, in the idea that ritual may facilitate psychological security and help people manage in disaster, by alleviating feelings of helplessness and despair. Postmodern social theorists are doubtless correct that we cannot fully appreciate the plight of those visited occasionally by the spectre of extreme hunger. Nor perhaps fully comprehend their cultural expression of these haunting experiences and efforts to wrestle control over the environmental forces responsible. But how many would be willing to give up the security of full stomachs to learn more? Anyone sophistical enough to try would be unlikely to continue to indulge in contemporary postmodern reflections that spawn these arguments, an ironic conundrum and suitably quixotic point on which to close. The Wola talk in awe about the caprice of the white woman's spirit, but believing that they can moderate her behaviour helps them to manage in hard times, by attributing the occurrence of these unforeseen events to unpredictable spirit-induced extremes in the weather.

Returning to this study's pedological theme, while periodic extreme fluctuations in the climate can cause human-beings acute hardship, from the perspective and time-scale of soil forming processes these events are unnoticeable. The contemporary weather regime of the Wola region is highly favourable to soil formation. While few soil scientists think that the climate is the dominant factor in determining the nature of soils — the zonal perspective — they agree that it is an important factor. The two most significant aspects of climate to the development of soil properties being moisture and temperature (Jenny 1941).

The majority of physical, chemical and biochemical processes that occur in soil involve moisture, and the volume of water arriving on the soil surface influences weathering and leaching activity. The wet climate that characterises the Wola region gives a water balance, after evaporative and run-off losses are deducted from precipitation, that ensures high soil moisture conditions all year round, except for the relatively infrequent and usually short-lived drier periods when soils, particularly topsoils, may to an extent dry out. These high soil moisture conditions promote rapid weathering and leaching processes, doubly so given the geology of Wolaland. Rainfall has a marked influence on the rates of weathering of volcanic ash, which is one of the region's primary soil forming materials, high precipitation promoting the alteration and rapid breakdown of tephra minerals (Mohr *et al.* 1972). Rain entering soils also contains appreciable amounts of dissolved carbon dioxide, which results in a weakly acidic, reactive solution, particularly significant where calcareous bedrocks occur, as they do across large areas of the region, comprising the other predominant source of regolith for soil formation.

The rate of chemical processes in the soil is in turn influenced by temperature, notably the hydrolysis of primary minerals and the biological breakdown of organic matter. The temperatures of the Wola region, although moderated by altitude, are reasonably high year round, sufficient to ensure fair rates. While the montane Wola region is not characterised by the intense,

deep weathering and leaching of lower altitude tropical forest regions —
where iron-and-aluminium-oxide-rich and high-kaolinite-clay-content soils
are common — its climate nevertheless encourages a vigorous rate of activity.
The outcome of this climatically promoted activity on soil processes depends
on a number of other factors, explored in the coming chapters, including:
parent material worked on; topographical influences on run-off, sunlight
interception and so on; vegetational cover, decomposition and organic matter
input; and soil related characteristics such as permeability, infiltration capacity, darkness affecting heat absorption, rate and depth of leaching, and so on.

SECTION III

THE LAND RESOURCES FACTOR

Nais g^enkden diy shongiriy haerob sin dorpind obokor. Nonk g^enkden diy onduw haerob sin dorpind obokor. Aenk diy wen boi sa sin dorpind obokor. Ol hombuwn luw hezaembakor ye. Hat ngo hombuwn tiybakor dorpind obokor. Iyb muw diy naelhund diy chay diy hat diy tibakor.

If small boys grow beards, *dorpind* will come. If small girls develop breasts, *dorpind* will come. If newly planted pandans bear nuts, *dorpind* will come. Everyone will be squashed, eh? All of the sky up there will fall down with the coming of *dorpind*. Gravel, charcoal, rain and sky will fall.

Pol Kot

CHAPTER 5

ETHNOGEOSCIENCE: TOPOGRAPHY AND GEOLOGY

Famine we commonly associate with adverse climatic occurrences, not geological events, but according to the Wola both may be implicated in causing crop failure and hungry times. In addition to unusual weather occurrences like periodic "droughts" (sometimes with potentially devastating frosts), and extended rainy spells, they cite tectonic events like earthquakes and volcanic eruptions as famine related phenomena. They maintain that an earth tremor in the morning foretells a hungry time ahead with sweet potato prolifically growing foliage at the expense of small and stringy tubers; an afternoon tremor indicates the reverse, a time of plenty. The association of some of their region's mineral oil and gas leaks with the flight of the mythical white-skinned woman who occasionally visits hardship on them, is another geological connection with famine and climate. A fall of hot volcanic ash from the sky is another geological threat that potentially menaces food supplies, blanketing gardens and killing crops, as legend recalls has happened in the past, with death on a plague-like scale, animals suffocated, streams choked and plants smothered.

The geophysical resources of a region, its topography and geology, both supply, and influence interactions between, the inorganic raw materials that feature in soil forming processes. The rocks that underlie any location, and their susceptibility to the weathering actions of the prevailing climate, contribute fundamentally to the soil resources that occur there, their development and agricultural potential. In the New Guinea highlands these involve the breakdown not only of rocks originating long ago in previous geological eras, but also relatively contemporary volcanic contributions. And these processes are fairly vigorous in Wola country, with its young and rugged landscape.

The field of ethnogeoscience, like ethnometeorology, is relatively neglected, although it too recommends itself, offering scope for intriguing investigations. It enquires into local perceptions of land resources. The interest that people have in these, how they classify them, their explanation and understanding of related natural processes, are again central to how they manage their region's resources. The topography of any place sets limits on the activities of the human population living there. Cultivating steep mountain slopes like those found in the New Guinea highlands demands different strategies to farming on level plains. Rocks outcropping at the surface, or impervious bedrocks which inhibit ready drainage, and so on, likewise present problems to farmers where they occur on their land. And the terrain of any region bears the stamp of its underlying lithology, influencing soil and other resources. The land plays a prominent part in both the creation and loss of soil resources. Different locales demand varying conservation strategies, with contour cultivation, cover crops, windbreaks, and so on. Soil erosion is more of a problem on some terrain than others, as people like the Wola are well aware; the steeper a slope the potentially more serious run-off losses, the flatter and more exposed an area the more vulnerable to windborne losses, and so forth.

I. TOPOGRAPHY

The countryside of the Wola region, like much else in their lives, is difficult to evoke in mere words, one needs to experience it, to appreciate truly its natural beauty and sometimes hair-raising awesomeness. It is necessary that you have some feel for it, sitting in your chair, to appreciate Wola attitudes to, and their management of their natural resources. You should imagine large, unspoilt and unpolluted tracts of countryside where panoramic views are commoner from the heavily settled parts of valleysides, where gardens and secondary grassy regrowth are more extensive, opening up scenes, than they are in the forest. But this cultivated country has a tamer quality to it, with evidence of wide-spread human occupation and activity. It lacks the raw wilderness feel of forested tracts, where gigantic trees and undergrowth hem you in and you rarely see the splendour of the landscape. If you can transport yourself above this forest screen, as memorably I was airlifted by helicopter during the filming of a documentary, you will see below the ruckled landscape, like a three-dimensional *National Geographic* map, with forested mountains rolling away into the distance, furrowed by steep-sided valleys along which run the silver threads of what are turbulent rivers.

Occasionally, a truly spectacular vista rewards forest travel, giving the weary walker a deep regard for this country, for it is only by slogging across its switchback terrain that you really come to understand something about it. You have to picture yourself climbing a steep, apparently never-ending slope, vegetation obscuring any prospect of its summit. You are travelling by way of a path which probably comprises sticky, slippery clay on which it is difficult to obtain a sure purchase and all too easy to slip over. Or it may be a path on a thick bed of dead leaves which you cannot recognise as a trail at all but your local guides assure you is the route followed by everyone in this part of their country. Or it may be a path across broken terrain where you are scrambling over rocks, paddling ankle-deep across waterlogged patches and balancing along greasy moss-covered tree trunks over ravines and other obstacles. Whatever, it is likely to be a strenuous journey, trying the skills and patience of even a committed walker and explorer. But wet from sweaty exertion, or rain showers, or both, you breast a ridge top and a breathtaking view of the precipitous countryside across which you have been struggling rewards you. It is at such heightened moments that one becomes intimately aware of the place, lungs heaving and heart pounding, as you labour at this altitude to obtain enough oxygen from the thinner atmosphere, lacking the physiological adaptations of local companions, with their increased red corpuscles and so on, who can happily chat and even smoke walking over the same route.

The valley bottoms are no less awe-inspiring nor easier to traverse. The larger valleys have sizeable rivers flowing along them. They are not tranquil watercourses but raging and boiling mountain torrents cascading over boulder-strewn beds and churning through narrow steep-sided rocky defiles. When adjacent to them it is sometimes difficult to make yourself heard above their thunderous roar. And their strength is occasionally borne home by the sight of a large tree trunk tossed along matchstick-like in the surging current. The largest are death-traps, even for strong swimmers, especially when swollen after heavy rains. At various locations, local residents may point out

Plate 5.1 Turbulent mountain rivers like the Was flow along valley bottoms.

Plate 5.2 Looking down on a vine bridge suspended across a narrow gorge in the Was river's course.

large water-splashed boulders on the edge of the white-water which are spots favoured by suicides who, in tormented desperation, throw themselves into the maelstrom to be battered unconscious and drowned.

The crossing of these rivers is a nerve-racking experience, even for someone with no vertigo hang-ups. They are spanned at irregular intervals by vine suspension bridges, frail-looking structures comprising V-shaped cradles which, as they often appear, are rather too insubstantial for comfort. When the vine guys supporting a walkway snapped with a crack and an entire bridge lurched with me swinging above the whirlpooling water below, I knew what it was to experience hair-standing-on-end fear. Another memory that always comes vividly into my mind is the moving sensation one experiences when looking down at the fast flowing water, as if the bridge and yourself are travelling rapidly sideways downstream; the best strategy is not to look down but to trust to your feet finding the tightrope-like walkway, even if it does occasionally drop down alarmingly as you transfer your weight along it and the odd vine strand snaps.

Local Orography

The topography of the Wola region bears the stamp of recent geological events. During the Quaternary era, folding and faulting occurred, which was responsible for the north-west/south-east orientation of today's landscape, evident in the strike of the region's mountain ranges and the course of its sub-parallel rivers (Maps 5.1 and 5.5). The relatively recent occurrence of these orographic events accounts for the rugged and sharp relief, which current vigorous geomorphological processes are continuing to maintain.

Overall the country has a youthful mountainous aspect, raw and precipitous, rivers running along valley floors. The Wola, living on the sides of the valleys, leave intervening rugged watersheds largely unpopulated. The majority of their gardens and most settlement occur between 1600 and 2000 metres (the white valley areas on Map 5.1), although occasionally they garden up to 2400 metres (notably in the Lake Egari region and the headwaters of the Kolpa river). They leave considerable areas unoccupied, some of them substantial mountain ranges; particularly in the east where Mount Giluwe rises to an inhospitable 4367 m., and in the west where Mount Kerewa and Mount Imila ascend to 3235 metres and 2998 metres respectively.

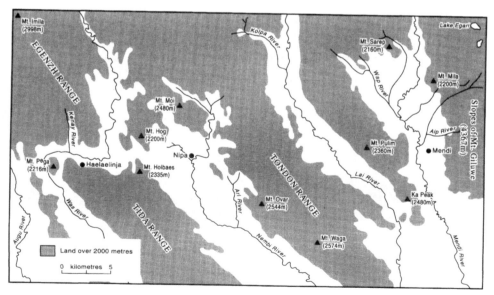

Map 5.1 Topography of the Wola region

The Wola have a considerable vocabulary to describe the topographical features found in their region but, although acutely aware of current geomorphological processes, they have relatively little to say about their origins. They refer to any area of country as *suw*, from large tracts to garden-sized plots (see Chapter 10). The two principal features of their region, mountains and rivers, they call *hat* and *iyb* respectively.[1] The cross-section in Figure 5.1 both illustrates the terrain and gives some further Wola topographical terms (compare Clarke 1971:35).

These people distinguish a mountain summit, called a *hat maenda*, *hat chem* or *hat nay* (lit. mountain tooth), from a ridge top which they call a *kuwng*, although sometimes they use all these words interchangeably. Mountains also have individual proper names, prefixed by the word *hat*, such as *Hat Pega* or *Hat Ebera*. These are not names of single peaks necessarily, they may refer to several points on some feature (for example, *Hat Kongon* refers to two crests at either end of a route called *Kongonhaeret* (lit. Kongon path) across a karst interfluve — Map 5.2). A small hill feature is an *aluw* or *muwtiy*. The large valleys dissecting the countryside are *hat tongowlow*, whereas smaller ravines and steep-sided stream channels are *iyb kaen* or *iyb hegben*. An extensive low-lying valley region of relatively flat terrain is *shuwol* and a very large level area (like Nipa) is *buwkbuwk suw* or *aiben pay* (lit. straight goes). A sinkhole bowl-like feature in karst terrain without a pothole evident at its centre is a *hat shuwol*. A steep mountain slope which is negotiable on foot is a *baeray aez hae* (lit. steep driven stands — i.e. vertical

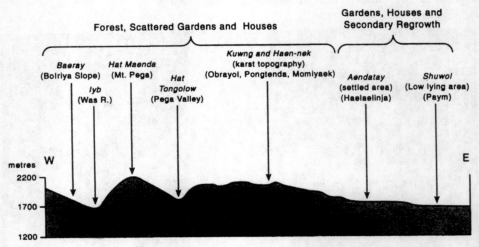

Figure 5.1 Topographical cross-section with some Wola terms for features (*from P.N.G. 1:100,000 Topographic Survey Series, sheet 7585 Kutubu, transect between grid refs. 570210 & 64021; no vertical exaggeration*).

---

[1] The word *suw* also refers to soil, and *iyb* to water generally.

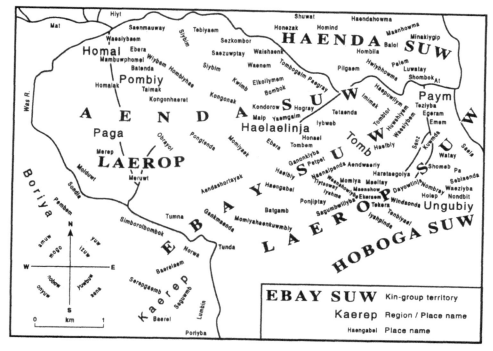

Map 5.2 Local place names in one part of the Was valley

like a fence stake) or *yin pay* (lit. upwards goes), whereas a near-vertical or vertical one which is unnegotiable is a *hat bagiy*.

Rivers, like mountains, have individual proper names, prefixed by the word *iyb*, the general term for any watercourse whatever its size, for instance *Iyb Was* (River Wage) or *Iyb Nemb* (River Nembi). The upstream direction is *iyb ma* (lit. water sprout) and downstream is *iyb tay* (lit. water base), and a confluence is an *iyb taip* or *iyb shoba* (lit. water fork).[2] The bank of a river is *iyb sowlow* (lit. water edge) or *iyb maeray* (lit. water carries). A stretch of particularly wild water with rapids is *iyb liy* (lit. water hits), and a river surging along in flood is *iyb day* (lit. water boils). A waterfall is an *iyb sond*, and a dry gully which fills with a torrent in time of heavy rain is a *h^{e}gben*. A stretch of relatively calm water is an *iyb mael*, as is any pool or pond of water, for example in a flooded sinkhole. A pothole or cave, common features in karst areas, are *haen-nek*. If water flows into such a feature it is called an *iyb bombok* or *iyb hondai* (lit. water enter), whereas if it flows out it is an *iyb komb deray* (lit. water opens out). A cavern that has eroded into a rock arch, across a river for instance, is also called an *iyb bombok* (as at the place Simborolbombok — see Map 5.2).

---

[2] The Wola language clearly intends to draw an analogy between rivers and plants, the three words sprout, base and fork all referring to equivalent botanical features too.

Place and Direction

In addition to the foregoing glossary of topographical terms, the Wola have an extensive vocabulary of place names, which can complement, even substitute in a synoptic sense for words that describe particular types of landscape. Communities divide their territories up into many small named locations (for example Boriya, Obrayol and Paym on Figure 5.1; see Map 5.2). There are two levels of place name. Some names refer to small areas (e.g. Tombem), and others to larger regions (e.g. Kaerep) which are in turn subdivided internally into smaller named places (e.g. Saguwmb). The larger region names commonly apply to forested areas away from main concentrations of settlement, although people unfamiliar with the small place names of heavily settled grassland areas may use the name of one widely known place there like a region-level-name (for example, the place names Haelaelinja, Tomb, Paym and Ungubiy are used in this manner on occasion). When they use place names in a less precise sense for a region, speakers may signify this by suffixing the place name with *naen* or "direction" (e.g. Tomb *naen*, meaning "at or in the vicinity of Tomb").

Any of these locale names, referring to places intimately known to residents, can equally well stand for particular landscapes and their associated topographical features, as they can for their geographical locations. The mention of a place's name can conjure up its attendant physical geography. It can serve to represent a landscape type and substitutes for a technical description. Reference to the place called Paym in a topographical context for example, will suggest a relatively level valley pocket and all that this implies physically, whereas citing the locale of Pongtenda will suggest karst terrain, and so on. When they relate events and tell stories, the Wola frequently depict the terrain featuring in the narrative by referring to somewhere like it known to all those listening, sometimes suffixing the place name with the word *nonbiy* or "like" (e.g. when relating the white-woman myth recorded in the previous chapter, storytellers often cite rockfaces on their territories, such as "Siybim *nonbiy*" — Map 5.2).

In short, indigenous topographical knowledge has a local reference.[3] People are keenly aware, as one might expect, of the terrain of the territories they occupy and cultivate, and think in local terms of places they know. While they have a technical vocabulary relating to features of their landscape, they feel no compunction to group together features that define different landforms. The naming of a place brings to mind its associated relief and other physical factors, obviating any need to categorise features into named landscape classes, equivalent to karst, volcanic plain or whatever. While they may not label different landforms, the implication is not that the Wola are unaware of them, only that they make no overt acknowledgement of them. They have

---

[3] It contrasts with the way Ok speakers to the west of the Strickland river conceive of their homeland, which they map through arrays of animal bone hunting trophies (Hyndman 1991).

no need of a landscape classification, they know the topography of their territories and their constituent locales intimately, living on and cultivating them. Their topographical knowledge is part of a practical tradition, relating to the exploitation and management of specific natural resources, not an abstract intellectual tradition worrying about geophysical theories and landscape processes and origins.

The intention is not to suggest that Wola geographical knowledge is restricted only to the territories that they occupy, although their appreciation of terrain is most acute here. While the extent of their traditional geographical knowledge is miniscule by our worldwide, remote-sensing satellite perspective, they were aware of the geography of quite distant places considering the limited sphere of their personal travels (Sillitoe 1979:26–30), and their horizons have increased dramatically in recent years with migration to find work elsewhere. The Wola were, and are becoming evermore aware of the gross topography of the central cordillera running spine-like through the New Guinea highlands. They refer to regions at lower altitudes for example as *waibsuw*. It is noticeably warmer here, malarial and people can grow some crops uncultivable higher up; it includes locales such as Salaenda and Kopa adjacent to Lake Kutubu.[4] Higher altitude regions they call *ibil*. It is cooler here, frosts are more likely, low cloud-mist is more persistent and trees somewhat stunted; they commonly cite Kandep as such a place.

If you travel anywhere to the north through to the east from where Wolaland is located on the central cordillera of New Guinea, in the direction of Margarima round to Mendi, they say that you go higher in altitude. They consequently refer to these directions as *yin* or up. On the other hand, if you go anywhere to the south through to the west, in the direction of Lake Kutubu and the Hegigio river, you move lower in altitude. They refer to these directions as *onyin* or down. They refer to the remaining two directions, from the north through to the west and the south through to the east, as *amuw* and *aena* respectively. These four directional forms refer to distant places and correspond broadly to the quadrats into which our cardinal directions divide the compass, approximating to the S.W. (S–W quadrant), N.W. (N–W quadrant), N.E. (N–E quadrant), and S.E. (S–E quadrant), as follows:

Wola term	quadrant
onyin or *onyuw*[5]	S – W
amuw	N – W
yin or *yuw*	N – E
ae or *aena*	S – E

---

[4] The Huli speakers living in these places (e.g. Bogbalay and Powa), they call *waib ol* (lit. *waib* men). The Foi of Kutubu are *waisem*, after Wasemi island in the lake.

[5] *Onyuw* also means down and *yuw* up, like *onyin* and *yin*.

It is noteworthy that the four indigenous cardinal directions parallel the region's axes of folding (cf. directional rose Map 5.2 and Map 5.5). They relate to local topography, referring to directions either along the line of folding or transverse across it.

The terms *yin* and *onyin*, when used locally to designate their relevant opposed directions, may not literally be true; the people living at Haelaelinja for example, refer to Nipa as *yin* whereas it is at a lower altitude to them and to Boriya as *onyin* when it is at a higher altitude. They have another set of terms which they use for more nearby places which prevent any confusion. These are not specific direction markers but relate to local topography with respect to the place named, hence two places in the same direction having different terrains will be prefixed by different markers. These relatively nearby topographically related designations are as follows:

Wola term		Topographical designation
nearby	**more distant**	
nobow	onyuw	somewhere lower in altitude than the place where the speaker is (i.e. down a valleyside)
yuwbuw	yuw.	somewhere higher in altitude than the place where the speaker is (i.e. higher up a mountain side)
izuw		somewhere at the same altitude relative to the speaker with no valley between (i.e. along the same valleyside)
mobo	mogo	somewhere at the same altitude relative to the speaker separated by a valley (i.e. across a valley)
or *mor*		

The Wola use these place reference terms when yodelling messages to one another, called *kort bay*; by prefixing a place name with one of these readily audible and euphonious prefixes they attract the attention of those there. The terms are reciprocal (i.e. someone at a place referred to as *nobow* by a caller will refer to their location as *yuw* in response, or at an *izuw* place will reciprocate with *izuw*). These location terms occur in pairs, which also gives added specificity to bearings when used in conversation, indicating if the speaker is referring to a nearby or a slightly more distant location.

The location markers reflect the local focus of indigenous geographical knowledge and perception of direction. They are initially an extension of geographical awareness, dependent on knowledge of nearby physical terrain, until regions of unknown topography are reached, when the four distant general direction markers take over. As two men put it to me when we were discussing these location markers, they know what the topography is nearby and so can use the descriptive markers relative to the place where they are, whatever the direction, but the terrain of far off places is largely unknown to them and hence they resort to their four general direction markers; nonetheless two of these latter (*yin* 'up' and *onyin* 'down'), reflecting general altitudinal changes, also double up as more distant topographical location markers.

The implication of the foregoing discussion of Wola landscape recognition and locale-centred knowledge is not that they are unable to appreciate the topographical features that characterise different landform associations found in their region, only that they have no use of any landscape classification with the specific neighbourhood focus of their concerns, which relate to limited domains. The people living in the western Wola region, for example, can comment on the topographical differences between limestone and volcanic country. They point out that the terrain of the former is more precipitous, with steep slopes and keen ridges, some vertical rock exposures and potholes, and sheer-sided, sharp-summitted mountains. The topography of the latter on the other hand is more rolling, they point out, not so broken and fragmented, with longer and gentler slopes, and rounder, more dome-shaped mountains. The boundary between the two is obvious they say, following the Ak river (this is a major geological boundary between limestone and volcanic lithologies, see Map 5.4). But they have no names for these two different landscapes. They have no need of them. Everyone in the region knows not only where any named locale is but also what the terrain is like there, that Pontenda for instance is limestone karst and Sabim on rounded volcanics.

Nevertheless, as outsiders, we have to resort to landform classifications to try and make sense of what is observed and to explain how we think people relate to their environment, for our knowledge is not locally centred, and the names of locales such as Pombiy, Kaerep or wherever bring to mind no landform systems for the reader. The up-shot is that any representation which we make of the topography of their region will differ from that of the people who live there (Brookfield & Brown 1963:29–37). The extent to which this distorts our understanding of their appreciation of their countryside is difficult to gauge, but for me to write in terms of local place names would defeat my anthropological objective to convey something about Wola country and their life there to English readers. There is no way around this conundrum (Hirsch & O'Hanlon 1995). The topographic knowledge of the Wola is too piecemeal and unsystematic for our purposes, depending on their intimate knowledge of restricted territories. They live their landscape whereas we wish to conceptualize it, their concerns are practical, ours intellectual. We need to arrange and classify data on the physical geography of their region to discuss and make sense of it, but in so doing we need to remember that we achieve another, somewhat alien understanding of their country compared to their appreciation of it.

Landforms

The landscape of the Wola region falls into two main structural and relief zones: 1) the strike ranges of folded sedimentary rocks, and 2) the mountains and slopes of volcanic origin. The folded zone predominates, sandwiched between two smaller regions of volcanics, one in the east and the other in the

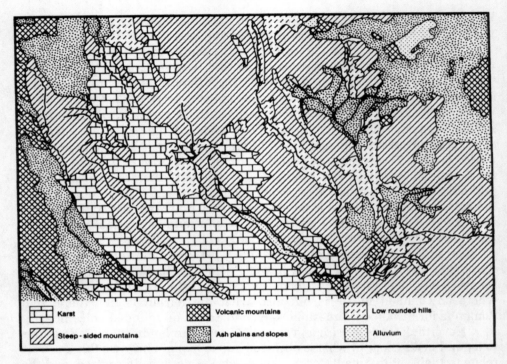

Map 5.3 Landform units of the Wola Region

west. The following landform units[6] can be distinguished within each of these two zones (Map 5.3):

Volcanic Zone:
- Volcanic mountains, hills and ridges (including scoria cones, necks etc.).
- Volcanic ash plains and footslopes.

Folded Zone:
- Steep-sided mountain ridges and escarpments (both limestone and non-calcareous sedimentary rocks).
- Polygonal karst and dolines.
- Low rounded hills and ridges (mainly non-calcareous sedimentary rocks).
- Alluvium (ash derived terraces and piedmont fans, flood-plains and peaty swamp plains).

---

[6] These units are largely an amalgamation of the C.S.I.R.O. 1965 land system classes, amended according to field observations and Radcliffe's 1986 work in Upper Mendi.

The **volcanic mountains, hills and ridges** comprise the rugged mountainous summit areas and ridges of volcanoes, and associated relatively gentle upland lava plateaux. A radial pattern of streams drains these areas. This landform includes younger volcanic features too, such as small adventive ash and scoria cones, necks, dykes and small valley lava flows, which occur scattered on both ash plains and sedimentary rock country.

The **volcanic ash plains and footslopes** comprise broadly concentric zones around volcanoes, built up from the vast quantities of tephra and other pyroclastic material ejected, both locally and from sources elsewhere, and deposited on the surrounding landscape. The depth of tephra blanketing these areas varies but it is typically several metres thick (up to 20 metres). The piedmont slopes are long and gentle (slopes 5° to 20°), comprising lava flows,

Plate 5.3 A limestone pinnacle in the Ak valley, typical of karst country.

scoria and ash accumulations. The plains, on the other hand, are gently undulating to rolling (slopes 1° to 5°). The drainage pattern is again radial, although the plains are poorly drained in places. The streams that dissect the footslopes and plains have carved V-shaped valleys, with local relief varying from 30 to 100 metres. The lower sections of the long slopes have been dissected into low hill ridges in parts, notably in the vicinity of lake Kutubu.

The **steep-sided mountain ridges and escarpments** comprise the strongly folded and faulted ranges. They occur on both limestone and sandstone, and also where both of these parent materials occur interbedded together with mudstones. All are characterised by steep slopes (20° to 50°) having an internal relief commonly ranging between 300 and 600 metres. The ridges on the limestone are perhaps sharper crested and more prominent than else-

Plate 5.4 A doline in limestone terrain (the sinkhole is near the many-crowned pandan at the centre).

where, with cliffs more common. Although structurally little influenced by Quaternary volcanic events, these rugged regions are covered in part with a mantle of tephra deposits, between 1 to 5 metres thick, commonly on gentle to moderate slopes (<35°). A series of major subparallel rivers run through these areas, typically along fault lines or beds of less resistant rocks, giving a strong element of structural control over drainage (watercourses flowing largely along the axis of folding, from north-west to south-east). These rivers carry surface run-off from neighbouring volcanic areas and non-calcareous sediments, with a few short tributaries off the limestone. Organised surface drainage is largely absent on the limestone where drainage occurs through percolation and subterranean passages.

Polygonal karst and dolines comprise fields of enclosed depressions, interspersed with pyramids and towers, having little organised surface drainage. The depressions vary in diameter from 50 metres to over 2 kilometres, and in depth from 20 to 150 metres. They are largely either bowl-shaped dolines, or polygonal-shaped catchment depressions. The towers and pyramids vary in size and shape, from cones to hemispheres to spectacular sharp-edged pinnacles (Jennings and Bik 1962: Williams 1972; Löffler 1977). The depressions frequently have sink holes roughly in their centres, down which disappear the streams that flow in heavy rains. This broken karst topography comprises a considerable part of the interfluves between major rivers on the limestone. A mantle of volcanic ash deposits covers it in places, like it does the steep-sided valleys and ridges that divide it up, largely sealing it off from further karst processes.

The **low rounded hills and ridges** comprise rolling or hummocky hills with subdued local relief up to 75 metres. It often occurs in summit areas and in some places it may represent the remnants of an uplifted mature land surface. Areas of round-crested low hills and ridges are developed mostly on sandstone, siltstone, mudstone and shale. Even with their relatively gentle slopes these areas can be very unstable and their surfaces subject to continuous slumping and earth movement.

Landforms on **alluvium** comprise deposits ranging in age from Pleistocene to Recent. The former, which cover a small area north of Mendi, are valley fills and colluvial fans associated with the disruption of drainage by Pleistocene vulcanism, now deeply dissected by narrow streams (Löffler 1977). The latter younger deposits comprise piedmont fans, and lake and river plains of Recent fine textured alluvium. These areas have characteristic flat or gently sloping topography, sometimes incised by recent streams. Where drainage is poor, organic matter accumulates and peat swamps have developed. These swampy plains, which are common at high elevations, may have peat deposits several metres thick and are traversed by meandering streams with levees.

The steep-sided mountain landform unit predominates across the Kerewa-Giluwe region, comprising over two-fifths of the area (Table 5.1). The other two units that cover substantial areas are karst landscape and volcanic piedmont slopes and ash plains, each making up about one-fifth of the region. The three remaining landform units cover small areas in comparison. Alluvial

Table 5.1
The areas and relative proportions of the Kerewa-Giluwe region covered by different landforms

	Entire Kerewa-Giluwe region		**West of region** (area west of diagonal line following the karst boundary to the west of the Nembi river and east of the Was river)	
	Area (km^2)	% of area	Area (km^2)	% of area
Steep-sided mountains	1060.6	42.8	174.6	18.4
Karst	517.9	20.9	418.5	44.1
Volcanic plains/footslopes	498.1	20.1	182.2	19.2
Rounded hills and ridges	190.8	7.7	34.2	3.6
Volcanic mountains	175.9	7.1	135.7	14.3
Alluvial landforms	34.7	1.4	3.8	0.4
TOTAL	2478	100	949	100

landforms are particularly small in area (although there are many narrow areas of flood-plain and terrace along the major rivers, too small to show on Map 5.3; their inclusion with other landforms does not substantially distort the calculations). The relative proportions of the different landforms varies in different parts of the Wola region. In the west, the predominating landform unit is karst, comprising over two-fifths of the area, with volcanic footslopes and plains and steep-sided mountains each covering a little under one-fifth of the area respectively (Table 5.1).

II. GEOLOGY

The Wola may be characterised in popular parlance as living, until relatively recently, in the "stone age", although somewhat malapropos, for stone has never featured prominently in their lives. The stone age epithet derives from the polished stone axes that men previously carried and used widely, made from imported, fine-grained, hard metamorphic rocks. Regarding the use of local geological resources today, a visitor would notice in many houseyards piles of blackened stones adjacent to shallow charcoal-stained holes. These are earth ovens, where people, to use the Pidgin term, *mumu*-cook food using hot stones. It is the most common use they make of local rocks, and before large communal pig kills and pork exchanges communities amass astonishing numbers of stones to use in long pit ovens (Ryan 1961; Sillitoe 1979; Lederman 1986), individuals sometimes struggling back with heavy boulders to break up and add to the piles.

While the Wola may not exploit the rocks of their region to any great extent, its geology has impressed itself in other ways on their culture. They associate some of the countless potholes that occur throughout the limestone for example, with spirit forces, particularly those containing a deep pool of

Plate 5.5 Stratification evident in a limestone exposure at Bombok (solution weathering is evident in the calcareous surface deposit).

water. They occasionally tossed offerings of pork into these dank and eerie fissures. Some are of considerable depth to judge from the sound of stones dropped down them. At a place in the Was valley called Simborolbombok (Map 5.2), there are some caves high above the river where local inhabitants say the spirits of the deceased reside, those who suffer violent deaths going to the left and others to the right hand caves.

Some geological formations add to the awesomeness of the mountainous Wola homeland. The foregoing spirit caves for instance are adjacent to a massive rock arch across the river which is used as a bridge, the crossing of which involves a thrilling climb up notched logs until you are perched tens of metres above the thundering river below, when you scramble using root handholds up a steep slope high onto the opposite bank. Other geological features associated with limestone include rock pinnacles and cliffs weathered to weird shapes, standing up like enormous jagged teeth. An image that frequently comes to mind is of a chipped and partly dissolved old conical sugarloaf, as occurred to an early Highlands explorer who so named one of the mountains beyond Mendi. In the damp climate, solution processes often result in the rocks having smooth surfaces, sometimes featuring a cake-icing-like white calcareous deposit.

The moss-clad rockfaces of Wolaland, called *haen bagiy*, and overhangs, called *haen abaenda*, have an eldritch and remote atmosphere, particularly in the forest, reminiscent of film-sets for the buccaneering archaeologist Indiana Jones. They are occasionally smoke-stained in part, evidence of fires kindled

Plate 5.6 Peering down a cleft in the limestone into a pothole.

close under them when people have stopped to shelter from the rain, or to rest and enjoy a tobacco smoke, or to cook and consume secretly an animal caught nearby; these locations, at the base of rockfaces are called *haen tay* (lit. rock base). Others may have served as a necropolis and today have bones and pieces of grave platform-wood littered around them, they are frightening and spooky places to the Wola, associated with dangerous ghosts.

Historical Geology and Parent Materials

The contemporary landmass of New Guinea is the result of a series of complex events associated largely with the movements of two crustal plates, the Australian plate moving northwards and the Pacific plate moving westwards. The island occupies an unstable, mobile zone adjacent to these two plates, which currently contact one another along New Guinea's north coast. Subject to the intense geotectonic forces associated with plate margins, young folded and faulted mountains characterise the region, together with contemporary volcanic and seismic activity (Davies & Smith 1971).

The early geological history of New Guinea is largely unknown, evidence being too fragmentary, comprising a few scattered pre-Permian formations. It is not until the Mesozoic that reliable inferences can be made (Pieters 1982). During this period a shelf-like extension of the southern Australian Palaeozoic basement block is thought to have reached to current day New Guinea, a proto-island forming on it. This margin experienced extensive terrestrial deposition, sedimentation continuing throughout much of the Tertiary, with thick deposits laid down, except on a few high fault-bounded anticlines which stood proud. Along the mobile belt, periods of volcanism alternated with periods of deposition, both of continentally and volcanically derived sediments. They filled the geosynclines and basins that developed, reaching their full extent in the Cretaceous through the Oligocene, with thick deposits of shale, shelf limestones and marine volcanics.

In the Neogene, extensive folding and large throw faulting occurred, due to the movement of the adjacent plates (Bain 1973), together with isostatic uplift, which continues today (as evidenced by the staircase-like raised platforms along parts of the north coast of New Guinea). During the intense tectonism of the Miocene large plutons were also emplaced, and island arc volcanoes, also evident today, emerged. In summary, although the present island land mass took shape in the lower Miocene, with the emergence of the New Guinea mobile belt (which was subsequently subjected to intense erosion), it was not until the upper Pliocene that the present landscape became evident, with recent folding, faulting and uplift, together with volcanism.

The geology of the Wola region comprises, in gross terms, an extensive area of folded sedimentary rocks, limestone predominating (together with mudstones, shales and sandstones), between two areas of volcanic rocks, comprising in the main basalt lavas, breccia, tuff and ash (Map 5.4). These two contrasting lithologies, evident in the two major relief zones described, also reflect the geological history of the region (Perry 1965; Brown and Robinson 1977). Throughout the Mesozoic era and into the Cainozoic, the Wola region was part of the Papuan Geosyncline, a subsiding belt beyond the north-eastern margin of the Australian continent. Although geologists distinguish some ancient periods of uplift, submergence and movement during this one hundred million plus years of Earth history, in summary, throughout this era great thicknesses of marine sediments were laid down continuously, the deposited material derived from the land mass to the south, with continental shelf and slope deposition (the Maril shale, through to Urubea sandstone series — Table 5.2). In the Tertiary period the seas covering the region favoured the building up of thick deposits of reef limestone (the Darai and

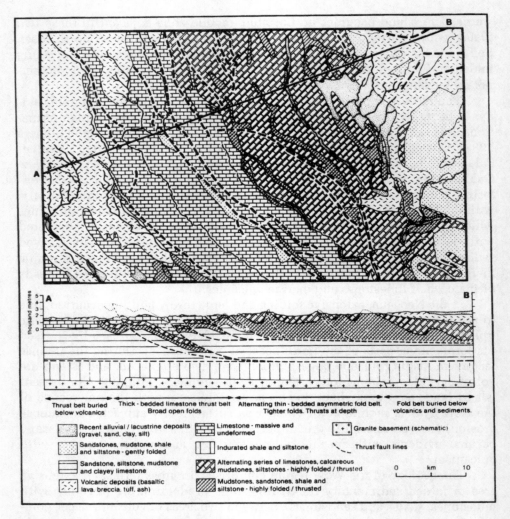

Map 5.4 Geology of the Wola region

Nembi limestone series — Table 5.2). And in the Oligocene and Miocene epochs uplift and erosion of adjacent high lands provided increased amounts of fine sedimentary material and the facies became more clastic compared to the earlier pure calcareous ones, with muds and shell-sand banks giving limey muds, sandy and shelly limestones, siltstones and mudstones (Ka mudstone to the Aure beds series — Table 5.2).

Regarding the present day land surface, it is the later Tertiary limestones and mudstones that comprise the greater part of the region that is underlaid by sedimentary rocks, the majority of Miocene age. The end of the Tertiary and the beginning of the Quaternary periods marked an orogenic phase when these thick sedimentary deposits were massively deformed (Hill 1991). The

Table 5.2
Geology of the Wola region: detailed lithology, local formation names and their geological periods

Mapping Unit	Geological Period/Epoch	Local Formation	Lithology
Recent alluvial/lacustrine deposits	Holocene	—	Angular rock fragments, clay matrix; chaotic scree deposits and talus cones.
	Holocene to Pleistocene	—	Gravel, sand, silt, mud, clay, peat; fluviatile, lacustrine valley fill deposits, alluvial plain deposits, volcanic ash deposits.
	Pleistocene	Kagua clays	Light grey carbonaceous non-calcareous clay, volcanic and cherty lithic sand and tuff laminae, gravel bed; lacustrine deposits.
Volcanic deposits	Pleistocene	Giluwe volcanics	Basaltic (shoshonitic) lava, ash, tuff and agglomerate.
	Pleistocene	Kerewa volcanics	Fine grained basaltic lava, lava breccia, andesitic agglomerate, tuff — with areas of volcaniclastic andesitic and basaltic breccia; reworked agglomerate, tuff intercalated volcanically derived conglomerate, sandstone; laharic deposits.
Highly folded/thrusted series of limestones, calcareous mudstones and siltstones	Early to Middle Miocene	Ka mudstone	Grey calcareous mudstone, interbedded siltstone.
		Mala limestone	Yellowish to grey-brown argillaceous micrite.
	Early Miocene to Late Oligocene	Lai siltstone	Grey calcareous siltstone, silty mudstone.
		Nembi limestone	Pale grey to brown foraminiferal micrite, fine grained calcarenite, rare chert nodules.
	Early Oligocene to Early Eocene	Mendi formation	Dense foraminiferal micrite, fine grained calcarenite, chert nodules, minor grey siltstone, silty mudstone.
	Late Palaeocene	Moogli mudstone	Soft, grey foraminiferal calcareous mudstone, minor quartz sandstone, siltstone interbeds.
Gently folded sandstone, mudstone, shale	Pliocene to Late Miocene	Orubadi beds	Blue-grey calcareous mudstone, shale; interbeds siltstone, sandstone.
		Aure beds	Greenish grey greywacke sandstone, dark grey mudstone, siltstone, minor conglomerate.
Limestone	Late Miocene to Late Eocene?	Darai limestone	Massive to thick bedded limestone; white to cream algalforaminiferal biomicrite, calcarenite, calcirudite, blocks of coralgal biosparite, minor dolomite, rare black chert lenses.
Highly folded/thrusted mudstone, sandstone	Late Palaeocene	Urubea sandstone	Grey to green calcareous, glauconitic sandstone, siltstone.
	Palaeocene to Late Cretaceous	Chim formation	Massive dark grey calcareous shaley mudstone, fine grained laminated sandstone, siltstone, shale, minor calcarenite.
Sandstone, siltstone, mudstone	Early to Late Cretaceous	Kerabi formation	Fine grained feldspathic, glauconitic sandstone, siltstone, mudstone, interbedded silty argillaceous limestone.
Indurated shale & siltstone	Middle? to Late Jurassic	Maril Shale	Indurated shale and siltstone; some sandstone, limestone, shale, arkose, breccia and conglomerate at base.
Granite	Late Permian	Strickland granite	Pink granite, chlorite, calcite and sericite alteration.

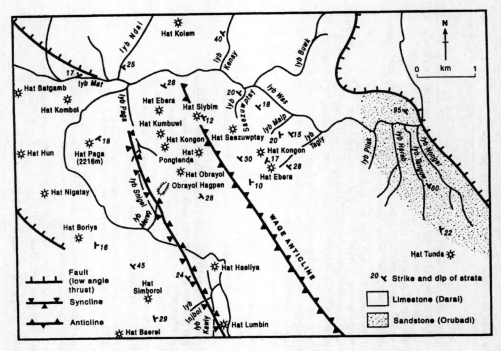

Map 5.5   Orogeny in a localised region of the Was valley

folding, accompanied by thrust faulting, has its north-west to south-east trend directly reflected in the orientation of today's strike ranges (Map 5.5 indicates the extent of the orogeny, giving strike and dip of strata, in a localised region of the Was valley). The present drainage pattern was broadly established following uplift and vigorous river erosion gave a relief approximating to present dimensions. It is from this time that processes start that contributed to the soils of today. Although derived in considerable part from Tertiary sediments, these have only been subject to pedological processes since the Pleistocene epoch over much of the region. The soils are consequently young ones.

The earth movements and dissection of the Pleistocene epoch were accompanied by intermittent volcanic activity which has had a profound effect on both the region's geology and soils, with important consequences for indigenous agriculture. Two centres of volcanic activity have predominantly affected the Wola region: in the east Mount Giluwe, which is a lava shield volcano dotted with numerous craters and scoria cones; and in the west Mount Kerewa and the Doma Peaks, which are respectively a steep-sided conical strato-volcano and a complex volcanic centre of numerous craters and cones (Löffler 1977:72–74). They have deposited thick layers of volcanic rocks and agglomerate. The volcanoes themselves consist of basalt lava flows, pyroclastic rocks and ash deposits which thicken downslope and cover wide piedmont zones (Table 5.2). The volcanic products are spread way beyond the cones and surrounding foot slopes. Extensive ash showers have blanketed a large part of the area and even where ash does not noticeably cover the landscape it has probably contributed in some measure to, and been incorporated to varying extents in, the soil, affecting fertility and local farming prospects.

Local Lithology

The Wola have no equivalent of the foregoing scientific explanation of their region's geological history, with its millenia-long epochs. They are pragmatic not speculative in their approach to their land resources, accepting their existence without any apparent need to explain how they came to be there. Nevertheless, while disinterested in the origins of rocks and lacking to my knowledge any theory or myth-equivalent about how they came to be or how they relate one to another, the Wola are fully aware of the different kinds of rocks that outcrop across the surface of their region, and their properties (compare Brookfield & Brown 1963: 34–35 on Chimbu geology).

The generic Wola term for all rocks is *haen*. The most common they maintain is limestone, called *hathaen*. A wide variety of limestones occur across their region, some markedly fossiliferous and others less so and more fine-grained (Table 5.2). Specimens analysed in section[7] include: a bioclastic limestone containing prominent fossil fragments (including pieces of bivalves, gastropods, echinoids, corals, Bryozoa and foraminifera, notably Globigerinidae) in a sparry cement, probably deposited on a shell bank; and a recrystallised limestone of fine-grained structure with numerous pelagic foraminifera (including Globigerinidae), probably deposited in a quiet lagoon. The Wola are aware of the range of limestones that occur. When fossils are prominent in them, they even comment on them, but they offer no traditional explanation of how they come to be where they are, the terrestrial snail-like appearance of some fossils notwithsanding. Even today, when several men have experienced migrant labour to coastal regions and have seen similar shells on the seashore, they are unforthcoming on the possible origins of limestones.

Regardless of their variety, the Wola make few terminological distinctions between limestones. The exceptions relate to the state of the rock, not its type. They distinguish rotten powdery limestone, found in damp locations (for instance, limestone buried in wet soil) or where pieces of fractured limestone have ground pulverulently together (for instance along a fault line), calling it *haen hok*. They use it occasionally as a pigment to paint face and beard. The soft and friable calcareous mudstones and siltstones are also *haen hok*. Any recent conglomerate-like deposit comprising friable limestone fragments cemented together by calcium carbonate or packed together in a hard clay matrix is called *haen kolkol*. The only use the Wola have for *hathaen*

---

[7] I thank Dr. C. Forbes and Mr. W. Manser, previously of the Sedgwick Museum in Cambridge and the Geology Department of the University of Papua New Guinea respectively, for looking at, and discussing rock specimens with me.

limestone is in earth ovens, although it is second-best to igneous rocks for this purpose. It is superabundant and may be collected from anywhere without restriction.

Occasionally lenses and nodules of chert outcrop in the limestone. This flinty, siliceous mineral occurs in a range of colours, from black and brown through various grey shades to dull white. The Wola call it *aeray*. It is a popular raw material for tool blades, producing a razor-sharp edge when knapped, the darker stone giving the more durable and sharpest-edged tools (Sillitoe 1988:58). It is fairly common, surface concentrations typically occurring along the sides and in the beds of streams, where they are exposed, and eroded fragments are deposited. Individuals are free to collect chert from anywhere, no persons exercising exclusive rights to sources on their territories. They distinguish between fresh chert, with its glassy surface, called simply *aeray*, and chert covered by a dull accretion of lime carbonate, which they call *araytol*, the name meaning literally 'dirty chert'. The Wola refer to any carbonate concretion on the surface of a rock as *tol* or dirt. Stones in the soil frequently occur as 'dirty chert' (see Chapter 10), and any stony soil is designated *suw araytol* (lit. earth dirty-chert, or stony).

The other common sedimentary rock occurring in the region is sandstone. The Wola call it *naenk* if it is consolidated and hard, or *iyb muw* if it is unconsolidated and sandy-soft (they distinguish reduced pale grey sands from oxidised browny yellow sands, calling the former *iyb muw tongom*, [after *tongom* their name for any grey-coloured gleyed soil], and the latter simply *iyb muw*). Sandstone outcrops less often than other sedimentary rocks, although it frequently occurs close to the surface, and is soon exposed when digging, it supporting only a thin soil. Detached rock fragments, often sizeable sandstone boulders, also occur along river courses, where transported and deposited on the banks, notably where they are flatter. Some of these locations are known locally as good sources of sandstone, which is useful intelligence, for pieces of this rock are sought as whetstones for sharpening axe blades (Sillitoe 1988:50). Again there is no restriction on access, anyone may help himself to pieces of rock from any location regardless of territory.

Regarding the igneous rocks that occur in their region, the Wola refer to any dark coloured extrusive or intrusive rock as *huwbiyp*, whether a fine-grained basalt of volcanic origin or a coarser-grained augite porphyry from a dyke intrusion. The lava and dyke sources of these rocks rarely outcrop to their knowledge, the rocks occurring largely at the surface as water-borne pebbles and boulders. The Wola refer to any such rounded water-borne stones as *haen h^{e}gben* (lit. stone water-channel), whatever their size or lithology. Men collect conveniently sized hand-held *huwbiyp* pebbles for use as hammer stones, basalt being the only locally occurring stone that is hard enough for this purpose; other rocks, such as limestone or chert, would shatter if used against any hard material, such as previously the metamorphic rock of stone axe blades. They also favour *huwbiyp* rocks above limestone for earth ovens because they hold more heat, although where they are uncommon people have resort to limestone. Again, there are no restrictions on the collection of this rock, sources are common and open to all to exploit. Igneous rocks also occur sometimes as varying sized stones buried in the soil. Other volcanic debris found in the clayey subsoil include reddish tuff and pyroclastic cinders, called *kolbatindiy*, and a softer, often grey-bluish through to red, ashy agglomerate, called *tiyptiyp*. These extruded materials, occurring buried in the subsoil, vary in hardness from friable masses, called *kolbatindiy* or *tiyptiyp kolkol*, to unbreakable rock called *kolbatindiy* or *tiyptiyp haen*.

The Wola perceive a connection between the rocks that predominate in any area, notably whether sedimentary or volcanic, and its landscape, as mentioned earlier. They point out that in the precipitous Was valley, *hathaen* limestone is the bedrock, it outcropping frequently there, sometimes with large cliff-like exposures. *Huwbiyp* basalt on the other hand is relatively uncommon here and found largely adjacent to rivers which transport it, together with pieces of limestone of all shapes and sizes, depositing them along their banks. In the more rolling Ak valley, limestone outcrops are infrequent, and virtually unknown to the west of the river, and basalt stones are very common both along watercourses and elsewhere. The Wola maintain that they are unsure what the bedrock here is because it is mantled in thick clay containing *kolbatindiy* tuff and *tiyptiyp* ash. Nevertheless, they appreciate in some measure that the different geologies, although imperfectly understood, exert some control over the landforms that they can see.

The other major class of rocks distinguished by the Wola includes those from which their stone axes were made (Sillitoe 1988:45). They represent in the indigenous scheme the third main rock class of geological science, being metamorphic in origin. They do not occur in the Wola region to their knowledge, they imported axes ready-made from sources to the east and north. The majority of axe heads originated from sources located in the Western Highlands Province to the north-east. These are a dark stone, ranging from black through to various dark blues and greens, sometimes mottled together. They consist of thermally metamorphosed greywackes or basalts varying geologically in character according to the quarry, although all petrographically similar. A few blades originated from river bed sources located in the Enga Province to the north. They are grey, varying in hue from dark to light and sometimes mottled (as a result of iron oxide concentrations), and consist of a low-grade metamorphosed fine-grained chert.

The Wola distinguish between blades originating from these two sources. The nigrescent ones originating from the north-east they call *aiben* (or *mumung* or *haelboi*), and the less common grey ones coming from the north they call *paym*. Although colour, and consequently source, are a fair indication of the class to which many blades belong, these are not the only criteria by which the Wola classify them. They have a third somewhat residual, category they call *haez*, which comprises both pale and dark coloured stone. It includes axe heads that are not highly polished in appearance. They consist of glaucophane schist, a metamorphic rock of somewhat more fragile structure than other axe stones. Practical considerations, relating to the hardness and durability of the stone, determine the classing of these axes. The quality of the stone determined a blade's value, together with its size. Men preferred small *aiben* blades to large *paym* ones, up to an indeterminate point where the difference in size tipped the scales the other way. The value of an axe depended on the size of its blade. A fair-sized one changed hands in return for highly valued wealth objects, such as pearl shells and pigs; indeed stone axes were so valuable that men occasionally gave them as wealth in ceremonial exchanges (Sillitoe 1979). Smaller axe blades changed hands for less valued items like cowrie shells, feathers and so on.

All other rocks the Wola call simply *haen*, regardless of the fact that they may differ markedly from the above named rocks and from one another. All of them occur infrequently. Rocks which they lump together as undifferentiated "stones" include crystals of calcite, found precipitated occasionally in the limestone and, somewhat similar in appearance, the odd piece of whitish quartz. They also include metaliferous nodules of iron pyrites, and any conglomerate (one example of which, collected from Norwa, comprised fragments of *hathaen* limestone, *huwbiyp* basalt and *aeray* chert cemented together by calcium carbonate). These are all residual rocks, of little interest to the Wola except as curiosities.

Other non-solid geological resources that interest the Wola are mineral oil and gas, which emerge at seeps dotted across their region on the limestone. The soil at these places they point out is markedly black, presumably due to high deposition of carbonaceous matter. They refer to the gas as *eltaekis*. It sometimes emerges from the ground with a bubbling sound, and they are well aware of its explosive qualities saying that it is especially violent during periods of fine weather.[8] The mineral oil they call *dez* or *poborgaim*. They collect it occasionally and use it to smear on objects, to give them a lustre. At certain locations the oil seeps out into pools of water, floating in an iridescent film. According to the Wola it increases noticeably in dry *baen* spells, water levels falling and the flow through pools decreasing or ceasing, not flushing the oil away but allowing it to accumulate. This is the best time to collect oil, which individuals do by placing the palm of the hand on the floating oil and scraping off the film of oil that adheres to it on the rim of a vessel. But they dislike its smell and use it rarely. Lice dislike the oil too, and men use it as a repellent when suffering from infestation. It takes a considerable time and some patience to collect a reasonable amount of oil, another reason why it is not popular. When collected, men place no transactable value on it.

These gas and oil reserves are currently the subject of intensive geological assessment and survey by oil companies (including Chevron, B.P., and Exon, among others), not only in the Wola region but also throughout the Southern Highlands Province. They are consequently of much greater interest to the Wola than ever before, particularly with the recent sinking of an oil well at Yagipu adjacent to lake Kutubu and drilling along the Mubi valley.

---

[8] Wenja Neleb, a friend of mine, burnt off his eyebrows and singed his hair in a demonstration with a glowing ember, igniting the gas with a flash of blue flame.

The Wola are particularly concerned that the tapping of reserves at Kutubu is draining the oil from under their region too, saying that its depletion is responsible for their soil recently showing signs of becoming more dry and hard, with a decline in yields, which is particularly irksome to them as they are receiving no share of the royalties.

While they apparently have no traditional account for the origins of many geological features, the Wola associate the genesis of *eltaekis* gas and *dez* mineral oil seeps with the mythical white-skinned women Horwar Saliyn. They link some seeps with her agonised flight, related in the previous chapter, with a hot stone thrust into her vagina. But possibly even more awesome and frightening than her capricious influence on the weather, causing crop failure and famine, is the legendary time of hunger and starvation associated with volcanic activity.

Tectonic Activity

While the local volcanoes in the Kerawa and Giluwe region are no longer active, with the exception of some solfataric emissions on the Doma Peaks (Taylor 1971),[9] volcanic ash has nevertheless continued to fall across the Wola area up to very recent times (Pain and Blong 1976; Blong 1982; Bik 1967). Several young volcanic ash layers (or tephra beds) have been deposited as a result of contemporary vulcanism several hundreds of kilometres away in the Bismarck Sea (e.g. Long Island which erupted some 300 years ago — Map 5.6). This distribution of ash in the atmosphere far beyond regions of

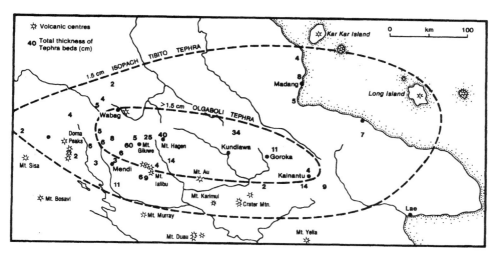

Map 5.6 The distribution of recent (<50,000 years B.P.) volcanic ash deposits across the Highlands (*after Pain & Blong 1979*)

---

[9] Mount Giluwe last erupted some 200,000 years B.P., Löffler 1977.

active vulcanism is significant for soil formation and rejuvenation, and hence agricultural potential and human activity.

It has also had an appreciable impact on people's perceptions of their relationship with, and ability to manage, their environment; it even prompts fears of possible future catastrophes. Legend tells of a "time of darkness", when an enormous ash fall turned days into night. Versions of the legend are widespread across parts of Papua New Guinea (see Blong 1982; Glasse 1963; Skeldon 1977), although people everywhere have not heard it; my friends in the Was valley, for instance, could not recall such a specific legend relating to an ash shower, yet many of their neighbours are familiar with it (for example the Kewa, Huli, Enga and Kaluli, and possibly even other Wola speakers in the Nembi valley). The gist of the legend, as reported in surrounding regions, is as follows:

> Long ago, the sky remained dark one morning, instead of as usual becoming light. There was like a storm, only not of rain but of a white sandy stuff. It rained like this for several days, the sun not coming up at all but everywhere remaining in darkness. The people stayed inside their houses. When the ash-rain stopped, and the sun appeared again, and they went outside, people found that the white ash had covered everything. It had destroyed unprotected crops in gardens and killed many trees, which had lost their leaves. It choked up streams, and smothered animals and birds. It was a time of hunger, food was short with gardens ruined. But following their replanting, when the white stuff had cooled off, crops grew quickly, and very large.

> At some places people mention various precautions to be taken in the event of another ash fall to prevent disaster. Among the neighbouring Huli (Glasse 1963: 271), they say that kinsmen should build communal houses at the first sign of an ash storm, called *bingi*,[10] and collect sufficient food and water to last them several days. No one should leave the houses during the fall, except sole surviving sons. People should spread grass over their gardens, and refrain from sexual intercourse, wives returning to their natal places. If they do not observe these prescriptions, considerable loss of life will ensue.

While those living in the Was valley deny any knowledge of such a legend founded on living oral memory that explicitly talks of a time of darkness when ash fell from the sky, they do have a brief foreboding myth that approximates many features of the legend, and may evidence their ancestors' experience of a volcanic ash fall long ago, recalled today somewhat fantastically in the myth. It talks about *dorpind* coming, although no one could tell

---

[10] Bilingual Wola friends tell me that the equivalent Wola word is *nenjiy*, which is an eclipse of the sun (someone who, when a boy, experienced an eclipse of about an hour's duration, while in the company of two older brothers, described to me how they sat indoors in fright holding their heads, and he was told that none of them could go outside again until the sun reappeared or else they would die, because only someone without any siblings could do so).

me exactly what *dorpind* would be, except that it would be a disaster falling from the sky. The premonitory myth is as follows:

> If casuarinas and pandans bear large numbers of seedlings, germinating prolifically like grasses, or if several women in a community give birth simultaneously, then *dorpind* will come. Or if a young girl prematurely develops breasts, or a small boy precociously grows a beard, then *dorpind* will come. Or if a juvenile pandan bears a cluster of nuts before fully grown, or seven generations pass in the descent of any *sem* from its acknowledged founding ancestor, then *dorpind* will come.
>
> When *dorpind* comes the sky will fall down on everyone, with heavy rain, clouds, gravel, hailstones, or whatever is up there, falling down on us. When *dorpind* comes, it will squash all those outside. It will kill us Wola up here in the highlands, but not the Foi down at lake Kutubu. When *dorpind* comes, those staying indoors will need strongly constructed houses that will not fall down and crush them. This is the reason why our ancestors, and we today, build strong houses with sturdy walls of beech wood and thick cane grass thatch, so that if *dorpind* comes they will not collapse on their occupants. While *dorpind* falls, all of us will have to stay inside our houses, except for only children, those with no siblings, who may go outside without *dorpind* injuring them. They will fetch food for their relatives confined indoors.[11]

People with whom I have discussed the *dorpind* myth maintain that they do not know what it is that will fall on them 'when *dorpind* comes'. Many maintain that it will be the sky, and as one man expressed it "we don't know what the sky is made of, do you?" When I related to my friends the time of darkness legend as reported elsewhere, they surmised that *dorpind* will bring a fall of black stones, ash clouds, and sandy gravel, commenting that their ancestors had overlooked to pass this information on, or were ignorant of it. It is noteworthy in this context that the forebodings in the myth relate to precocious fertility, acknowledgement perhaps of the otherwise unremarked soil rejuvinating properties of volcanic ash, that *dorpind* marks a possible increase in fertility.

It is unclear whether these legends refer to one widespread ash fall associated with a single catastrophic eruption, or several different lesser volcanic eruptions.[12] Whatever, they indicate that in the past volcanism has figured cataclysmically in people's lives. In all probability it will again, their apprehensions and precautionary lore may prove no mere mythical fantasy but life preserving advice, for the island of New Guinea, situated geographically adjacent to a destructive plate margin, on the so-called Pacific "ring of fire", can expect to experience further volcanic eruptions as subduction proceeds

---

[11] I am grateful to Pol Kot and Mayka Sal for relating the story of *dorpind* to me.
[12] Different writers have mooted various volcanic eruptions as possible candidates: Watson (1963) suggested distant Krakatoa; Glasse (1963) the local Doma Peaks; and Blong (1982), in a detailed and sophisticated analysis of the evidence, Long Island in the Bismarck Sea.

(Johnson 1976). The Highlanders regularly experience evidence of this plate movement with earthquakes and tremors, other phenomena associated with such tectonic activity (Denham 1969). These point to continued fault activity in the region (Map 5.5), other evidence for which includes rejuvenated streams.

The Wola refer to earthquakes and tremors as *duwlbaeray*. They say that small sub-human creatures who live under the ground cause them.[13] Every so often they gather at large stones that protrude at, or near the surface and shake them, so making the ground move under foot. When encouraged to describe the creatures and their behaviour further, my friends maintained that they know no more. They point out that the explanation of earth movements is only speculation or *konay waeray* (lit. thoughts made) because no one has ever seen the subterranean creatures nor found any evidence for their existence. It is all hearsay.

Earth tremors are of infrequent occurrence and not particularly strong usually, although they are a strange sensation, especially when experienced for the first time. When we first experienced a *duwlbaeray*, my wife thought that some people were playing a practical joke and shaking our house, and I thought that a large pig was scratching its back on one corner and causing the whole structure to sway! Avalanches of earth, rocks and trees, on the other hand, are considerably more formidable, and potentially life threatening, although fortunately fairly rare too. The sound of tons of material crashing down a hillside and the sight of the bare soil scar and tangled tree trunks, like a gigantic foot having slipped on the mountainside, tearing away the vegetation to expose the earth and rock deep below, is not one readily forgotten. It is a reminder of nature's fearsome power, which is somehow accentuated for me — an Englishman used to intensively farmed countryside and concrete dominated urban sprawls — by the relatively untrammelled and largely unmodified New Guinea highlands.

Contemporary Geomorphological Processes

The Wola are acutely aware of the destruction that landslides can cause. On occasion they witness the sweeping away of sweet potato mounds and crops, sometimes taking all the topsoil and exposing the sticky uncultivable clay below. In the forest they occasionally see a swathe of trees and shrubs uprooted and transported downhill on a wave of earth and stones. They associate landslides with heavy rainfall, not tectonic movements. Intense storms will not only wash soil particles off steep, exposed slopes, but also destabilise soil masses, acting lubricant-like when the soil approaches saturation point, triggering off avalanches. The Wola refer to a landslide as *suw engay may* or *suw mundit may* (lit. earth avalanche takes). They distinguish between small falls of

---

[13] The people living in the Hagen region have a similar idea, believing in 'Earthquake People' called *Kidlömbömb-wamb* (Strauss & Tischner 1962).

soil called *suw kombay* (lit. earth open), avalanches of rocks called *haen kombay* (lit. rocks open), and large slides that sweep over trees in their path called *iysh tam ebay* (lit. trees fall-and-cover come). The cracks that appear in the soil as a landslide starts are called *suw aevril*, they mark the line along which a mass of soil is detaching itself from the rest, to fall away in a landslide.

Besides catastrophic soil loss in landslides, other less obvious contemporary erosive processes are constantly active. They include mass movement on slopes, resulting from vigorous downcutting by streams, serving to maintain many steep slopes. Creep, which occurs on nearly all slopes, and rotational slipping due to alternate weathering and slumping, which occurs on the steeper ones, together with the occasional large scale movement. In addition to carving the landscape, rivers carry away a considerable volume of the eroded material, especially when in spate.[14] Weathering is also rapid in the relatively cool and wet climate, especially given the unresistant nature of many of the rocks (Haantjens & Bleeker 1970). On the limestone it occurs mostly through solution, which is fairly swift because water has a fairly high CO_2 content under the prevailing climatic and vegetational conditions. Experiencing these vigorous processes of change, any locale exists against time; these agencies have transformed some locations I have known unrecognisably in twenty or so years.

Depositional features are less evident in the Wola region. The fall-out from previous volcanic eruptions interrupted the drainage system to some extent and, overloading rivers with ash, led to the build-up of alluvial deposits of volcanic detritus. Alluviation, unlike the occasional intermittent ash shower, proceeds as a continuous contemporary process, although restricted largely to the narrow floodplains adjacent to parts of larger rivers. The local residents are familiar with river deposition, particularly on flooded flatter banks, when a veneer of sediment is sometimes left behind, and they point to rounded water-worn pebbles laid down in sedimentary laminae as visible evidence in some banks of previous deposition.

The Wola are aware of the contemporary geomorphological processes actively shaping their country, living in a youthful region experiencing vigorous erosion and obvious modification (compare Clarke 1971:37–38 on Maring landscape). Nevertheless, while familiar with short-term geomorphological processes, witnessing their effects in their daily lives, they do not extend this awareness to explain longer-term geological processes. The time scale and gradual cumulative impact of geological events appears to be too considerable for them to associate the two together, to make the geologists' uniformitarian jump. But they can extend their observation of contemporary erosive and depositional processes forward into the future, if not backwards into the past, to explain aspects of soil formation.

During their transport, water-borne *haen h^{e}gben* stones are worn round and smooth, and at a certain indeterminate point the Wola relate how they are

---

[14] Humphreys (1984) cites a sediment transport rate from the Purari catchment of 79 cm/1000 yrs.

Plate 5.7 A *mundit* landslide that has torn a gaping hole in a hillside, sweeping away trees in its path.

ground down beyond pebbles to gravel and sand-sized particles which they call *iyb muw*. This is a size class not a lithological one, any rock can be abraded to *iyb muw*, whether quartz grains, tuff grit or limestone gravel. The particles occur in a range of colours. When asked if *iyb muw* sand/gravel is *haen* rock or *suw* soil, the Wola maintain that it is neither but *tuwguwn* or 'between the two'. They similarly classify *kolbatindiy* and *tiyptiyp* volcanic debris as between rock and soil, pointing out that while some pieces occur rock-hard, others are ped-like and soft. When they crumble them up, they can sometimes see dark grey patches which are, they say, rotten fragments of *huwbiyp* basalt (some people are of the opinion that buried for a long time in wet soil *huwbiyp* basalt rots into such ashy and tufaceous material). Limestone too rots under the right conditions to produce soft *haen hok*. This can accumulate to give a pale soil-like deposit called *haen paenj*, which at the surface, on rock ledges and so forth, can support some plant life like mosses. It is, the Wola indicate, almost soil, like *iyb muw* sand, and pale silty sediments laid down on river banks, and soft *kolbatindiy* or *tiyptiyp* ashy blocks too. The addition of organic matter to these materials from rotting vegetation will, they say, eventually produce a fertile soil.

The Wola, in associating rock abrasion, deep weathering and depositional processes with the formation of new soils, raise again to the issue of food security. According to them, some geological events can not only rather unexpectedly, cause hunger and famine, but they can also supply some of the raw materials necessary for soil genesis, and hence help ensure full or, as they express it, 'standing bellies' *tombow hae*.[15] They readily acknowledge, as accomplished agriculturalists, the importance of the continuous replenishment of fertile soil resources, not only from familiar organic sources, an issue explored more fully in coming chapters, but also from longer term mineralogical ones. But while some primary material and much organic matter finds its way into the region's soils, some quantities are lost through erosion, together with a certain volume of soil particles. The rate of erosional losses is important, particularly compared with rates of new soil genesis: overall are they losing or gaining cultivable soil resources?

---

[15] When they have eaten sufficient, the Wola talk of their extended stomachs as standing, that is feeling firm and pushed out when full up with food.

CHAPTER 6

LIVING WITH LAND LOSS: THE COSTS OF EROSION

The loss of soil through erosion is of concern to farmers the world over (Hudson 1981). Some think that soil loss through erosion ranks as a major environmental issue like global warming, ozone depletion, rain forest destruction, and such like. They calculate that the world is sustaining massive soil losses; various figures are bandied around — it is estimated, for example, that 10 hectares of arable land are being lost globally to erosion every minute (Huypers et al. 1987). It is another race against time, to effect tolerable conservation measures and protect soil resources.

Water is the dominant agent of erosion throughout New Guinea. In the Wola region, water erosion seems potentially a particularly serious threat. The combination of high, and sometimes prolonged rainfall, with precipitous terrain, as described in the previous chapters, suggests a region particularly at risk. People regularly cultivate steep slopes, periodically exposing soil to the incessant rainfall when tilled and newly planted. Yet Wola attitudes seem strangely off-hand regarding the dangers of erosion, and they take few apparent conservation measures. It appears as if these skilful subsistence cultivators are recklessly risking their most valuable productive resource. The rate of soil loss anticipated with the wet climate and rugged topography could threaten land degradation, the local population deforesting areas and quickly reducing them to barren expanses of scoured sub-soil and scrubby vegetation (Woods 1982, 1984; Allen & Crittenden 1987). How can it be that they are foolhardy enough to follow cultivation practices which dice with losing down river the topsoil on which their crops and livelihood depend?

The impression of recklessness given by the Wola commonly prompts informed visitors (geographers, agriculturalists, teachers etc.) to voice their concerns, even to recommend changes to local cultivation strategies to conserve soil resources. An expatriate teacher at the newly established high school in Nipa for example, acted on this urge to intervene and 'improve' peoples' cultivation practices through demonstration, deciding that the school's food garden should have contour ridges to reduce topsoil loss. The pupils and their relatives advised strongly against the strategy warning of serious soil movement, but the teacher, convinced that they were too conservative-minded and hide-bound to experiment with improved cultivation techniques, pressed on regardless — after all, he was there to teach them! The consequences, unexpected by the expatriate, were very instructive regarding soil erosion in the region and the soundness of local attitudes to conservation, whatever first impressions.

I. SOIL EROSION

Geo-scientists distinguish several types of water erosion (Morgan 1986; Barrow 1991:199–202), of which mass movement, soil creep, rainsplash and surface wash, and rilling and gullying are particularly active in the New Guinea highlands (Löffler 1977; Humphreys 1984). Mass movement includes landslides, slumps and mud flows, which events like intense or prolonged rainfalls, earthquakes or other destabilising processes like stream undercutting or soil creep usually trigger, as related in the previous chapter. It gives rise to thick, poorly sorted colluvial deposits and, regarding volume of soil moved, ending up in rivers to be transported out of the region, it is the dominant erosion process occurring in humid mountainous areas like that of the Wola. In contrast, soil creep is a slow process affecting soil to depth, espe-

Plate 6.1 Evidence of surface wash erosion from and between sweet potato mounds where soil exposed by recent cultivation.

Plate 6.2 Incipient rill developing on soil exposed after pulling up remaining crops and weeds prior to recultivating a garden.

cially on steep slopes, the effects of which are not always readily apparent, unless they initiate mass movement.

Rainsplash, which detaches soil particles, and surface wash, which moves them downslope, although responsible for only removing small amounts of soil at any time, are cumulatively important processes regarding topsoil loss, especially under cultivation; they result in the thicker deposits of topsoil that accumulate against downslope obstacles in gardens, notably along fence lines, which the Wola recognise as locations of potential high fertility, producing the soil they call *suw hemem* (see Chapter 11). Rills are ephemeral small channels down slopes which may disappear between storms, whereas gullies are permanent, sometimes deep channels, along which water flows in rainstorms; rills occasionally develop naturally in the Wola region, notably on bare surfaces during cultivation and are evidence of vigorous localised surface transport, gullies on the other hand tend to originate from human activity, notably the digging of ditches to enclose some gardens which, where they run downhill, subsequently become conduits of, and may be enlarged by, waterflow during rain. The rooting of pigs in forest and garden areas, following the harvest of sweet potato, and during recultivation and fallow intervals, may significantly accelerate surface and rill erosion; the effects of other animals and soil borrowing fauna, like earthworms, crickets and ants, is probably minimal (Humphreys 1984).

The Wola are fully aware of the occurrence of topsoil loss through erosion. The movement of soil particles downslope generally they call *suw silsil pay* (lit. soil *silsil* goes), whereas the rolling of larger destabilised aggregates and stones downslope is *pendet pay* (lit. downhill-rolls goes). The transport of soil by water, often visible as a browny suspension, is *suw iybom tok pay* (lit. soil water carry goes). They are also awake to the movement of soil promoted by pigs rooting for earthworms, referring to the turning over and downhill transport of the soil by these animals as *showmay pil suw tombow pen kayor* (lit. pigs earthworms soil dig send away); they do not apparently consider this a serious hazard leading to substantial soil losses, allowing their pigs to roam and root at will during the day, women sometimes putting them inside gardens before recultivating them to root out any remaining tubers and turn over the soil.

The Universal Soil Loss Equation (USLE) summarises the principal factors affecting rate of soil loss through water erosion under cultivation. It is the most widely used of various equations to predict soil losses (Wischmeier & Smith 1978), and its use has been advocated in Papua New Guinea (Bleeker 1983; Humphreys 1984).[1] The equation, in the form $E = f(RKLSCP)$, specifies the following indices as determining mean annual soil loss through surface flow (E): rainfall erosivity (R), soil erodibility (K), slope length and steepness (L & S), crop cover (C), and conservation measures practised (P).[2] The equation provides a useful framework around which to structure a discussion and assessment of erosion in the Wola region.

Rain Erosivity

The erosivity factor relates to the ability of rainfall to erode soil, and is a function of its energy to dislodge and move soil particles. The critical issue is not necessarily the amount of rainfall received as its intensity. Data on rainfall intensity across the Wola region suggest that soil losses may not be of the order feared given its high annual total rainfall (Table 6.1). The chances of receiving rainfall in a 24-hour period, of sufficient intensity to cause serious erosion (taken as over 50 mm in 24 hours), are low at about 1.4% or 5 days a year. The monthly distribution is broadly even, as with the mean annual rainfall statistics (see Chapter 3), although intense rain is perhaps less likely in the middle of the year. Catastrophically heavy falls are rare, rainfall being driven largely by a diurnal cycle of relatively even convectional precipitation, although when they do occur they are memorable — a storm on the night of 7th July 1977, in which 151 mm of rain fell at Haelaelinja, caused many large and terrifying landslides throughout the region. The sparse data on short-term

---

[1] Alternatives have been proposed to the USLE, which was developed largely from temperate experience in the USA, in response to frightening erosion experiences in the mid-West (e.g. Elwell 1981), but the balance of informed opinion is that the USLE is suitable for use in Papua New Guinea (Bleeker 1983; Humphreys 1984); it also has the advantage of wide use and acceptability.

[2] I have written the equation as a function to underline the point that there are significant interactions between the parameters (as recognised by the use of nomograms to evaluate some of them), although E is calculated by simply multiplying the values together.

Table 6.1
Probability of rainfall intensities
(average no. days per month with rainfall within specified classes)

	CLASS (mm)	Jan	Feb	Mar	Apr	May	Jun	Jul	Aug	Sep	Oct	Nov	Dec
HAELAELINJA (Field Base)	0	4.7	2.7	7.0	5.0	4.5	7.3	8.0	19.0	8.0	5.0	8.5	6.5
	1–4	9.0	4.7	12.5	9.7	10.0	9.7	12.0	5.0	6.0	6.0	10.5	9.5
	5–24	14.3	17.0	9.0	11.7	11.5	10.3	10.0	7.0	16.0	15.0	10.0	11.5
	25–9	2.7	3.3	2.0	3.3	5.0	2.0	0.0	0.0	0.0	4.0	1.0	3.0
	50–99	0.3	0.3	0.5	0.3	0.0	0.7	1.0	0.0	0.0	1.0	0.0	0.5
	100	0.0	0.0	0.0	0.0	0.0	0.0	0.0	0.0	0.0	0.0	0.0	0.0
TILIBA (Mission station)	0	14.0	9.0	4.0	15.0	6.0	15.5	9.5	6.5	3.0	3.0	7.5	9.0
	1–4	7.0	7.0	7.0	7.0	8.0	5.0	8.5	10.5	11.0	4.0	5.5	8.5
	5–24	9.0	10.0	17.0	5.0	15.0	9.0	9.5	13.0	13.0	18.0	12.5	9.5
	25–49	1.0	2.0	3.0	3.0	2.0	0.5	3.5	1.0	2.5	5.0	4.0	2.5
	50–99	0.0	0.0	0.0	0.0	0.0	0.0	0.0	0.5	1.0	0.5	1.5	
	100	0.0	0.0	0.0	0.0	0.0	0.0	0.0	0.0	0.0	0.0	0.0	0.0
MENDI (Provincial (H.Q.))	0	9.9	7.1	6.5	9.9	10.4	12.1	10.9	8.6	7.8	9.1	12.5	10.2
	1–4	8.1	6.9	7.9	7.1	0.6	6.9	6.9	8.1	7.6	7.7	5.7	7.3
	5–24	10.6	12.0	13.9	11.0	10.3	9.7	10.1	12.1	12.2	11.0	9.8	11.8
	25–49	2.2	2.1	2.4	1.9	1.4	1.0	2.5	1.7	2.1	3.1	1.9	1.5
	50–99	0.2	0.2	0.3	0.1	0.3	0.3	0.6	0.5	0.3	0.1	0.1	0.2
	100	0.0	0.0	0.0	0.0	0.0	0.0	0.0	0.0	0.0	0.0	0.0	0.0

Sources: Haelaelinja, readings made in the field. No. years records each month same as Figure 3.1.
Tiliba, records kindly from Ed Staich
Mendi H.Q. records McAlpine et al. (1975:105) cover fourteen years' readings.

intensities, of heavy rain falling in an hour or so, which is particularly erosive, support the above observations, for they are uncommon (Clarke 1971:41; McAlpine et al. 1983:75–81).

Several indices of erosivity have been proposed relating kinetic energy and intensity of rainfall (Morgan 1986). They require pluviometer data, whereas only daily rainfall readings are available for the Wola region. Nonetheless, a tolerably accurate value for the R erosivity index can be estimated using these rainfall data, sufficient to indicate the force with which rain falls and the likelihood of it causing significant erosional losses across the Wola region,[3] by employing the information on rainfall intensity collected by Humphreys (1984) in the Kundiawa region of Simbu[4] Province (on the assumption that rainfall across the Central Highlands of Papua New

---

[3] When daily rainfall data only are available, Bleeker (1983:177) advocates the use of a modified Fournier index, which correlates with the USLE rainfall factor, to obtain an indication of the magnitude of R; on the basis of such a calculation the Wola region has a modest rainfall erosivity index.

[4] Another spelling of Simbu, dating from the time of the Australian administration, is Chimbu.

Guinea has similar intensity patterns relative to totals in any period, as the meteorological evidence and experts suggest — McAlpine et al. 1983).[5] By arranging average annual rainfall into daily range classes (5 to 25 mm hr^{-1} etc.) and multiplying these by their kinetic energies (calculated using the appropriate equation), we obtain the kinetic energy per rainfall class, and summing these gives the total annual average kinetic energy of the rain falling in the Wola region (Table 6.2).

Two equations are used here to calculate kinetic energy, one derived from temperate rain in the U.S.A. (Wischmeier & Smith 1978 — used to calculate the R index of the USLE), and the other derived from work on tropical rain in Zimbabwe (Hudson 1981). Neither of the equations applies unambiguously to the New Guinea Highlands, where the rainfall regime falls somewhere between tropical and temperate; they indicate the range within which rainfall energies

Table 6.2
Kinetic energy and erosivity calculations for rainfall across the Wola region

Rainfall Class (mm 24 hrs^{-1})	Average Annual Rainfall per Class (mm yr^{-1})	Total Kinetic Energy per Rainfall Class (ft-tons acre^{-1})[1]	I_{30} (ins)[2]	EI_{30} (col. 3 × col. 4)[3]	Total Kinetic Energy per Rainfall Class (J m^{-2} mm^{-1})[4]
WESTERN WOLA REGION (HAELAELINJA + TILIBA DATA):					
<5	1570.8	45256.4	0.07	3168.0	17083.0
5–25	1253.6	38586.3	0.41	15820.4	17864.9
25–50	237.8	7648.4	0.74	5659.8	3953.6
50–100	40.3	1357.2	1.44	1954.4	761.5
>100	1.6	54.9	1.74	95.5	30.7
TOTALS:	3104.0	92903.2		26698.1	39693.7
EASTERN WOLA REGION (MENDI DATA):					
<5	1562.4	45017.8	0.07	3151.3	16993.8
5–25	1030.4	31717.6	0.41	13004.2	14685.4
25–50	182.0	5853.9	0.74	4331.9	3026.1
50–100	25.2	845.1	1.44	1216.9	474.0
>100	0.0	0	1.74	0	0
TOTALS:	2800.0	83434.4		21704.3	35179.3

[1] After Wischmeier & Smith 1978, where kinetic energy (KE) equation is: KE = 916 + 331 log$_{10}$I (where I is rainfall intensity in ins hr^{-1}).
[2] After Humphreys 1984:23
[3] Rainfall erosivity index R is calculated as $EI_{30}/100$. For the western Wola region R = 267, and for the eastern Wola region R = 217.
[4] After Hudson 1981, where kinetic equation is: KE + 29.8 – 127.5/I (where I is rainfall intensity in mm hr^{-1})

---

[5] Humphreys (1984:78) is confident that his data can be used elsewhere in the Highlands of New Guinea to compute erosivity, after comparing his Simbu results with those from other regions with similar rainfall regimes and finding a fair level of agreement (see also Morgan 1986 and Lal 1990 for further comparative data on USLE calculations of erosivity elsewhere in the world).

may vary.[6] There is nevertheless some danger, which we need to bear in mind, that using equations developed in the U.S.A and Southern Africa may overestimate erosivity. Both the United States and Zimbabwe have climates which feature dry seasons and periods of intense rainfall, making them, for various reasons, considerably more erosive than Highland New Guinea.

The rainfall data, divided into the western and eastern Wola regions respectively, indicate similar patterns and intensities across Wolaland, with Mendi in the east having slightly less erosive rainfall (perhaps the result of some rain-shadow effect due to adjacent Mount Giluwe). The region's overall R index of 242 is considerably less than that for Kundiawa in Simbu Province at 431 (Humphreys 1984:78). While the Southern Highlands region has a larger annual rainfall than Simbu, it has fewer large storms of short duration which, possessing more kinetic energy, have greater erosive potential; some 92% of annual daily rainfall in the Wola region is <25 mm, whereas in Kundiawa the same statistic is 63%.

The difference between the R indices for these two Highlands regions points to the importance of the amount of rain falling in any time interval to its intensity and total kinetic energy, the more rain and the shorter the period, the greater the energy and erosivity potential. The I_{30} value of the USLE is intended to correct for overestimating the impact of less intense rain, although its appropriateness is open to question (Morgan 1986:46).[7] Indeed there has been some controversy over identifying the threshold at which rain is likely to become noticeably erosive. It is is argued that as low intensity rainfall causes relatively little erosion, dislodging and moving few soil particles, allowance needs to be made for this. On these grounds, Hudson (1981) argues that in sub-tropical Africa rainfall intensities less than 25 mm hr^{-1} are not important to overall erosion and proposes his index KE > 25 for predicting erosivity (i.e. only kinetic energy when rainfall intensity exceeds 25 mm hr^{-1} is used to compute the index). If we accept that only rainfall in excess of 25 mm hr^{-1} is going to cause noticeable erosion, this suggests further that the rain received in the Wola region, although annually larger in volume than that received in Simbu, is considerably less erosive (Table 6.3) — only 9.4% of Wola annual rainfall exceeds this threshold compared with 21.7% in Simbu, delivering 7420.9 J m^{-2} mm^{-1} of kinetic energy compared with 12996.3 J m^{-2} mm^{-1} (Humphreys 1984:79).

While the idea of a threshold is logically appealing, its application presents problems in regions with markedly sloping terrain like Wolaland

---

[6] These two equations are the most widely used in soil erosion studies, and programmes to predict losses and recommend conservation measures, although a range of alternative indices have been proposed, and demonstrated to have relevance in certain contexts, particularly in some parts of the tropics where rainfall has a different character to that assumed in the USLE equation (Lal 1977, 1990; Kinnel 1981).

[7] In response to criticism, Wischmeier & Smith (1978) have recommended that under tropical rainfall conditions maximum values be set for rainfall intensity, of 76.2 mm hr^{-1} in calculating kinetic energy per unit rain and 63.5 mm hr^{-1} for I_{30}.

because the trigger value may decrease with increasing slope, when it requires less energy to dislodge and move particles by raindrop impact. Humphreys (1984:67–69) has evaluated the efficacy of the differing allowances made by the EI_{30} and $KE > 25$ indices for rainfall of different intensities, on a series of experimental plots of varying slope and vegetation cover in Simbu Province, and concludes that both are equally good predictors of rainfall erosivity and soil loss (although as he points out this does not imply that appreciable losses are not occurring at lower intensities).[8] He also thinks that any threshold value should decrease with increasing slope.

The division of data into classes for analysis relates to the issue of thresholds, involving the imposition of boundary values for each class, which may in themselves introduce some distortions, resulting in possible overestimates of erosivity. According to Table 6.3, for instance, some 80% of the rain falling across the Wola region occurs in the <12.5 mm hr^{-1} intensity class, which suggests that a considerable proportion of the rainfall recorded in the 5–25 mm 24 hrs^{-1} rainfall class in Table 6.2 occurs in the 5–12.5 mm^{-1} range, making it considerably less erosive than the calculated mean value.

A further complicating factor is that the soil is frequently not saturated when rainfall begins. According to the evaporation data presented earlier (see Chapter 3), the daily average evaporation rate is 4.7 mm (with little annual

Table 6.3
Kinetic energy by rainfall intensity class across the Wola region

Rainfall Intensity Class (mm hr^{-1})	Average Annual Rainfall per Class (mm yr^{-1})	Total Kinetic Energy per Intensity Class (ft-tons acre^{-1})	(%)	Total Kinetic Energy per Intensity Class (J mm^{-2} mm^{-1})	(%)
WESTERN WOLA REGION (HAELAELINJA & TILIBA DATA):					
0–12.5	2414.8	67870.0	73.1	22699.2	57.2
12.5–25	382.2	13110.4	14.1	8790.6	22.2
25–50	269.6	10300.9	11.1	7117.4	17.9
50–75	24.9	1023.1	1.1	692.2	1.7
75–100	13.9	598.8	0.6	394.3	1.0
EASTERN WOLA REGION (MENDI DATA):					
0–12.5	2216.4	62293.9	74.6	20834.2	59.2
12.5–25	335.1	11494.6	13.8	7707.3	21.9
25–50	219.1	8371.4	10.0	5784.3	16.5
50–75	19.3	793.0	1.0	536.5	1.5
75–100	11.2	481.5	0.6	317.0	0.9

---

[8] These two indices, derived from different equations, cannot be substituted for each other. While the $KE > 25$ index is easier to calculate and involves metric units, the EI_{30} index is required for computation of the R index of the USLE (an example of a widely understood, extensively used and often quoted equation keeping the use of imperial units alive).

variation), which taken together with evidence presented here (that on average some 8.4 days a month experience <1 mm of rain and a further 7.8 days <5 mm), suggest that for some 16 days a month rainfall is less than evaporation. It is probable that during these periods the soil takes up a goodly part of the rain falling on it in wetting itself, leaving little to run off (assuming that the rainfall is not too intense, although data on infiltration rates discussed below indicate that these soils can cope with very heavy falls before runoff begins). Consequently, some water deficits must occur both within and between days. A great deal of rain therefore falls on soils that are frequently drier than field capacity, which consume at least part of the water in wetting themselves to saturation before any runoff and erosion occurs.

In conclusion, while it appears reasonable to accept the calculated R value of 242 as a fair estimate of rainfall erosivity in the Wola region, in calculating it we have erred on the side of generosity. It is consequently probably an overestimate. The erosivity of rainfall across Wolaland is, if anything, smaller than the index suggests.

Soil Erodibility

Soil erodibility relates to the susceptibility of a soil to erosion and is a function of soil physical properties[9] such as texture and structural stability, together with contributory factors such as organic matter, clay mineralogy, moisture and permeability (Greenland 1977). Later chapters, dealing with the classification and appraisal of soils in the Wola region, describe and analyse some of these properties further (see Chapters 10 & 11). This section covers analytical data particularly relevant to the susceptibility of soils to erosion. The results discussed come from a survey of sites in the western Wola region, analysed further in an appraisal of indigenous soil assessment (Chapter 11). Four horizons were identified as possibly present in profiles:

- horizon 1. organic horizon, either F (organic material partly decomposed, plant structures visible) or H (well decomposed organic material, no visible plant structures);
- horizon 2. surface or near surface, dark coloured mineral A horizon, featuring well incorporated humified organic matter;
- horizon 3. subsurface mineral AB horizon of transitional character, in which properties of A and B are mixed;
- horizon 4. subsurface mineral B horizon without rock structure, and no evidence of a weakly consolidated mineral C horizon retaining parent material structure.

The results of **texture** assessments indicate that it becomes finer with depth, sandy and silty textures being more evident in the top two horizons

---

[9] For detailed data on the soil physical properties discussed here, see *Land Degradation & Rehabilitation* Vol. 4.

(91% comprising silty clay loams or coarser), and clayey textured soils being more evident in horizons 3 and 4 (97% comprising clay loams). Laboratory particle size analyses suggest that clay contents are somewhat higher than estimated. The reason is the mineralogy of the volcanic ash or tephra component in these soils, allophane and halloysite imparting a loamy feel.[10] Andepts, contrary to feel, have higher clay contents than those soils derived more from the underlying sedimentary parent material, predominantly limestone (although these possibly have some tephra component too, particularly in their topsoils, which are usually volcanic ash influenced). The reason for the heavier, more clayey, sticky and plastic feel of these latter soils in the field is their domination by 2:1 crystalline clay minerals as opposed to the amorphous ones derived from tephra. The silt/clay ratios suggest that the surface horizons are the least weathered due to the presence of more recent tephra deposits (the presence of unweathered tephra also indicates slow rates of erosion). These will impart a sandier texture to the topsoils, accentuated by the preferential removal of fine particles in sheet erosion. The impression overall, sought for an indication of the region's general susceptibility to erosion, is one of fairly clayey soils, ranging from clay loams near the surface to clays at depth.

There was a noticeable trend in the way soil **consistence** — which relates to how soil reacts to applied forces and resists deformation — changes with depth, from surface horizons that are only slightly sticky to non-sticky (93% of the top two horizons), moderately plastic to non-plastic (87% of these two horizons) and loose to friable in strength (96% of these horizons), to subsurface horizons that are moderately to very sticky (88% of horizons 3 and 4), very plastic (83% of these two horizons) and firm to very firm (76% of these horizons). When measured in the laboratory using a drop-cone penetrometer, the plasticity index and plastic and liquid limits of volcanic ash soils in the Southern Highlands Province were high, the results not correlating with clay percentage (Allbrook & Radcliffe 1987; also Wallace 1971).

The **structure** of a soil depends on the aggregation of primary mineral and organic particles to form peds, separated by voids and surfaces of weakness. There was again a trend discernible in the survey between soil structure and depth, from surface horizons that had granular or composite granular peds (87% of the top two horizons), of fine to medium size (62.5% of horizon 1 and 2 peds), to sub-surface horizons that had angular shaped peds (67% of the lower two horizons), of coarse to very coarse size (71% of horizon 3 and 4 peds). This correlates with a predictable increase in bulk density (ρ) on average down the soil profile, although there was a considerable overlap between the upper and lower horizons (73% of horizons 1 and 2 had ρ of 1.1 g cm^{-3} or less, and 88% of horizons 3 and 4 had ρ of 1.2 g cm^{-3} or more). Bulk density — which measures

---

[10] The presence of volcanic ash minerals further presents some problems in particle size analysis, results varying with the method used to disperse samples, which accounts for discrepancies in clay content reported in the literature (Bleeker & Healy 1980; Radcliffe 1986:84). More vigorous dispersion of soil samples suggests even higher clay contents for soils derived from airfall tephra in the nearby Mendi valley (Allbrook & Radcliffe 1987).

the compaction of a soil and relates to texture, structure, clay mineralogy and organic matter content — was lowest in topsoils on volcanic ash, reflecting their allophane rich clay fraction and high organic matter contents, and highest in subsoils derived largely from sedimentary rocks (Radcliffe 1986).

Bulk density includes pore space in its measure of soil-mass-per-unit-of-bulk-volume, giving an index that relates to both the packing of aggregates and the proportion of voids to solid material. The porosity of a soil derives from this, depending on the packing and arrangement of peds, and relates inversely to bulk density. Fissures were uncommon in these soils, and macropores predictably became smaller and less abundant down the profile, correlating with observed changes in soil structure (100% of horizons 1 and 2 had macropores and 69% of these had >2% of their area covered with pores, whereas 22% of horizons 3 and 4 evidenced no macropores and only 4% of those with them had >2% of their area covered). The low densities indicate large porosity, part of which will be within the micropores. When measured in the laboratory, there is a slight decrease in total porosity with depth, tephra-derived soils have the highest total porosities between 65% and 85%, and sedimentary-derived subsoils the lowest between 55% and 60% (Radcliffe 1986:87), probably reflecting differences in pore continuity.

The porosity of a soil relates to the drainage of water through it. The field assessment of soil-water state indicates that for all horizons the predominant moisture condition is slightly moist followed by very moist. On the whole the soils are adequately drained. The dry weight percentage of water in sedimentary-derived soil taken at field conditions ranged from 30% to 45% for topsoils and was 33% for subsoil, and laboratory oven drying of soil samples previously sun-dried for transporatation gave moisture weight percentages of between 4.2% and 32.8% (mean 15.9%) for topsoils and 4.8% and 39.7% (mean 17.1%) for subsoils, indicating that aggregate micropores, which strongly retain water, make up a large part of soil void volume, as expected in clayey soils.[11] The results of infiltration measurements[12] indicate very fast vertical intake of water into these soils, with basic infiltration rates of 12 to 87 cm hr^{-1}, putting them into the rapid to very rapid infiltration categories (Landon 1991:69).

Using moisture retention curves, Radcliffe (1986:89) infers pore size classes, and he takes the volume of large pores (equivalent cylindrical diameters 50 μm to 500 μm) as corresponding to the amount of water held between field capacity and saturation, and indicating soil aeration and permeability; nearly all topsoils measured were above the critical level of 10%, which he takes as limiting root oxygen supply, but few of the subsoils exceed it, either tephra- or sedimentary-derived, suggesting imperfect or poor drainage. Hence,

---

[11] I am grateful to Jan Staich, one time of Tiliba mission near Nipa, for allowing me to use her oven to dry soil samples (at 105°C to 120°C), and the Department of Primary Industry's soil fertility laboratory at Konedobu for laboratory oven drying samples.

[12] Measurements were made using the double ring infiltometer method, with a metal inner cylinder and outer clay bund, on *hundbiy* tropept clay soils with fairly high stone contents.

regardless of high total porosity, the volume of transmission pores may be low and the development of anaerobic conditions a possibility (Allbrook & Radcliffe 1987). But there is rarely evidence of gleying in soils on slopes, which we should expect if the subsoil was saturated for extended periods. Gleying indicates an increased risk of erosion over that in a freely drained soil, and its absence suggests that these laboratory measures overestimate the drainage impediment, and consequent heightened danger of erosion. And anyway, on steep slopes water may move rapidly downhill within the porous A horizon, whatever the drainage status of the lower horizons.[13] The rapid infiltration rates into topsoils combined with the slower movement into subsoils accounts in part for the susceptibility of these soils to mass movements like landslides in times of very intense rainfall, when it acts as a lubricant between the two.

Moisture retention measurements on soils in the eastern Wola region, and on similar soils elsewhere in the Highlands (Radcliffe 1986; Allbrook & Radcliffe 1987) indicate that the tephra-derived soils may have somewhat higher available water capacities, with topsoil moisture content values of 19 vol% to 33 vol%, indicating good water holding capacity (although they evidence a sharp fall in the subsoil to between 13 vol% and 20 vol%), compared with the largely sedimentary-derived soils which have topsoil available water capacities of 15 vol% to 20 vol% (although they have a smaller decrease in the subsoil to 16 vol%). The predictable and high rainfall conditions that prevail in the Wola region result in soil moisture contents at, or in excess of, field capacity a deal of the time. The tephra-derived soils regularly have field moisture contents approaching 100%, above 80% of the liquid limit, and their somewhat anomalous structural stability has been attributed to bonding between discrete soil particles within a 'viscous gel' (Wallace 1971).

The soils of the Wola region, like those in the Southern Highlands generally (Radcliffe 1986; Wood 1984) have high organic matter contents for the tropics, as discussed further in the investigation of soil fertility under cultivation (see Chapter 12). The topsoils have a mean organic matter content of 23.5% (ranging from 9.0% to 47%), and the subsoils evidence a notable decrease with a mean organic matter content of 6.0% (ranging from 0.1% to 12.5%),[14] which parallels the fall noted in other properties like bulk density (Chartres & Pain 1984; Allbrook & Radcliffe 1987). Among the possible factors suggested responsible for these high organic matter levels are the region's high precipitation and moderate temperatures, which result generally in an excess of rainfall over evapotranspiration and give moist, cool soils that check bio-

---

[13] Regarding the assessment of soil permeability for the calculation of the K erodibility value of the USLE, Bleeker (1983:185) makes an estimate based on other soil properties and concludes that Andepts and Tropepts like those in the Wola region evidence rapid permeability (at 6 to 20 cm hr^{-1}).
[14] These figures are based on organic carbon analyses of 42 topsoil and 18 subsoil samples (Walkley-Black oxidation method), and the soil organic matter contents estimated by multiplying the results by a conversion factor of 1.724 (which assumes that average soil organic matter comprises 58% carbon.)

logical activity; and tephra-derived mineral complexing, which results in the formation of stable organo-mineral complexes with allophane (Bleeker 1983). The latter factor perhaps explains why some workers report that tephra-derived soils have higher organic matter contents than largely sedimentary-derived soils (Radcliffe 1986 gives topsoil means of 26% for Andepts compared with 16% for Tropepts).

The minerals dominating the clay fraction of a soil dictate both some of its physical and chemical properties. The presence of short-range order, tephra-derived minerals like allophane, and its weathered product halloysite, impart good structure to topsoils and promote high organic matter contents, as evidenced in the soils of Wolaland. The mineralogy of the relatively young, recently ash rejuvinated soils of this region is complex; the results of mineralogical assays suggest that it varies over short distances, with allophane, gibbsite and halloysite predominating (Bleeker & Healy 1980; Chartres & Pain 1984). Regarding the sedimentary-derived soils, it is probable that kaolinite, montmorillonite and quartz dominate, and that their topsoils also contain allophane and halloysite, being volcanic ash influenced to some degree (Radcliffe 1986).

The soil erodibility factor K for topsoil and subsoil can be determined by plotting the mean values for grain size distribution, organic matter percentage, soil structure and permeability, as discussed here, on the USLE soil-erodibility nomograph (Wischmeier *et al.* 1971). The extremely high organic matter contents of the soils, way beyond the nomograph range, make it difficult to extrapolate K factors (Morgan 1986:53), but it is possible to ascertain the likely compass of topsoil and subsoil values. If the plot goes to the highest organic matter content (4%) on the nomograph, greatly underestimating actual organic matter percentage values, the K factor for topsoil is ~0.05 and subsoil is ~0.10, whereas if some allowance is made for the much higher organic matter contents, the vertical line joining structure to permeability moves towards the limit of resolution for both topsoil and subsoil and their K values become exceedingly small. This low estimate of K parallels those reported elsewhere in the Highlands for similar soils (Humphreys 1984; Bleeker 1983:181), and other places with humus-rich Andepts (Utomo & Mahmud 1984; Lal 1990). The reason for it is the favourable combination of high organic matter and high clay contents, together with good soil structure; the suite of clay minerals present may also further favour the resistance that these soils evidence to erosion.

The thorough reworking of topsoil that features at each round of sweet potato cultivation among the Wola may further promote the expression of these favourable physical characteristics and assist in conservation. Little change occurs in soil physical properties over time on semi-continuously farmed sites in this region, with the soil thoroughly loosened and broken up with digging sticks for each sweet potato crop. This maximises their favourable physical properties, by reducing compaction and promoting porosity and infiltration, lowering bulk density and increasing aeration. But on the debit side, the exposure of bare soil to rain increases runoff losses; clearance and cultivation also lead to a marked loss of organic matter (see Chapter 12).

Slope Length and Steepness

The anticipated effect of slope on soil erosion, as already suggested in the discussion of rainfall intensity, is to increase losses (Hudson 1981; Morgan 1986). An increase in either the steepness or the length of slopes is expected to increase the rate of erosion, slope affecting both the shearing and transport capacity of flowing water, velocity increasing with steepness and surface-runoff-volume with length.

In the mountainous Wola region we should anticipate slope to contribute markedly to soil loss, given the precipitous and rugged terrain. Erosion experiments conducted in the Highlands indicate a clear relationship between rate of soil loss and site steepness,[15] particularly where there is little or no vegetation cover (Figures 6.1 to 6.3[16]). People here regularly cultivate gardens on sloping land, sometimes on very steep inclines, and their practice of at times abutting gardens one to another down a slope, often because collectively it reduces the work needed to enclose the site, can considerably extend the uninterrupted length disturbed by cultivation activities, exacerbating the dangers of erosion losses. In a survey of Wola gardens, excluding very small cultivated plots adjacent to houses, the average steepness of slope cultivated was 20°, ranging from 0° to 46° (n = 291 gardens), and the average length of slope was 47 metres, ranging from 7 metres to 175 metres (n = 178 gardens).

Geo-scientists have conducted considerable research to establish the impact of slope on erosion rates and to formulate equations to predict its effects on soil loss (Morgan 1986; Lal 1990). In the New Guinea highlands, it is possible to fit a range of regression lines and equations to soil losses on dif-

---

[15] Lal 1974 demonstrates a similar relationship for soils under shifting cultivation in Africa.

[16] The annual rates of soil loss for the Wola region I have calculated from erosion measurements made over only three months. I made the measurements using modified Gerlach troughs of plastic sheeting (Morgan 1986:145–150), sited along the bottom edge of enclosed 1 m² square plots. All of the experimental sites were on tropepts (*pombray* topsoils over *hundbiy* clay subsoils), and were located in sweet potato gardens. The number of sites = 2 per mean plot on the graphs. The sites under crops had a dense cover of sweet potato vines, which by the end of the experiment afforded ~100% cover on some plots. The bare soil plots were maintained clean weeded (with no working of the soil). On two of the bare soil plots there were events which increased soil movement into the troughs: on one plot (~40° slope) a woman crossing the garden higher up dislodged some soil which rolled into the trough, and on the other plot (~25° slope) a large rill initiated by work up slope cut across the plot and deposited some soil in the trough. Neither plot was discounted in calculating rates of soil loss because such events occur occasionally in everyday life, moving soil downslope. Furthermore, the period during which I made the erosion measurements was particularly wet; when compared with the average annual rainfall total for the western Wola region the rainfall was 1.16 times higher (also during this period Radio Southern Highlands reported landslides occurring throughout the Province, one in the Mendi region sweeping away some gardens). The reported rates of soil loss are probably on the high side as a result of these experimental conditions, particularly on the unprotected bare plots.

Living with Land Loss: The Costs of Erosion 153

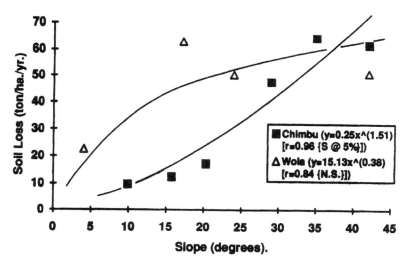

Figure 6.1 Soil loss according to slope on bare soil where vegetation cover 0% (Simbu data, Humphreys 1984 [no. sites {with increasing slope} = 4,3,6,2,2 & 3]; Wola data, this study).

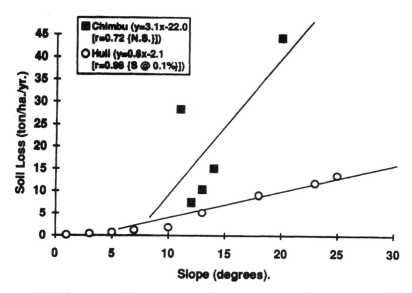

Figure 6.2 Soil loss according to slope where crop vegetation cover <50% (Simbu data, Humphreys 1984 [no. sites {with increasing slope} = 2,2,1,1 & 3]; Huli data, Wood 1984 [no. sites = 1 per plot].

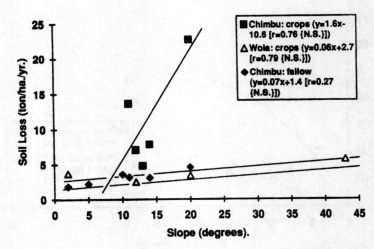

Figure 6.3 Soil loss according to slope where crop/fallow vegetation cover >50% (Chimbu data, Humphreys 1984 [no. fallow sites {with increasing slope} = 3,3,3,5,4 & 2, no. crop sites same as Fig. 6.4]; Wola data, this study).

ferent slopes under various vegetation covers (Figures 6.1 to 6.3).[17] In the USLE a combined factor for slope steepness and length is calculated to cover the L and S terms in the equation (Wischmeier & Smith 1978).[18] The LS factor for cultivated sites in the Wola region is 14.73, according to the mean garden length and slope steepness data. But the range of LS factor values is considerable given the large variation in the steepness and size of gardens; it goes from a mere 0.03 for small level cultivations adjacent to houses, to 17.43 for the area of combined gardens with the longest slope surveyed (175 m), and up to an enormous 85.76 for the site with the steepest slope (46°).

The topography of slopes further complicates efforts to predict soil loss (Lal 1990). The slopes cultivated by the Wola are rarely uniform, and frequently evidence markedly uneven micro-relief. Irregularly shaped slopes make it difficult to model erosion processes (Siebert 1987). The variation in shape implies that soil lost on a steeper section may be deposited lower down a site, where slope eases off or a concave feature like a fold promotes retention. In his study of erosion in the Tari Basin, Wood (1984), who found a consistent relationship between soil loss and slope in sweet potato gardens (Figure 6.2), draws attention to the redistribution of topsoil across sites rather

---

[17] It is common practice to express the relationship between slope and soil loss as a power function (Lal 1990), but a power curve only fits the bare soil data from Highland New Guinea better than any other plot, a straight line better matches losses from sites under vegetation.

[18] The LS factor can be estimated from a nomograph (Hudson 1981; Bleeker 1983:188), or it can be calculated using the equation LS = $\sqrt{L/22.13}$ $(0.065 + 0.045S + 0.0065S^2)$ where L is slope length in metres and S is steepness in per cent (Morgan 1986:117).

than its complete removal (he observed shallower soils on upper slopes, where mounding sometimes exposed bright brown subsoil, and found lower slope soils had higher levels of organic matter, nitrogen, phosphorus and exchangeable bases, evidence he thinks of the import of nutrients from up-slope); nevertheless he argues that the continued movement of topsoil, albeit without marked loss, contributes to seriously declining soil fertility and land degradation. The more level areas on the broken up slopes cultivated in the Highlands consequently act as terraces trapping soil and water locally, which may be beneficial even if the soil is thinned on some parts of the cultivated area during the process. This suggests that applying the USLE to the Wola data may markedly overestimate the contribution of slope length to potential erosion losses.

The assessment of the contribution of slope angle to erosion hazard demands qualification too. The serious threat of soil loss faced by Wola farmers obliged to cultivate steeply sloping land may not be so dire as the slope analysis suggests so far. There is some evidence that rate of soil loss may not continue to increase relentlessly with ever steeper slopes, but peaks and then declines (Figure 6.1 suggests that losses on bare plots may peak at between about 20° and 35°). A possible explanation is that rain striking a surface obliquely has less erosive potential than rain falling perpendicularly on to it, and the steeper a slope the more oblique the angle at which vertically falling rain will strike it (wind blowing off, or onto a slope will enhance or diminish this effect — Ahmad & Breckner 1974; Humphreys 1984; Bleeker 1983:188–89). It is possible that the positive relationship between increasing slope and rate of soil loss may only hold in the New Guinea Highlands for slopes up to 25° or 30° (Humphreys 1984:75–76). In this event, the LS factor values calculated for the Wola region over-estimate the potential contribution of slope to soil loss, particularly when taken together with the ameliorating effects of irregular surface topography. A further consideration is that on steep slopes water may not enter so deeply into the soil due to the rate of surface runoff, reducing the danger of slumping and mass movement.

II. SOIL PROTECTION

Vegetation Cover

The extent and type of vegetation cover on a site may also moderate the effect of slope, or any other factor that promotes erosion, like intense rain or highly erodible soil. Vegetation intercepts raindrops, reducing the kinetic energy of their impact on the soil (e.g. in his study of New Guinea montane forest ecology, Edwards (1977:987) found that about 80% of the rain falling during wet months reached the forest floor, declining to only 30% during dry months). Plants also present obstacles to running water, cutting flow velocity and transport capacity. The effectiveness of plant cover in reducing erosion

Plate 6.3 The dense ground cover of established sweet potato vines affords excellent protection of the soil.

depends on its density, the height and continuity of its canopy, and the extent of its root network.

The norm under forest is for lower rates of soil loss to occur because canopy interception reduces runoff volume, the presence of a thick litter layer and dense understorey protects the ground, and soils are better aggregated, with roots opening up macropores and promoting high rates of infiltration. These beneficial traits reportedly increase with altitude in Papua New Guinean forests, the intensity of slope wash decreasing with height above sea-level. A high rate of surficial wash occurs in lowland rainforests, with their high canopies, open undergrowth and patchy ground cover; whereas in montane forests, like that of the Wola, a subsurface wash of considerably reduced intensity occurs under the dense vegetative carpet mantling the forest floor, this thick ground cover of matted rootlets and moss, together with the lower canopy, affording considerably increased protection (Ruxton 1967; Bik 1967; Löffler 1977).

It is thought that dense grass growth may be almost as effective as forest in reducing erosion losses (Morgan 1986:61), if it extends over a sufficient proportion of the ground surface. And grassy regrowth features prominently in the management of some semi-continuously cultivated sweet' potato gardens in the New Guinea highlands (Floyd *et al.* 1988). The reduction in the rate of soil loss on land under fallow grasses (*Themeda australis* and *Imperata cylindrica*) is dramatic when compared to bare land fully exposed to the rain

Plate 6.4 A pig rooting over topsoil, moving it downslope and exposing it to rain erosion.

(Figure 6.3); even on very steep slopes the rate of soil loss is not markedly different to that on gentle ones (fallow vegetation affording >50% cover).

The protection afforded by crops varies, not only according to species and cultivation practice, but also with the changing cover offered during the cropping period. This complicates efforts to model soil loss under cultivation. Rates of erosion are anticipated to vary at different stages of the cultivation cycle, reaching a peak during tillage and immediately after planting with soil exposed, and subsequently declining as the crop establishes itself, cover perhaps peaking and then falling back somewhat as the crop matures, until it is disturbed at harvest. This approximates to the situation under sweet potato in the New Guinea Highlands (Figures 6.2 & 6.3), where cover increases from 0% following tillage and mounding to about 50% after three months or so, with further expansion subsequently as the crop matures. The disturbance of the crop during sporadic weeding, and with harvesting from about six months onwards, deranges the cover to some extent and makes likely some increased soil losses. Soil loss in the first three months is larger than in the following period (according to the Simbu data, the factor of difference in soil loss averages 1.7[19] — Humphreys 1984:77; and Lal 1990:333 states that runoff decreases by 82% during the first four months under sweet potato). The soil

---

[19] Soil loss difference factor calculated as: soil loss <3 months/soil loss >3 months.

loss recorded in the Simbu region during the first three month period fell within the range of loss from bare plots, and in the second period it fell between the losses occurring on bare and fallow plots.[20]

Sweet potato, which predominates on intensively cultivated sites in the Wola region, quickly establishes a dense cover close to the soil surface sufficient to reduce considerably erosion losses compared to other crops; Lal (1990:336) cites Brazilian research where, of all crops investigated, soil loss was lowest under sweet potato, with only 4.0% of rainfall flowing away as runoff giving soil erosion losses of 1.6 tons ha^{-1}. Different cultivars can have a marked influence however, slow-growing ones can give considerable erosion risks and substantially higher losses, particularly with increases in slope. Erosion losses in Nigeria increased from 1.7 tons ha^{-1} on a 3° slope to 6.5 tons ha^{-1} on a 9° slope, resulting in a high C factor value (Lal 1990:265,327). But these measurements were made on old infertile soils, whereas those cultivated by the Wola are young and considerably more fertile. The increased fertility means more rapid growth of ground cover, significantly reducing erosion risks.[21] Consequently, we can only use these data as a guide to possible C factor values under sweet potato in the New Guinea highlands.

The soil actually lost from any cultivated area will depend on several variables including crops grown and their planting density, and cultivation practices adopted, such as frequency and thoroughness of weeding, method and extent of harvesting, and so on (Lal 1974). The weed factor, difficult to assess, may be particularly significant, because weeds act as efficient ground cover and can help check erosion in the early stages by their rapid growth; the assiduity with which the Wola weed their gardens varies but it is uncommon to see people weeding during the first three months of so after planting, when the exposed soil is at greatest risk of erosion. The practice of intercropping further complicates the situation regarding erosion estimation in the New Guinea Highlands, as in other tropical regions. While the staple crop of sweet potato predominates in many gardens, it is infrequently grown as an exclusive monocrop, patches of cucurbits, green leafy vegetables and sugar cane, among other crops, usually occur dotted about gardens too. And newly cleared gardens, in contrast to ones under semi-permanent sweet potato cultivation, commonly have a wide range of intercropped plants mixed up irregularly across them (Sillitoe 1983).

The existence of diverse cropping patterns and cultivation practices makes estimates of the effects of vegetation cover on soil loss in the Wola region difficult. The tables available to estimate the average cover afforded by different crops at different stages of their cultivation are inappropriate

---

[20] The differences in the Simbu and Huli data for soil losses when crop cover <50%, and the Simbu and Wola data when crop cover >50%, are probably reflections of the different rainfall regimes in the regions, the Tari Basin, like the neighbouring Wola region, having less rain of high intensity and erosivity (McAlpine *et al.* 1975).

[21] It is noteworthy in this regard that serious erosion infrequently occurs immediately after forest clearance, and is usually worst when the soil loses its ability to grow plants quickly.

(Wischmeier & Smith 1978), particularly as developed for monocrops under temperate regimes (Bleeker 1983:190). Nevertheless, it is possible to make a fair estimate of the C crop index used in the USLE, which measures the amount of effective cover as the ratio of soil lost under a specified crop to that lost from a bare surface in a year (and may vary from unity, where the soil is bare, to zero, where the cover it so complete and effective that it prevents any soil loss).

We can draw up a standard accounts-like summary of vegetation cover. According to this procedure, when vegetation cover varies throughout the year, as it does under crops, the accepted way of accounting for the changing C factor is to divide the year up into periods, corresponding to different stages of crop growth when cover is fairly uniform, calculate a C factor for each, adjust for any seasonal variation in rainfall intensity, and sum them to obtain an annual C value (Morgan 1986:117–119). The span of the cropping cycle in the New Guinea Highlands, which regularly exceeds one year with intermittent harvesting of crops over many months, further complicates the computation of this ratio. An intensive cropping cycle in this region covers on average about three years, although longer periods between cultivations are common. If we assume a three year cycle, with a site under crops for half this time and grassy fallow for the remainder, it is possible to calculate a reasonable estimate of the C factor for sites under intensive semi-continuous cultivation. The predominant crop in such gardens is sweet potato:

Table 6.4
Cropping cycle in the New Guinea Highlands

Months	Cover	C value[22]	Adjusted C value[23]
1	0% bare	1.0	0.028
3	<50% crop	0.5	0.042
3	>50% crop	0.2	0.017
11	>50% crop (with harvesting disturbance)	0.3	0.092
2	10–30% grass (with declining crops)	0.3	0.017
4	50–70% grass (with few crops)	0.2	0.022
12	>70% grass	0.1	0.033
		TOTAL	: 0.251

The resulting C value of 0.25 is an over-estimate for all land under cultivation in the Wola region. If we reduce the intensity of land use, and extend the cultivation/fallow sequence to a five year cycle, as Humphreys (1984:85) assumes for Simbu, the C value falls to 0.19. It would be even lower for newly established gardens. The calculation of a C value for newly cleared gardens

---

[22] C values from Humphries 1984:80; Lal 1990:265.
[23] Adjusted by multiplying C values by: number months under cover specified/36 (assuming rainfall erosivity is the same in all months).

under a wide range of intercropped plants would be extremely complex given the almost infinite variety of cropping patterns, but the combination of quick growing leafy green vegetables and a multiple storied canopy when the garden is established (ranging from the creeping and thick foliage of sweet potato to the enormous leaves of banana plants), probably results in a markedly reduced C index value. Several gardens are also cultivated once only and then abandoned for decades to natural vegetation, when the C value falls greatly, especially as secondary forest establishes itself.[24]

Conservation Practices

The final factor in the USLE makes allowance for conservation measures taken to reduce soil erosion losses, and includes any practices employed to slow down runoff rate and reduce denudation, such as terracing, contour tillage and strip cropping, mulching, and so on (Lal 1974). The Wola traditionally employ no specific conservation measures. The value of the conservation factor P would appear to be 1.0. But their conservation practices may be more real than apparent.

While the Wola take no steps to control soil erosion as such, some of their agricultural practices have an influence on soil movement and loss. The enclosure of gardens to protect them against the depredations of foraging pigs, largely with fences and sometimes ditches, results in barriers downslope which, as noted, catch and trap moving soil, and act like widely spaced terraces, preventing transport of soil off cultivated sites. The natural breaks and irregularities in slopes, incorporated in garden areas, also act terrace-like. And the practice of heaping up mounds of topsoil in which to plant sweet potato, although primarily intended to give well-drained beds of friable topsoil well-suited to tuber expansion, also serves, with their arrangement in regular rows, to divert runoff into downhill channels, so effecting efficient drainage. They act like contour rows with the channels as overflows, preventing the ponding of water in depressions and reducing potentially damaging erosion around crop roots. Management practices that prevent soil compaction, encourage infiltration, maintain fertility, and so on — such as the regular breaking up of the soil during cultivation — all contribute to erosion reduction too. And the practice of leaving trees around and within gardens, such as pandans and casuarinas, and not removing the stumps of felled trees, further serves as an

---

[24] Since Papua New Guinea has a dense natural vegetation on the whole, Bleeker (1983:191) maintains that the cover factor is only significant under very intensive cultivation regimes and, assuming that vegetation cover then only falls below 30% for very brief periods, he estimates that serious soil erosion only occurs on average during one month each year under a four year cultivation cycle; he does not consequently expect the cover factor to affect soil losses greatly, calculated according to the USLE, given traditional farming methods and a climate that promotes rapid plant growth.

erosion control measure, as does the unhindered initial proliferation of weeds in newly planted gardens which afford rapid soil cover and protection.

Nonetheless, the absence of any specific conservation practices appears reckless, particularly given the high annual rainfall of the Wola region and the steepness of the slopes regularly cultivated there. It contrasts with the Highlands to the east where farmers often take extensive measures to reduce erosion. The Simbu erect low soil retaining barriers of short sticks called *giu* along contours, breaking up slopes and trapping sediment (Paglau 1982; Wood & Humphreys 1982); the Bena Bena and others dig grid-iron ditches to direct water flow and reduce erosion by dividing slopes into shorter sections, although they run the risk of gullies developing in heavy rain (Bleeker 1983:194); and in the Simbai valley they construct rough terraces from the trunks of casuarina trees (Burnett 1963).

The data presented here on soil erosion suggest however that, whatever initial appearances, the Wola are not reckless farmers. Rather they are farmers who through generations of experience know their region's environmental constraints and soil resources intimately, and realise that conservation measures beyond their prudent cultivation practices are unnecessary. They farm on soils that are very stable and resistant to erosion, they cultivate a staple crop which affords very good cover and soil protection when established, and they live in a region where intense and erosive falls of rain are uncommon compared to Highland regions to their east where increased rainfall intensity makes necessary the practice of traditional erosion control measures. On the grounds that people who take measures that prevent a problem occurring, even if fortuitous and unremarked, cannot be accused of failing to take any measures, I have decided to reduce the P factor value to 0.7.

Soil Losses

According to calculations made here, an average garden, predominantly under sweet potato and subject to the most intensive cultivation regime possible, could lose soil at something like the rate of 25 tonnes ha^{-1} yr^{-1}. The FAO (1978) classification of water erosion risk ranks this as a "moderate degree" of soil loss. It represents the extreme for the Wola region and occurs very infrequently, if at all. It is the slope factor that results in this high rate, whereas areas under intensive, semi-permanent cultivation are usually located on less steeply sloping land than average, which reduces rates of soil loss considerably; setting the average slope of these gardens at 10°, half the overall mean slope value, reduces the annual rate of soil loss to 5 tonnes ha^{-1}. On another tack, the value of the cover factor C falls with reductions in the intensity of cultivation; it approaches 0.1 with several years under grassy regrowth and annual soil loss falls to 9 tonnes ha^{-1} on mean slopes, and if woody regrowth occurs it may go below 0.1 and annual soil losses decline to 5 tonnes ha^{-1} or less. In conclusion, it is unlikely that soil losses in the Wola region exceed

10 tonnes ha^{-1} yr^{-1}, except on infrequent occasions. This rate of erosion for land disturbed by human cultivation falls into the FAO's (1978) lightest erosion category of "slight soil loss". The lack of evidence of noticeable erosion, other than localised slumping, supports this conclusion. It matches the empirical evidence too (Figure 6.3); and also complies with findings elsewhere — in the Simbu region, for example, Humphreys (1984:85) concludes that the overall annual rate of erosion is 11 tonnes ha^{-1}, just within the "moderate soil loss" range.

Regarding total soil losses, it is mass movements, not the effects of rainsplash and slopewash on cultivated sites, that are the major erosion processes in the Wola region. These are events over which the local population can exert little, if any control, triggered by unusually heavy rainstorms and earthquakes, and involving the movement of large volumes of soil in landslides, slumps, and so forth. In July 1977 I witnessed several such movements following twenty-four hours of exceptionally heavy rainfall, when luckily no fatalities occurred. Only a few gardens were affected, not sufficient to disrupt subsistence, and some tracks were buried under mudflows. A house was left precariously and comically balanced above a steep scar, a landslide taking away the earth up to its walls. Only a fraction of the region's land area was affected, and herbaceous plants and grasses soon colonised the scar sites. Sometime later these locations may become good garden sites, the changes in slope morphology (e.g. headscarp depressions), may act to trap sediment and eventually develop deeper and more fertile soils than average, although they are often wet locales suitable for only a few crops (Humphreys 1984:83).

It is difficult to estimate the volume of soil lost in these events. The rainfall intensity thresholds at which substantial mass movement losses are likely under different slope conditions and vegetation cover are unknown, and would be difficult to assess accurately. Exceptional storms that cause short-term processes of high-intensity, with slumping, landslides and mudflows, are superimposed on long-term low-intensity processes, of slopewash, creep and so on. Nevertheless the overall annual rates of loss appear to be modest; according to Löffler (1977:127) the rate of denudation on Mount Giluwe is 0.5 mm yr^{-1} which, assuming a mean soil bulk density of 1.2 g cm^{-3}, computes to an annual rate of erosion of about 6 tonnes ha^{-1}, which is similar to that for runoff erosion on cultivated sites.

It will come as no revelation to those who believe in the soundness of indigenous practices that the Wola have evolved an efficient agricultural regime which does not, first impressions notwithstanding, court massive erosion losses. These findings lend support to those advocating close consultation with farmers, and understanding of their practices before recommending changes (Richards 1986; Warren 1991; Chambers *et al.* 1989). Techniques of soil conservation developed in drier and less precipitous environments, which aim to conserve water for plant growth while allowing any excess to run away without needless erosion, are unlikely to prove suitable under New Guinean conditions of steep slopes and humid climate. Structures like terraces and ridges are more susceptible to slumping in such environments where rainfall is high. Recommendations for interventions, for example of contour-based or biological systems (Gagné 1977; Williams 1982; Huypers *et al.* 1987), will demand rigorous evaluation before promoted as ways of reducing erosion, which is anyway not high (Humphreys 1984). The result of the high school teacher's innovations to reduce erosional soil losses graphically illustrate this point. His interventions resulted in a considerable movement of topsoil, as predicted by local people. The school garden was alright until the first large rainstorm, when water ponding up behind the contour ridges breached them and swept everything before it into the ditch at the bottom. A better demonstration of the soundness of local cultivation practices would be difficult to conceive.

While the calculations made here demonstrate that the Wola do not face serious soil loses from cultivated sites, and are not agents of serious land degradation, regardless of their wet climate and steep terrain, nevertheless any loss of topsoil is undesireable, potentially reducing their agricultural productive capacity. Ideally, an agricultural regime ought to maintain, or better improve on, the stability and productivity of its soil resources through appropriate land use and management (Wood & Humphreys 1982). It needs at least to keep soil losses below that threshold at which natural soil formation processes can no longer keep pace with them, if it is to sustain productivity. The calculation of maximum acceptable rates of erosion, so-called soil loss tolerance levels, is not really feasible, unlike measuring rates of soil loss, because rates of soil formation are so slow and currently poorly understood (Morgan 1986). But it is possible to give an indication of the magnitude of replenishment rates.

Under the Wola agricultural regime, the replenishment of lost topsoil resources depends on the period under natural fallow vegetation, when the soil is largely protected from erosion and the input of organic matter contributes to the build-up of new topsoil, as it is incorporated together with the simultaneously weathering mineral fraction of the soil (see next Section). The average annual litter production under montane rain forest in the New Guinea Highlands is of the order of 7.5 tonnes ha^{-1}, and while the annual rate of decomposition is relatively slow for the tropics at $k = 1.2$, it is sufficient to ensure notable humus build up (Edwards 1977). The Wola themselves acknowledge that this is an important process, for the replenishment of what they call *suw iyba* or 'grease', which they judge as a soapy-silty feel imparted to the soil by decomposing vegetation, and which they take as an important indicator of fertility. According to Bleeker (1983:198), a humic topsoil nearly 15 cm thick can form within 25 years under the relatively cool and wet conditions of the Highlands, if left undisturbed. The return of organic matter under grass will be less, but still significant, even on sites under long-term cultivation. The Wola practice of building mounds over the composted vegetation pulled up from sites when reworked ensures the return of fair amounts of organic matter, with composting rates of up to 85 tonnes ha^{-1} at each recultivation (see Chapter 13). This further promotes the conservation of soil, additions of organic matter increasing the resistance of soil to erosion.

The evidence suggests that the Wola practice a conservative, not degrading, agricultural regime. Perhaps an unexceptionable conclusion. After all, archaeological evidence suggests that the Highlands of New Guinea was possibly among the first places on earth where human-beings cultivated domesticated plants, something like 10,000 years ago, contemporarily with the beginning of settled farming in the so-called fertile crescent of the Middle East (Golson & Gardner 1990; Golson & Hughes 1976; Golson 1982). They are the inheritors of a long-lived farming tradition, accordant with the natural environment, not discordant agents of degradation, whatever initial impressions. The lessons learned over millenia have been passed down to today's farmers, albeit changes having occurred in their agricultural practices, notably attendant upon the inferred fairly recent arrival of sweet potato. The archaeological evidence on agricultural and environmental change supports the contemporary perspective, pointing to the long-term sustainability of the agricultural tradition. The palaeolimnological information suggests increased soil losses with the beginning of the current day sweet potato mounding agricultural regime. Palynological evidence from montane lakes indicates an acceleration in sedimentation over the last 300 years or so, interpreted as due to increased human activity and deforestation consequent upon the arrival of sweet potato with upward extension of the altitudinal limits of subsistence farming, more intensive land use and population growth (Oldfield et al. 1980, 1985a & b). Soil erosion is worse today. The highlanders' ancestors were perhaps less of a force for land degradation.

SECTION IV

THE BIOTIC FACTORS

'*Obun tay ma hombuwnja ombes way buway iyba misor, iyba pombray chem bort nokor. Ngubiy tomb g^enk mon taebuwgiy non biy, iyba g^enk mon, ereb iyba waek waerakor pombray hombuwn bort.... Den hombuwnja aend sokor tomb, shor tiybaem tiybaem pung bukor, ebay sokor. Obun waip waem waem tomb, bundun bort waip suw pombray menkoroga; waip obuw ora pung bukor ngo tomb obuw pombray menkoroga.*'

'The reason [for abandonment] is that the taro and other crops planted have taken the 'grease', they have eaten the 'grease' in the black topsoil. Now it is somewhat tired with only a little 'grease', but later it will make 'grease' again throughout all the black topsoil.... When grass and everything grows, leaves fall and fall and rot, and it will become alright. When vegetation litter accumulates, underneath it changes into black soil; when litter rots completely, then it transforms to black [soil]'.

Maenget Pes

CHAPTER 7

ETHNOBOTANY: PLANTS AND VEGETATION COMMUNITIES

The vegetation and soils of any region are closely related. Living on the soil, plants use it physically to anchor themselves and draw from it the nutrients they require for growth and development. And they in turn feed the soil, returning to it when they die to decompose and create organic matter, improving soil structure and replenishing nutrient supplies. It is part of the grand ecological cycle which gives us today's arable resources. Farmers attempt to fit themselves into this cycle where they rely on the regeneration of natural vegetation to manage the fertility of their soils. The Wola, like other cultivators, are quite aware of the importance of natural vegetation to their management of soils, readily pointing out, as in Pes' opening comments to this section, how rotting vegetative matter 'builds' topsoil, adding to cultivable soil resources, giving it 'grease' and restoring its fertility.

The growth and decay of vegetation is the third factor featuring in soil formation (see Chapter 1). While the climate of any region determines the conditions for soil processes, and geological resources weather to supply the inorganic fraction, the cycle of plant life is critical too, comprising the organic phase. Although the plants that grow in any region depend in considerable part on the climatological and geological preconditions, in an established ecological community they play a prominent role in soil genesis and processes, influencing the edaphic environment for future plant generations. It is to this topic, and the part uncultivated plants play generally in the agricultural system, that this section turns. The awareness that local populations have of plant ecology, vegetational successions and related edaphic processes will condition their attitudes towards, and management of, their natural resources.

In relation to ecosystems, nutrient cycling and natural resources generally, a very wide range of plants contributes to mineral and organic matter interactions between vegetation and soil in the New Guinea Highlands. Countless complex micro-ecological systems occur across the Wola region, with a myriad of differingly composed communities at the species level. Regarding the plants themselves, the Wola are keenly aware of many of the species that comprise the different vegetational communities of their region. Their extensive system of botanical classification reflects their considerable acquaintance, it including several hundreds of names familiar to all as everyday knowledge (Straatmans 1967).

The study of ethnobotany, as opposed to ethnometeorology and ethnogeoscience, is well established in New Guinea as elsewhere, with many excellent studies in the field (e.g. Hays 1979; Haberle 1991; Kocher Schmid 1991;

Hide *et al.* 1979; Powell 1976a&b; Sterly 1974, 1977). The emphasis in this work is on local plant classification, together with the documentation of the uses to which people put plants. The aim here is to explore other issues in addition, by placing plants in their ecological context. The relation of plants to soil depends on the supply and cycling of mineral nutrients. The botanical composition of the different plant communities found in a region determines the nature of organic matter inputs. Indigenous perceptions of communities, and of ecological processes going on within them, are relevant, as are their attitudes to management and conservation. Local manipulation of plant successions, attempting to manage soil conditions and fertility, influences organic matter inputs. People's activities impact on decomposition processes and soil formation.

Plate 7.1   The *iysh* family: a Southern beech (*Nothofagus*).

Plate 7.2 The *henk* family: a tree fern (*Cyathea*) growing in *gaimb* cane grass (*Miscanthus*).

I. THE PLANTS

The folk classification of the Wola, indicating how they think their plants relate together, is relevant to an ethno-environmental study, for it conveys something about those qualities perceived locally to be ecologically important. The key features used by the Wola in classifiying plants are overall morphology and habit. Some of their categories consequently cut across certain botanical ones — scientifically grounded classification sometimes grouping otherwise outwardly dissimilar species according to quite foreign, technically defined characteristics, not always obvious to others nor easily observed. Disagreements between them are particularly evident on the family level and

Plate 7.3 Plants of no family: an *aenk* screw pine (*Pandanus* sp.) growing in *gaimb* cane grass (*Miscanthus*).

above: the Wola, for instance, unequivocally classify certain climbing palms and pandans as *ya* or vines, excluding the free-standing members of the Palmaceae and Pandanaceae families, for which they have no overall class names.

The Wola name six major life-form categories, as follows:

WOLA FAMILY NAME	ENGLISH LIFE-FORM GLOSS
Iysh	Trees, woody plants.
Henk	Tree ferns
Ya	Vines and climbers
Munk shor	Large-leaved herbs, some epiphytes
Den	Grasses, herbaceous plants
Kwimb	Mosses and liverworts

Plate 7.4 A homestead: a bamboo (*Nastus*) overhanging a women's house.

Some of these primary taxa terms, where they exist, also have wider connotations in certain contexts; the word *iysh*, for example may also refer to firewood and timber generally, the word *ya* to string or rope, and *den* to weeds when used in relation to cultivations.

But some plants belong to no named life-form class. While the Wola call all ligneous plants *iysh* and all climbing ones *ya*, they have no named categories covering palms, pandans, ferns, bamboos or canes. Here they assign names to individual species but do not group them together into higher named taxa. For them there are not palms or canes but only such-and-such a palm or cane. These differences raise questions about the nature of Wola plant classification compared to that of either our folk system or that of botanical science (Berlin 1992). There are similarities evident, but also some significant differences. Not only do the Wola for instance have no mid-level named taxa equivalent to life-class, order or family covering all plants, but they have no word equivalent to plant or vegetation. They do not in speaking group together all plants into a named taxon equivalent to our upper level concept of a plant kingdom.

Plant Taxonomy

The Wola refer to life-form categories, where they exist, as *sem* (lit. family), talking for instance about the *iysh sem* (lit. tree family) or *henk sem* (lit. tree-fern family). When naming plants the Wola may, but need not, use primary taxa where available to form composite names (Bulmer 1974), for example people may talk of *iysh pel*, which is equivalent in English to referring to 'beech tree'. The use of binomials varies with context, emphasis, danger of confusion if not used, and so on. There is also an element of customary usage with some plants; people often talk of *den leb* (*Acorus calamus*) for instance, but rarely speak of *den bol* (*Ischaemum polystachyum*), referring instead to just *bol* — when both are classed as *den* grasses. In some contexts, polysemous binomial taxa terms may not relate to plants at all, for example someone referring to *ya hung* is talking about string made of *hung* (*Pipturus* sp.) tree bast fibre, not a vine of any kind. The use of binomials, featuring uninomial

Plate 7.5 The *den* family: a garden site overgrown with herbaceous invaders (*Arthraxon, Polygonum, Viola, Crassocephalum* etc.), some sweet potato (*Ipomoea*) and Highland pipit (*Setaria*) among them.

secondary taxa labels, is more common at the lowest tertiary taxonomic level, although not invariable.

It is the next taxonomic level, which occurs below the *sem* where present, that is central to Wola plant classification and nomenclature. It equates in most cases with the genus and species taxa levels of scientific botany and the oak or primrose terminological level of the common English system. The equation of Wola names at this level with scientific ones throughout this study firstly requires qualification; the latter are relatively constant, experts defining categories carefully and applying specified criteria consistently to specimens when making identifications. Wola categories are not necessarily so rigid, individuals disagreeing on occasion over the naming of plants, sometimes displaying a surprising degree of dissent (see Sillitoe 1983). While they may be almost unanimous over the naming of common plants, their non-literate classification system has an intrinsically flexible aspect and they may disagree considerably over uncommon ones, calling into question the entire enterprise of matching some Wola names with those of natural science. Regardless of the extent of variation, there exists a majority opinion on the correct name for any plant, and this study gives these consensus identifications, so far as they are determinable using a few respondents.

The extent of disagreements over plant identification depend on the level involved in the classification hierarchy, the commonness of the plant concerned, and the fineness of the distinctions made in naming it. There are

Plate 7.6 Some cultivated food plants: a mixed vegetable garden (*em g^{e}mb*) with taro (*Colocasia*), sugar cane (*Saccharum*) and pumpkin (*Cucurbita*), among other crops.

few disputes at the upper, life-form taxa level of classification, people largely agree over when a plant is an *iysh* tree or a *ya* vine or whatever. Nonetheless, the ascription of some plants to primary life-form taxa is not invariable, and sometimes people place a plant in more than one superordinate category on different occasions (Healey 1979); for example, they may sometimes refer to *shaenshuwril* (*Pennisetum macrostachyum*) as *den* grass but on others talk about it as more akin to *gaimb* sword grass (*Miscanthus floridulus*) which they never think of as a *den* grass. Other plants are anomalous regarding class ascription and their status open to dispute, the epiphyte *borok* (*Microsorium punctatum*) some think belongs to the large-leaved herb taxon of *munk-shor* because people sometimes use its leaves in earth ovens, while others maintain that it belongs to the tree-fern taxon of *henk* because it bears spores on the undersides of its leaves. The aroids *mondba* and *dedwal* (*Alocasia* spp.) are also irregular, for while some say they are *ma sem* (lit. taro family) others disagree because they are inedible and uncultivated; their ascription to the *munk-shor* taxon is similarly open to dispute because no one uses their large leaves in earth ovens, which only allows the *den* taxon, for which they are really overly large to be members.

Few individuals disagree over the identification of common plants, for example crop plants like sweet potato, taro and so on at the species level, or frequently encountered trees or shrubs like casuarina and cordyline. But less often seen plants, for instance of remote forested regions, may provoke denials of others' identifications or claims of ignorance about names at all. People are also more likely to dispute identifications where the discriminations required in naming plants are particularly fine; for example to differentiate between the ferns called *saezuwp* (*Dicranopteris linearis* var. *altissima*) and puwt (*D. linearis* var. *montana*) demands the making of some particularly acute distinctions (B. Parris pers. comm.). It is at the lowest indigenous taxonomic levels that disagreements over the naming of plants are most probable. The Wola continue to classify some plants on below the level equivalent to genera or species, to one approximating to discrimination between either very closely related species or lower still between varieties and cultivars. They distinguish, for example, considerable numbers of cultivars of some crops (Sillitoe 1983), and they differentiate between similar species and/or varieties of several plants, like four *pel* southern beeches (*Nothofagus* spp.) and four varieties of *muwnaen* bracket fungus (*Grifola frondosa*).

It is understandable that disagreements over naming plants is most likely at the lowest taxonomic levels, with identifications frequently depending on micro-morphological variations in plant shape, size and colour, together sometimes with other small differences in habitat and growth. Nevertheless, the extent of aberrations between individuals can be disconcerting at times, leading one to ponder the nature and significance of differences between our notion and their's of what a classification system comprises. In short, to what extent are the Wola, who are socialised into an entirely alien cultural tradition, doing something analagous to classifying in

Western thinking when they order plants, or any other natural phenomena, into classes? They appear to conceive of plant ordering in a way that is both familiar to a European and yet, at the same time, different too. This impression, of familiarity mixed with strangeness, is commonly encountered in accounts of how other cultures classify the natural phenomena found in their regions. It is commonly explained by reference to the varying scopes that different taxonomic levels afford for cultural elaboration, that at higher levels there is more opportunity for cultural innovation and invention, whereas at lower ones the closely observed differences and discriminations leave little room for variation between cultures.

The absence of any term at the kingdom level analagous to 'plants' immediately marks off Wola plant taxonomy as somewhat different to ours. They do not taxanomically conceive of all plants being collected together at the apex of a classificatory hierarchy. But, leaving aside the absence of such a supra taxonomic level, the manner in which the Wola classify *some* plants, although not so elaborate regarding numbers of classes, parallels the hierarchical classification of botanical science, with up to three taxonomic levels, which equate with:

WOLA CLASS TERM	**ENGLISH CLASS EQUIVALENT**
sem	life forms (primary taxa)[1]
↓	↓
semonda	genera/species (secondary taxa)
↓	↓
semg^enk	species/varieties (tertiary taxa)

It is noteworthy that the Wola refer to these classes as *sem* or 'families'. They use the word *sem* widely for groups of phenomena, including local community groupings of human-beings, down to the extended or nuclear family called a *sem* (see Chapter 1). They frequently qualify the *sem* term as either large (*onda*) or small (*g^enk*).[1] Regarding plant classification, there are, at the highest indigenous taxonomic level, life-form *sem* (e.g. *iysh sem*, *ya sem* etc.), followed by mid-level classes that have *semonda imbiy* (lit. family-large names), and at the lowest class level there are the plants' *semg^enk imbiy* (lit. family-small names). The use of the same terms by the Wola for taxonomic classes as they employ in their classification of social groups (Ryan 1961; Sillitoe 1979; Lederman 1986) suggests that they conceive of them as analagous categories, with plant taxa organised hierarchically, one descending from the other in the same way as local groups represented on genealogies. When they classify some plants, they think of them as arranged in nesting hierarchies, in a manner similar to botanical science, as follows:

---

[1] The terms in brackets are those suggested by Bulmer 1974.

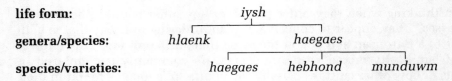

The classification of other plants, on the other hand, appears to reinforce the strangeness intimated in the Wola taxonomic system by the absence of any kingdom-like 'plant' term. Many plants do not fit into the above hierarchical scheme. They have one name only and are assigned to no higher nor lower level classes. Several of the pandans and palms are classified in this way; for example the large multi-crowned forest pandan *aendashor* (*Pandanus antaresensis*) and the tall stately palm *goiz* (*Gulubia* sp.) are not members of any higher level named taxa nor do they divide up below into different named varieties, they are all just *aendash*or and *goiz*.

It is difficult to understand why the Wola should gather together some plants into life-form classes and not others. If those not so classified were grouped together, they would comprise relatively small life-form classes because there would be relatively few plants eligible for each. These unclassed plants are also quite distinctive and stand out from others. Regarding the lowest *semg^{e}nk* taxonomic level, it is easier to appreciate why the Wola may divide up some plants more than others. Although some anthropologists eschew pragmatic explanations, there is a relationship apparent between the extent to which the Wola classify plants at this level, if at all, and their utilitarian importance to them, either as food or raw materials (for example, they distinguish at least sixty-four cultivars of their staple crop sweet potato — Sillitoe 1983).[2] There is also some relationship between the extent of *semg^{e}nk* level classification and the commonness of plants, how often people see them, and the occurrence of notable differences between those classed together in the same mid-level *semonda* taxon.

The Wola further fly-in-the-face of generally assumed tenets of classification by giving the same name to two or more plants which although similar they acknowledge are not identical. They frequently refer to such plants, if a pair, as such-and-such *oliy* and *weray* (lit. husband and wife) — for example *haeraedaepon oliy* and *haeraedaepon weray* (*Pipturus* spp.), which they say are 'one leaf' regarding shape but different with respect to colour, the husband being white on the underside. At other times they refer to such plants, if there is a marked size difference between them, as *onda* and *g^{e}nk* (lit. large and small) — for example *woluwmsaeren onda* (*Cypholophus* sp.) and *woluwmsaeren g^{e}nk* (*Elatostema* sp.) to distinguish between woody and soft-stemmed plants of the same name. On other occasions they may distinguish between such plants, if they vary in height, as *sol* and *hiy* (lit. tall and short) — for example they refer to three types of sedge grass as *dunguwlumb* and distinguish between them as *dunguwlumb sol* (*Kyllinga melanosperma; Junus*

---

[2] Haberle 1991 makes the same point for the neighbouring Huli.

effusus), *dunguwlumb hiy* (*Kyllinga brevifolia*) and, *dunguwlumb tuwguwn* (*Eleocharis* sp.), the latter being the middle-sized one (the qualifying adjective *tuwguwn*, meaning middle, may also be used where three plants of 'one name' are distinguished by overall size).

Distorting Indigenous Knowledge

The up-shot is that no single classificatory scheme can adequately represent Wola ordering of plants, whatever framework we adopt will be somewhat distorting, which is perhaps to be expected because writing down any oral scheme misrepresents it. The approach taken in this study is to use indigenous names for all plants together with botanical identifications.[3] This equation of Wola plant names with the family, genus and species labels of botanical science risks misrepresenting their views. There is a danger that the idea might be conveyed that the Wola not only classify but also identify plants in a way similar to us, seeing the same objective specimen 'out there'. When asked how they identify particular plants, informants usually point to morphological features as differentiating between them. They give no standard responses. Different individuals may point out varying features, suggesting that when the Wola identify a plant they see it in its entirety and do not customarily search for specific cues as grounds for naming it. They simultaneously consider a range of observable criteria, viewing a plant as a distinct entity and not as something distinguished by having a certain number of distinctive, limited features. Those characteristics which seem to figure prominently in the configuration seen by the Wola cover plant form primarily, particularly the shape, size and colour of a plant's organs. Occasionally scent features too, and habitat. When making identifications at the *semgenk* level, the points people look for become narrower, with micro-morphological variations and colour changes important. They look for these narrow cues more systematically, in a manner familiar to botanists, because of their fineness. The problem here is the considerable level of disagreement encountered between informants about the use of these diagnostic criteria to name particular plant specimens, which again takes us away from any close parallels that we might assume with scientific botany.

The names which the Wola give to some plants are also descriptive of their habits and may assist in identification. The moss *shononpep*, for instance, grows only on the *shonon* (*Acalypha* sp.) tree, and any plant name including the words *showmay* (lit. pig) or *iyb* (lit. water) have associations with pigs or water respectively. Similarly, plants incorporating the words *towmow* (lit. ghost) or *deraen* (lit. outside) are emphatically wild species, associating them with ghostly propagation away from human habitation. The sedge grass *wesaembowshoba* (*Cyperus distans*) (lit. head-fork) is so called

---

[3] For a detailed catalogue of the plants identified by the Wola, see the *Journal of Ethnobiology* Vol. 15.

because of its large dense panicle reminiscent of a hairy head, and the bulb-bearing orchid *tombshombiy* (*Spathoglottis parviflora*) is so named because it is an ingredient used in the concoction administered to those thought to have been *moktomb* poisoned, to make them vomit. The name *kuwmkuwm* (*Blumea arnakidophora*) relates to the soft leaves of this plant which people use to wipe sweaty faces, making them *kuwmbiy* (lit. cool), and the word *aegop* (lit. tail; *Cordyline fruticosa*) makes clear allusions to men's tail-like sporting of Cordyline leaves to cover their buttocks. The name *muwmonhuwshiy* (*Viola arcuata*) (lit. Brown-Quail's hibiscus-spinach) recalls that the *muwm* quail eats this plant's fruits and leaves like humans eat those of *Hibiscus manihot*, and the newly arrived plant *cowaden* (*Desmodium* sp.) (lit. cow small-plant) reflects the introduction of this plant for cattle grazing by agricultural officers. The name *pibiytaeztaez* (*Isachne arfakensis*) refers to the way in which large beads of dew (*pibiy*) collect on the panicles of this grass, and *mondkaend* (*Pouzolzia* sp.) relates to the fact that this plant frequently germinates in the channels (*kaena*) between sweet potato mounds (*mond*). And so on.

It is pertinent here to note briefly how I learnt about the way the Wola identify and classify the plants of their region. This information derives from botanical enquiries, supported with plant collections made largely in the Was valley, and some in the neighbouring Nembi and Ak valleys, all west of Nipa. The collections were made from a Wola viewpoint, noting the local names for plants, together with other related information such as habitat where found, relative abundance and any uses to which people put them. The collected specimens were pressed for scientific identification later, sometimes several years later.[4] The botanical specimens were gathered in a variety of ways. Many of them I collected personally, not always on special plant collecting trips but frequently when engaged in ethnological research, walking from one place to another. Always accompanied by one or more Wola friends, I regularly enquire about plants we pass, and they, aware of my interest, frequently volunteer information. And to guide collection, I have compiled a check-list of Wola plant names (an open-ended list to which I continue to add further names as learnt), informing friends of gaps so that we might fill them.

This way of learning is quite foreign to the Wola, who normally pass on knowledge in a casual and piecemeal manner. Asking them to find plants to fit names on a list, how they classify them, why they have life-form classes for some and not others, why those plants that have life-form names are so labelled, and so on, are odd questions demanding contrived answers. Nevertheless, the principles the Wola use in ordering plants became familiar

---

[4] I acknowledge with gratitude the assistance I have received from several scientists and institutions in identifying the plants collected from the Wola region. In particular I thank Barbara Parris, Bob Johns, David Frodin, Peter Edwards and other staff at the Herbarium of the Royal Botanical Gardens at Kew for their matchless assistance in facilitating the identification of many plants; the late Ben Stone one-time Malaya University; also the staff of the following herbaria: the Botanical Gardens at Edinburgh, the Rijksherbarium in Leiden, the Department of Forests at Lae, and the Manchester University Museum.

to me, and it is these that feature in this study, giving it a Wola focus. Even so, it inevitably distorts their ideas to some extent, presenting them as more formalised than they are, although no more so than any anthropological account of aspects of another culture — contemporary post-modern criticism of this unavoidable misrepresentation notwithstanding.

II. THE VEGETATION COMMUNITIES

The understanding that people the world over have of their natural environments extends further than just naming the flora and fauna that occur in their regions and arranging them according to some classificatory scheme or another, although the emphasis which much ethnoscientific writing gives to classification could lead one to assume otherwise. It is necessary to go beyond a discussion of taxonomic schemes I argue, interesting as these are, to explore more fully the understanding people have of their environments — which they achieve in part, of course, using their classifications — to enquire how their perceptions influence their behaviour relative to the ecology of their homelands. In documenting how the Wola classify the plants that occur in their region, we have taken only the first step to some appreciation of their environmental understanding and the exploration of their interactions with, and their influence upon, their place.

The next step involves the recognition of different plant communities and habitats, and an investigation of their structure and dynamics. The Wola recognise several different vegetational communities, comprising varying populations of plants, both named and unnamed. Their awareness of these communities and understanding of their dynamics influences their attempts to manage their natural resources and their consequent impact upon the environment. Different plant communities have differing relations with soils, as they know full well, notably in terms of organic matter inputs, with all their consequent knock-on effects. This knowledge informs their cultivation strategies, although again it is more evident in their practices than in their verbalised accounts of these, which as related to me, feature nothing remotely similar to our concepts of nutrient cycling within ecosystems and so on.

The Plant Communities

The vegetation of the Wola region relates grossly to topography and altitude, notably as these influence human settlement patterns and land exploitation.[5] In the majority of valleys, between 1,600 m. and 2,000 m., where people live and cultivate most of their gardens, dense cane grass regrowth predominates, interspersed with the short grassy clearings of recently abandoned gardens

---

[5] For further information on the various vegetational communities described here and a finer botanical classification of the different communities see Robbins and Pullen (1965), Paijmans (1976:84–97), and Johns (1976;1982).

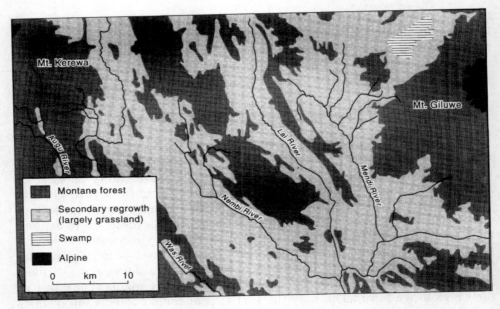

Map 7.1 Vegetation of the Kerewa-Giluwe region

and the dark soil and crops of current cultivations. On steep and uncultivable land, pockets of undisturbed forest occur. In the unpopulated areas of river valleys, on the mountains and watershed ridges and dolines, and above 2,000 m. generally, lower montane rainforest predominates, with a few patches of regrowth and occasional gardens (Map 7.1). The cane grasslands, besides having an abundant cover of cane or sword grass, support a limited range and number of secondary regrowth trees and a relatively meagre wildlife population, consisting primarily of small rodents and birds. The forest, on the other hand, is notably richer, supporting many hundreds of species of trees and other plants, together with a teeming animal population of marsupials, rodents and birds, some of them large and colourful.

While this prelusive description of the vegetation of the Wola region according to its two major plant successions, of forest and grassland, serves broadly to characterise it, particularly as it first strikes the visitor, it overlooks some noteworthy smaller plant communities and fails to do justice to Wola conceptions regarding their region's floristic ecology. They distinguish the following eight vegetational communities:

- *iyshabuw:* lower montane rainforest
- *pa:* wetland vegetation
- *haenbora:* rocky vegetation
- *yom:* alpine vegetation
- *em* and *aendtay:* gardens and houseyard environs
- *mokombai:* recently abandoned garden successions
- *gaimb:* cane grass regrowth
- *obael:* secondary forest regrowth

The lower montane forest and cane grass regrowth are the two vegetational communities that predominate across the Kerewa-Giluwe region, covering over ninety eight per cent of the area (Table 7.1). The other communities are small in comparison. These statistics may however under represent the area covered by some of these smaller communities, being calculated from large scale maps and aerial photographs,[6] supplemented by my own limited observations (notably they will omit small-sized areas under some of these less extensive vegetational communities which, occurring in one of the two major communities, are 'lost' because below the minimal size represented at the gross scale of the reconnaissance).

Thorough and closer work in more restricted areas indicates nonetheless that these overall figures are of the correct order. The swampy area to the east of Haelaelinja for example (Map 7.2), in the the Was valley region where I conduct fieldwork, covers about 0.6 km² which, while a noticeable part of the local territory on which it occurs, is too small to show on the map of the entire Wola region and makes no difference to the percentage given for this vegetational community overall, comprising only 0.02% of the entire region.[7] Detailed data on areas under cultivation on the territories of two neighbouring *semonda* communities in the Haelaelinja region (population of approximately

Table 7.1
The areas and relative proportions of the Kerewa-Giluwe region covered by different vegetation

	Kerewa-Giluwe Region		Aenda and Ebay *semonda* territory*	
	Area (km²)	% of area	Area (km²)	% of area
Lower montane rainforest (inc. some tree regrowth & a few gardens)	1295.3	52.2	32.8	83.4
Cane grass regrowth (inc. some tree regrowth & short grass areas)	1143.6	46.2	6.0	15.3
Swamp vegetation	17.4	0.7	~0.1	~0.3
Gardens	included under cane regrowth largely (some in forest)		0.4‡	1.0
Alpine vegetation	21.7	0.9		
TOTAL	2478	100	39.3	100

* Area within sweeping bend of Was river & slightly smaller area to the south of the bend, together with territory around northern area of regrowth in Ak valley & part of the forest in the south between the Ak and Dorwael rivers.
‡ 395,396 m².

---

[6] C.S.I.R.O. (1965) Forest Types map; Radcliffe (1986:28); Papua New Guinea 1:100,000 topographic maps; and R.A.A.F. (1959) 1:35,000 aerial photographs.
[7] This swampy area occurs on the territory of a *semonda* neighbouring those documented in the right hand column of Table 8.1 and hence is omitted from it.

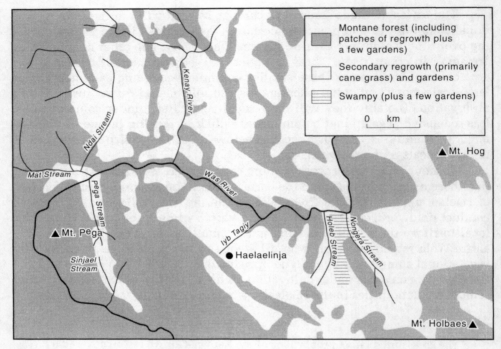

Map 7.2   Vegetation of the Was valley study area

300 persons) indicate that they only cover small areas too, at one per cent of these communities' territories (Table 7.1). The local territories in the Was valley region include considerably larger areas of lower montane rainforest than average, other territories elsewhere have more grassland, swamp and so on. In his study of LANDSAT imagery for example, Radcliffe (1986:28–29) found that only 39% (125.4 km^2) of the Upper Mendi region is under rainforest, whereas 15% (45.7 km^2) of it is under wetland vegetation (the region surveyed including the large Kombie swamp within it), and cane grassland comprises the balance, covering 46% (144.6 km^2) of the area. In summary, while closer study of more limited areas reveals predictable variation between territories across the Wola region, the broad picture is one of forest and cane grassland communities predominating, with patches of other vegetational communities dotted about them.

Montane Rainforest

The lower montane forest or *iyshabuw* (lit: wood-*abuw*) is not what I had expected rainforests to be like from boys' adventure stories, with impenetrable green walls of rank jungle vegetation. In places on the ground it can be of unexpectedly open aspect, with the sky visible through the canopy overhead. It is difficult to evoke its feel on flat paper, as it is to depict the rugged aspect of the Wola countryside. It is grand, and cathedral-like it inspires humility. It can frighten too, particularly those unfamiliar with it, by its size and extent. It is easy to lose one's bearings here. It recalls childhood fears of being lost in the woods. It sometimes intimidates the Wola too, who project their fears in forest-dwelling demon spirits.

The Wola maintain that the *pel* southern beech (*Nothofagus* spp.) predominates in the forest, with many gigantic mature trees and a considerable scattering of younger ones plus the occasional dead or dying tree with bare stagheaded crown. Quadrat survey data indicate that some 12% of the large trees in the forest are *Nothofagus* beech.[8] While beech is dominant, other *iysh* trees sometimes occur in considerable numbers in the forest, rivalling the beech in places, giving a mixed aspect to large tracts. They include *kaeriyl* and *pay* oaks (*Lithocarpus* and *Castonopsis*), which are members of the Fagaceae family like the southern beech and have retained similar ecological habitat compatibilities; the *Lithocarpus* oaks are particularly common in heavily disturbed forested pockets on lower valley slopes. Other trees include *ponjip, pil, soiz* and other figs (*Ficus*), *hok* colas (*Sterculia*), *haezuwmb* white magnolias (*Galbulimima*), *munk* and *wok* gamboges (*Garcinia*) and *ongol, shwimb* and other elaeocarps (*Elaeocarpus*), among others. The trees form a c.80% canopy cover at about thirty metres, with some emergents above it, and sometimes with a secondary or diffuse layer at twenty metres or so, all competing upwards for a share of the light. It is difficult to identify any clear stratification in the forest because of the overlapping extents of the crowns in the various 'layers'. All the trees are shallow rooting on the whole, with most of their roots fairly evenly distributed in the top 20 to 30 centimetres of the soil (Edwards & Grubb 1977, 1982). The ground cover is frequently heavy, often restricting visibility, and varies from dense stands of saplings and shrubs (including *gwaigwai* and *hombom* [Melastomaceae], *kol* [Phyllanthus], *haeraedaep* [Pipturus], *polpol* [Cyrtandra], *kongol* [Piper], *laenjlaenj* [Symplocos], and *penden* [Daphniphyllum]), to impenetrable tangles of slender stemmed climbing *mael* bamboo (*Racemobambos congesta*), to masses of fleshy leafed herbs, notably *munk* gingerworts (Zingiberaceae, Urticaceae) and ferns, both *henk* tree ferns (Cyatheaceae) and in places numerous low *meshmesh* clump ferns (*Cyclosorus*).

Frequently enveloped in low cloud and subject almost daily to considerable rainfall, the forest has a wet aspect, dripping much of the time. The wetness is one of the most irksome aspects of forest travel because it renders everything slippery underfoot and can make walking along some parts of paths akin to paddling across ponds. Mosses (*Frullania, Meteorium, Bazzania, Dicranoloma* and *Lepidozia*) thrive in this environment, and thick mats festoon trees and shrubs. They also cover the ground in a springy bed, together with a thick layer of rotting vegetation and fallen leaves, giving a raw

---

[8] For detailed data on the species composition of this, and the following vegetation communities, see the *Journal of Ethnobiology* Vol. 16.

or partially decomposed litter through which surface roots run. The forest floor is criss-crossed with surface and stilt roots, and fallen timber too which, together with the uneven ground, pitted by hollows and crevices, and the standing vegetation, make travelling awkward off of the established paths, and progress slow.

The trees also support a variety of *ya* climbing plants, from woody lianas like *toben* (*Alyxia*), *paerelya* (*Cayratia*), and *wolaya* (*Dimorphanthera*) to *haeluwp* palms (*Rattan*) and *gaiya* pandan vines (*Freycinetia*), twining strongly around towering trunks like sinewy arms trying to force their way into canopy gaps and seize the last chinks of light overhead. Trees act as hosts for many epiphytic plants too, all of which thrive in the humid atmosphere of the forest. They range from bryophtes on trunks, to vascular epiphytes and ferns on branches, to sooty moulds on leaves. On the crown branches of some trees there is a peat-like accumulation comprising the remains of epiphytes and litter which forms a substrate for larger epiphytic plants, and occasionally some tree species too which are more usually found growing on the ground, including *haegaes* (*Schefflera*), *nemb* (*Pittosporum*), *komb* (*Timonius*) and *sabhul* (*Gardenia*).[9] The older and larger trees predictably support more climbers and epiphytes. The epiphytic *pondiyp* orchids (*Dendrobium*), together with some other herbs, sometimes relieve the dank grey-green aspect of the forest with a medal-ribbon-like splash of colour to a uniform background.

In this forest lives an abundant and varied wildlife which, while less vocal than that found at lower altitudes with its noisy bird species, is nonetheless present in considerable numbers. Common furry animals include cuscuses and possums, together with tree kangaroos. A wide variety of rodents, including giant rats also occur, as does the egg-laying quilled echidna, a rare animal of remote forested regions. Birds are numerous, varied and colourful, ranging from the large flightless cassowary to various small flycatchers, from colourful parrots and lories to soft hued pigeons, and from numerous honeyeaters to some of New Guinea's renowned birds of paradise, among many others. Although not frequently seen in the dense vegetation of the forest, they can often be heard singing near and far. The reptile and insect populations are also numerous and varied.

A marked feature of the forest is the similarity and continuity of its structural form and floristic composition through a range of stands. While there are doubtless environmental gradients which account for species distribution, with nodal structural types occurring, they feature complex interactions and are difficult on the whole to specify. Nonetheless, there are locations where obvious environmental variations occur and give rise to particular plant asso-

---

[9] One way the Wola explain that these trees germinate in the crowns of larger trees is by birds depositing seeds there in their droppings; they even name the Red capped flowerpecker (*Dicaeum geelvinkianum*) after the *mondiyt* shrub, which it helps to propagate by distributing its sticky seed via its droppings (having disgested the outer fruit casing), calling it *mondiytiylkaelenj* (lit. *mondiyt*-shrub seed *kaelenj*).

Plate 7.7 The *iyshabuw* lower montane rainforest, seen from a garden, Southern beech (*Nothfagus*) predominating (including the gaunt, moss-clad trees).

ciations. These are well known to those familiar with any territory, occurring for instance in wet or waterlogged pockets, on bare steep slopes and along watercourses, sometimes varying with aspect. Landslides and slips, if extensive, can also change the floristic community, although the most pervasive interference and consequent modification of succesions results from the actions of human beings. While the forest is predominantly primary, human activity has disturbed considerable tracts (Flenley 1969 suggests that human disturbance is one of the main factors controlling vegetation, the other is altitude). In some places the interference is minimal, a hunter perhaps having felled a tree or cleared some undergrowth. In other places the disturbance is extensive, a man having maybe established a clearing to allow the sun access to nut-bearing *aenk* screw-pines (*Pandanus julianettii*) he has planted. These areas sometimes becoming quite large, developing into pandan groves.

Plate 7.8 An expanse of *pa* wetland, swamp grass (*Leersia*) predominating, giving way to a cane grass (*Miscanthus*) community.

Elsewhere in the forest wild pandans, with their prominent prop-roots, grow singly here and there. Nearer to settlements people have considerably disturbed the forest and altered its floristic composition, with fewer beech and more faster growing, softer-wooded species evident.

Swamp Vegetation

Wetland vegetational communities occupy varyingly sized areas throughout forest and grassland, occurring in poorly drained locations where bog conditions prevail, called *suw pa* (lit: place bog). The wetness of these areas varies, depending on the closeness of the water table to the surface, from spongy damp swards to waterlogged swamps. The vegetation consists of water loving *den* grasses like *obol* (*Leersia*), *bol* (*Ischaemum*), *pibiytaeztaez* (*Isachne*), and *mombiltay* (*Panicum*), which dominate swards, and herbs, such as *huwguwp* (*Cyperus*) and *dunguwlumb* (*Kyllinga*) sedges, *ungwem* horsetails (*Equisetum*) and *taziy* parsley (*Oenanthe*), growing in relatively low tussocks, together with a scattering of wild *mokombez* sugar (*Saccharum robustum*), and *gaimb* (*Miscanthus*) and *holor* (*Coix*) cane grass clumps, plus the occasional *iysh* tree and dwarf shrub, like *pak* water-gums (*Syzygium*), *iybkol* icacinads (*Rhyticaryum*) and *naep* she-oaks (*Casuarina*), particularly on the swamp margins.

Swamp forest occurs in some waterlogged locales (Johns 1980). Conifers like *tibil* are common here (*Dacrydium, Podocarpus*), together with a range of other *iysh* trees like *mul* (*Glochidion*), *paerep* (*Maesa*), and *timbol* (*Homolanthus*), giving a variable canopy beneath which occurs a dense layer of shrubby vegetation. The forest floor is fairly open with pools of water separated by irregular hummocks. Botanists have suggested that these conifer dominated swamp forests may have arisen due to extreme frosts killing off broadleaved trees (Robbins & Pullen 1965), or they may represent an early stage in mixed montane swamp forest development (Johns 1980). The wildlife populations supported by these areas depend on and reflect the surrounding vegetational community, plentiful if forest, more meagre if grassland.

The presence of hydrophytes is indicative of locations with gleyed soils, well suited to taro cultivation. These gardens, which the Wola call *ma em* (lit. taro garden), may support a range of crops in addition to taro, which tends to predominate, particularly in mature gardens. They often plant these other crops on drier hillocks and around tree stumps; they include various cucurbits and green leaved crops like crucifers and acanths. The early weed colonisers of these sites are similar to those found in other gardens (Compositae, Gramineae), often with several tree seedlings, these gardens frequently being sited in forested locales that supply abundant seeds.

Rocky Vegetation

Rocky vegetational communities occupy limited areas throughout the Wola region where the underlying geology outcrops at the surface sufficiently massively to gave areas of thin skeletal soil. Rocky locations that support some plant life are called *haenbora*. They vary in extent, but are generally small. The vegetation consists of hardy plants capable of colonising thin regolith, notably *kwimb* mosses initially (*Frullania, Meteorium*), followed when a suitable soil-like deposit has accumulated, by some ferns like *pukuwmb* (*Pteridium*) and *meshmesh* (*Cyclosorus*), hardy *pondiyp* orchids (*Spathoglottis*), stunted *den* grasses including *senz* (*Imperata*) and *gaimb* (*Miscanthus*) and the occasional dwarfed *iysh* sapling of *gwai* (*Dodonaea*), *komnol* (*Piper*), and *shonon* (*Acalypha*), among others. It is on the whole a monotonous vegetational succession with hardy ferns and grasses predominating. These locales are of no horticultural use, although they are sometimes disturbed by humans setting fire to them in dry periods.

Plate 7.9 A steep-sided limestone pyramid clothed in *haenbora* rocky vegetation of fern (*Pteridium*), orchids (*Spathoglottis*) and kunai grass (*Imperata*).

Alpine Vegetation

Alpine heath and grassland vegetation occurs on the high volcanic summits that flank Wolaland to the east and west, and is called *yom*. It is of little concern to the Wola, few people ever having reason to climb to these high inhospitable altitudes, with few resources of use to them. One exception is the hardy pandan called *dalep* or *tuwmok* (*Pandanus brosimos*) which grows in these regions, prompting some men to visit them occasionally to collect their nuts when in season. But they are dangerous locales because of their forest demon denizens, and best avoided (see Chapter 8). The flora comprises mainly tussock grassland (*Danthonia, Poa, Deschampsia, Festuca*), with small heath-like shrubs (*Rhododendron, Coprosma, Styphelia*), and low herbs (*Ranunculus, Gentiana, Lycopodium*), plus gaunt stunted *henk* tree ferns (*Cyathea*). The festucoid composition of mountain grasslands contrasts with the panicoid one of lower regions. Mires are common, with hummock plants and shrubs prominent in stagnant acid bogs (*Gleichenia, Trochocarpa, Astelia*), and sedges and grasses in fens with moving ground water (*Carex, Brachyposium*).

At high altitudes the forest adjoining these alpine grasslands may be of a quite different aspect to that lower down. Cloud envelops it daily, influencing its floristic composition. It is a single-tree-layered forest with *shina-aenk* (*Rhododendron*) and *hobaen* (*Rapanea*), of stunted and crooked aspect, with a low (10 m.) canopy, except for some emergent trees which may include *morowa* (*Papuacedrus*), *haegaes* (*Schefflera*), *gun* (*Dacrycarpus*), *oljomb* (*Saurauia*) and *tuwn* (*Podocarpus*). Thickets, particularly of fern, are common in this high-

Plate 7.10 Moss-clad stunted trees of high altitude cloud forest adjacent to the *yom* alpine community.

montane or cloud forest. The Wola refer to them as *pletbok*, a term which covers any dense and impenetrable stand of vegetation that requires the cutting of a path to pass through it. Although *pletbok* thickets are common at higher altitudes, they are not restricted to these locales. They commonly comprise a tangle of *saezuwp* ferns (*Dicranopteris*), *mael* climbing bamboo vine (*Racemobambos*), and *gaimb* sword grass clumps (*Miscanthus*), together with low bushy saplings.

Mosses and liverworts are also common in the cloud forest, sometimes festooning lower branches, exposed roots and crooked trunks; they may even cover the tangled roots and decaying forest debris on the forest floor in a thick, wet spongy carpet. Sprawling *den* shrubs are also common (*Vaccinium*, *Coprosma*, *Dimorphanthera*). A dynamic transition zone exists between forest and high altitude grassland, shrubs colonising grassy areas by creeping vegetative propagation, and grasses encroaching on shrubbery after its occasional

Plate 7.11 Sweet potato (*Ipomoea*) mounds in newly cultivated *em* garden with some sugar cane (*Saccharum*) and Highland pipit (*Setaria*).

destruction in fires (Gillison 1969, 1970). These regions are of no horticultural significance, being beyond the altitudinal range of crop plants.

Gardens and Homesteads

The Wola call their gardens collectively *em* (lit: gardens), distinguishing between several kinds. Here they cultivate a wide range of crops (Sillitoe 1983). Newly established gardens, particularly those adjacent to homesteads, feature a wide range of intercropped plants. They appear to be a confusion of crop species, and Clarke (1971:76) likens walking across such cultivations to wading in a green sea, pushing waist-high through irregular waves of crop plants, with some taller species rising above the flood overhead to provide some scattered shade. Longer established cultivations have less crop variety, *hokay* sweet potato (*Ipomoea*) vines predominating over large areas, together perhaps with some pumpkin (*Cucurbita*), *wol* sugar cane (*Saccharum*), *shombay* and *kot* (*Rungia* and *Setaria*) green leafy vegetables, and other crops interspersed among them. The occurrence of crops in gardens largely under sweet potato does not differ greatly with the time they have been under cultivation. All of these gardens support large numbers of sweet potato plants, usually cultivated in mounds, and while there is an evident increase in sweet potato plants per unit area as gardens age (from 16 to 18 to 21 plants per square metre, in gardens planted once, twice to four times, and five or more

Plate 7.12 Wenja Muwiy stands in an established sweet potato garden with a quadrat frame used in the survey of vegetation community composition.

times respectively), the range of crops cultivated does not change noticeably. Regarding changes in garden fertility over time this is particularly intriguing, for it suggests that no dramatic changes occur, corroborating the evidence of little variation in sweet potato tuber yields from gardens of differing ages. The Wola often cultivate a range of other crops in small fertile pockets that persist on sites, such as along downslope fence lines, in surface dips, and so on.

A range of cultivated plants also occurs around houseyards, called *aend bort* (lit: house at), and to a lesser extent around adjacent *howma* ceremonial grounds, where people may tend a limited range of plants, sometimes overflowing from neighbouring gardens. They cultivate long-term crops around the edges of these clearings, such as *taembok* bamboo (*Nastus*), *aegop* palm lilies (*Cordyline*), *diyr* bananas (*Musa*) and *aenk* screw-pines (*Pandanus*), and sometimes a few ornamental plants to enhance their outlook like *shonon* (*Acalypta*), *niysh* (*Laportea*) and *gonkliyp* (*Graptophyllum*), plus others on occasion that supply useful materials, including *maen* hoop-pines (*Araucariq*), *naep* she-oaks (*Casuarina*), *duwk* spurges (*Euphorbia*), *ongol* bead ashes (*Elaeocarpus*), and *hweb* marants (*Cominisia*).

Some of the early wild plants that invade garden sites are edible too, and on occasion people cultivate them, like *hombiyhaem* (*Commelina*), *taziy* (*Oenanthe*), and *suwtaguwt* (*Solanum*). Others are truly weeds and of no apparent use to the Wola, although they may serve important purposes, such as affording early protection against soil erosion (see Chapter 6). There is no correlation apparent between the weed species that colonise sites, and

Plate 7.13 A sweet potato garden in the later *puw* yield stage with a dense cover of invading weeds (*Bidens*, *Polygonum*, *Paspalum* etc.).

portend future natural regrowth if the site is not recultivated, and the range of crops cultivated or time under cultivation. The occurrence of different associations of weed plants seems to relate more closely to the natural vegetation adjacent to sites than any other factor and hence seed supply; a range of tree seedlings for example is commoner adjacent to forest, and certain herbs and grasses near grassland.

Recent Regrowth

When initially abandoned, gardens go through a rapid series of overlapping plant successions before either tree or sword grass regrowth become established. These sites are called *em mokombai* (lit: garden immature-regrowth). They are botanically varied, with quickly changing communities. Although easy to walk across, particularly during the early regrowth stage, with knee high, easily trampled vegetation, I find these sites irritating to traverse in footwear because of the several herbaceous plants that colonise them which rely on spikey and sticky burrs to disperse their seeds (e.g *Bidens pilosus, Adenostemma lavenia, Cynoglossum javanicum* — called generically *kobkob* by the Wola), for these invariably adhere in large numbers to socks and irritate my legs for the remainder of the day. They are particularly common on recently abandoned garden sites.

The species of plants occurring and their prevalence varies from one site to another, depending on the conditions that prevail, but in broad terms

Plate 7.14 The early *mokombai* phase of recent regrowth with early herbaceous invaders (*Arthraxon, Bidens, Polygonum, Viola* etc.) well established.

the pattern is the same everywhere (Spenceley 1980). By the time people abandon gardens certain rapidly growing pioneer *den* grasses like *taengbiyp* (*Arthaxon*), *holigiyn* (*Paspalum*), *bol* (*Ischaemum*), *dikiytagot* (*Setaria*) and *pibiytaeztaez* (*Isachne*) and herbs, which besides *kobkob*, include *waembuw* (*Crassocephalum*), *ngat* (*Polygonum*), *muwmonhuwshiy* (*Viola*), and *momoniyl* (*Rubus*) will have established themselves on the site, flourishing at the expense of the few remaining crop plants, before finally displacing them largely. This is the early *mokombai* phase, which the Wola may refer to as *taengbiyp* after one of the grasses that characterises it.

Some crops, like sweet potato, *kot* Highland pitpit (*Setaria*), bananas, sugar cane and the dye plant *komnol* (*Plectranthus*), compete successfully against the invading weeds and maintain their position on the site for some time. Eventually, robust and vigorous *den* grasses (notably *Ischaemum*, but also *Paspalum, Arthraxon, Isachne* among others) take over, possibly with *obol* swamp grass (*Leersia hexandra*) in wet depressions, and supersede both any remaining crop plants and many of the early weed colonisers too. The occasional tree seedling or juvenile cane grass clump may also occur. The Wola commonly refer to this later *mokombai* phase as *bol*, after the coarse *Ischaemum* grass that predominates on it.

Some garden fallows never advance beyond one or other of these stages of regrowth, people pulling up the herbaceous regrowth or coarse grasses and re-cultivating the sites. If natural regeneration proceeds, either saplings or cane grass, or a mixture of both, begin to grow up in the grass (see Walker 1966 for a schematic representation of various possible sequences). Perennial

Plate 7.15 The later *bol* phase of recent regrowth with coarse grasses (*Ischaemum* predominating) well developed (the smoke in the background comes from fires lit in recultivating such an adjacent fallow area).

short grassland like that in the Eastern Highlands is uncommon, probably because of the higher year round rainfall (Henty 1982). In drier eastern regions, burning is more frequent and destructive, helping maintain a continuous cover of short grasses; it is not necessarily more mature than a cane grass cover nor an indication of earlier settlement and longer disturbance (Robbins 1960).

Cane Grassland

In the long-term one of two major floristic successions will establish themselves on abandoned cultivation sites: cane grass or secondary forest. The Wola call communities dominated by sword or cane grass *gaimb* after the *Miscanthus floridulus* cane that predominates in them. Cane grassland, like secondary forest, occurs predominantly as garden regrowth, although it sometimes colonises sites disturbed and deforested for other reasons. *Miscanthus* is an erect cane-like grass with robust culms. It grows in dense clumps. Its lanceolate leaves, 2 to 3 centimetres wide, have finely serrated margins and taper to sharp points. And its inflorescence is a large, open, branching panicle bearing many paired spikelets on a continuous rachis. It produces large amounts of fluffy wind-borne seed, well adapted to travel distances and colonise disturbed locales (Figure 7.1). It grows prolifically even when cut right back, and eradication from a site, for example when clearing it for a new garden, demands levering out its clumpy rootstock.

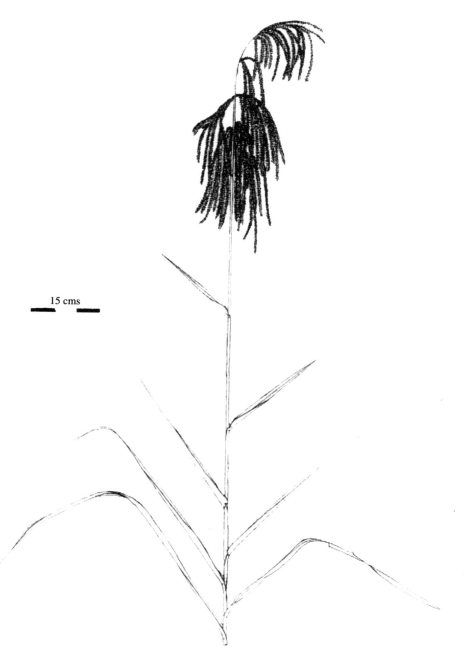

Figure 7.1 Cane grass (*Miscanthus*).

Dense clumps of this sharp-leaved cane grass, two to three metres or so high, cover large areas in valley basins. Quadrat survey data indicate thirty or so large clumps to every 100 square metres where it predominates, sometimes comprising 90% or more of the cover. The cane gives a dense *pletbok* 'thicket' cover where clumps grow closely crowded together, frequently impenetrable without a bush knife to cut a path. Thick brakes of *saezuwp* fern (*Dicranopteris*) also occur in some locations among the cane, and clumps of low fern (Thelypteridaceae) are common where the cane is less dense. The cane is more open of access where less mature and more widely spaced, particularly near homesteads where the rooting activities of foraging pigs give patches of exposed earth around clumps. While pig rooting may keep the spaces between tussocks in grassland near settlements reasonably bare, a fair layer of leaf and cane stem litter still builds up immediately around clumps, supporting local assertions that under cane grass fallow a good layer of dark topsoil suitable for recultivation soon accumulates. *Miscanthus* appears less readily to colonise sites close to the forest edge or other shaded places, these seem to reduce its competitive ability (Walker 1966).

Other tall grasses occasionally occur in amongst the *Miscanthus*, including clumps of *holor* Job's Tears (*Coix lachryma-jobi*), brakes of *mokombez* wild sugar (*Saccharum robustum*), particularly along stream banks, and patches of *shaenshuwril* grass (*Pennisetum macrostachyum*) on wooded margins. Where the cane grass is of more open aspect and the soil not turned over too frequently by rooting pigs, a considerable range of low *den* grasses (*Isachne, Paspalum, Ischaemum, Sacciolepsis, Setaria*) and various herbaceous *den* plants (Compositae, *Desmodium, Selaginella, Oenanthe, Plectranthus, Rubus, Viola*) may form a ground cover between cane clumps and trees.

Cane grassland is a monotonous looking vegetational succession, second only in extent to montane forest, which it has gradually replaced as the human population has expanded. Although seemingly less varied to walk through than the forest, forming almost mono-specific stands, one has to be attentive because the razor-sharp leaves caught edge-on can inflict painful cuts. When established, cane grassland supports a few *iysh* trees scattered here and there, notably lower-statured, soft-wooded species such as *haeraedaep* nettles (*Pipturus*), *op* ochnas (*Schuurmansia*), *bort* silkwoods (*Cryptocarya*), *oljomb* dillenias (*Saurauia*) and *waen* woolly cedars (*Trema*) together with *naep* she-oaks (*Casuarina*), *shuwat* figs (*Ficus*), *gwai* switchsorrels (*Dodonea*), *nemb* parchment barks (*Pittosporum*), *haegaes* umbrella trees (*Schefflera*) and others. Stands of cultivated *aenk* screw-pines (*Pandanus*) are also common, often remaining over from previous gardens, sometimes growing together in rows with *aegop* palm lilies (*Cordyline*), marking old fence lines. The graceful *henk* tree fern (Cyathaceae) is common too, sometimes growing in considerable numbers, giving rise to distinctive vegetational successions.

Any vegetation community is subject to gradual change, pitted against time to exist. In grassland areas, a *Miscanthus* succession generally replaces shorter grasses if a site is left undisturbed, the tall cane out-competing its shorter cousins, even vigorous and persistent *kunai* grass (*Imperata conferta*). The conditions that promote *Miscanthus* in competition over *Imperata* are an absence of extensive burning, which the wet Southern Highland's climate generally assures, and the presence of foraging pigs, which local herding practices ensure (Walker 1966). Not all grassland vegetational changes are due to human interference. Earth movements and so on can disturb communities, sometimes permanently if they result, for example, in locations with altered drainage status. And it is thought that some of the cane grassland that has replaced the region's original climax vegetation of lower montane rainforest established itself during climatic fluctuations associated with Pleistocene glaciation (glacial evidence, such as moraines, existing on high volcanic peaks — Walker and Flenley 1979).

Human activities have undoubtedly extended the area under grassland. Grasslands occur where neither climate nor soil would preclude the growth of forest, for example forest-grassland boundaries frequently occur independently of changes in soil types, having no apparent relationship with them. This is taken as evidence that grassland is largely anthropogenic, resulting from human destruction of the forest (Robbins 1960). When the rotation period in cultivation is relatively short, woody plants have less chance to regenerate in any numbers and to any size, and grassland becomes established (Henty 1982). Repeated cultivation, occasional grass fires, and other disturbances contribute to the maintenance of cane grass fallows following forest clearance, together with environmental factors that assist seedling death, such as soil-plant nutrient imbalances and sub-optimal drainage conditions. The Wola themselves acknowledge that they are agents of the forest's destruction; when digging a ditch once to enclose a garden in a cane grassland area, we uncovered some intriguingly gnarled pieces of beech root and tree stump (some of which I took home, to everyone's amusement, as natural sculptures) and those with me readily explained that they were evidence the forest at one time extended across the area.

The change in vegetational communities with human interference, in which cane grass features prominently, does not necessarily spell irreversible degradation, with permanent reduction in species diversity and biomass, at least in the long run. While there is piecemeal destruction of forest to establish new gardens, this may be seen as the start of a long term chain involving garden and houseyard sites, which may pass through a series of successions upon abandonment to become cane grass and/or secondary forest regrowth, even mature forest if left long enough, although more likely they are disturbed again at some time relatively early in their progression towards maturity. Whatever, subsistence gardening does not appear to be overly destructive of the forest environment, at least in the short-run; comparison of aerial photographs taken over part of the Wola region by the U.S.A.F. in 1948 with more recent ones shows that no marked increase has occurred in forest destruction over the last forty years or so (Radcliffe 1986:29), during which time more efficient steel tools have become available, making clearance somewhat easier, and administrative control extended over the area, with the introduction of

Plate 7.16 An expanse of *gaimb* cane (*Miscanthus*) grassland adjacent to the Uwt stream in the Nembi valley.

health services resulting in a spurt in population growth. The dramatic increase in rates of labour migration out of the region over the last decade also suggests that pressure on forest resources is not increasing markedly, and may even decline.

Secondary Forest

Secondary forest is the alternative long-term floristic succession to cane grass following the clearance of any area for cultivation. When plots cleared for gardens in the rainforest are abandoned they rapidly regenerate into patches of secondary forest, which the Wola call *em obael* (lit: garden mature-regrowth). Tree regrowth occurs too in pockets throughout the cane grassland zone. It has a markedly different floristic composition to montane forest. It is altogether of a softer aspect and less formidable.

The *obael* secondary forest has a considerably lower canopy than the *iyshabuw* montane forest at ten up to twenty metres and comprises fast growing soft-wooded *iysh* trees primarily, such as various spurges like *timbol* and *en* (Euphorbiaceae), *kongol* pipers (Piperaceae), *haeraedaep* nettles (*Pipturus*), several figs like *ponjip*, *pil* and *pakpak* (*Ficus*), *maenget* umbrella trees (*Schefflera*), *oljomb* dillenias (*Saurauia*), *bort* silkwoods (*Cryptocarya*), *waen* woolly cedars (*Trema*) and *gwai* switchsorrels (*Dodonaea*), among others. The *henk* tree fern (Cyatheaceae) is also common, often occurring as an understory tree. Clumps of *gaimb* and *holor* cane grasses (*Miscanthus*,

Plate 7.17 The *obael* phase of secondary regrowth, soft-wooded trees, pandans and cane grass.

Coix) are common too, and when they exceed a certain indeterminate number, the regrowth becomes more akin to cane grassland; there is no sharp distinction between these two vegetational communities, nor any others that pass from one to another, they gradually merge, as the Wola acknowledge, with no abrupt change.

The ground cover is on the whole considerably less dense under secondary woodland than primary forest, consisting of various coarse and creeping *den* grasses like *burumbol* (*Paspalum*), *dikiytagot* (*Setaria*) and *bol* (*Ischaemum*), and a range of *den* herbs and shrubs including *kobkob* (Compositae), *torwatorwa* (*Desmodium*), *hungmaenk* (*Impatiens*), *taziy* (*Oenanthe*), *kaerobkaerob* (*Plectranthus*), *ngat* (*Polygonum*) and *kuwmkaes* (*Selaginella*), sometimes growing to waist height. Ferns are also common, notably in sprays across the forest floor (Thelypteridaceae), and sometimes in tangles of *pukuwmb* (*Pteridium*) and *saezuwp* (*Dicranopteris*); tall leafy herbs, notably *munk* gingers (Zingiberaceae) are frequently seen too. The wildlife inhabiting secondary forest depends on its location. When surrounded by montane forest, where wildlife is abundant, it is likewise plentiful. But when situated island-like in a sea of cane grassland, where wildlife is limited, it is sparse too; although the fruits of some trees growing in these wooded islands are popular with birds and attract them here in considerable numbers when ripe.

These areas rarely develop into mature wooded stands, those with rights to the land usually clearing them again for gardens before they reach this stage, or otherwise hinder their development by repeatedly interfering with them in collecting firewood and raw materials. Sometimes, where near homesteads, they are disturbed by pigs rooting for food, with patches of churned sod and vegetation across them. Nonetheless if left undisturbed, the Wola maintain that *iyshabuw* forest would eventually establish itself in these areas. Some men told me that if they abandoned their valleys, montane rainforest would eventually replace both *obael* secondary forest and *gaimb* cane grassland, to cover them just as it used to before their ancestors cleared it. They spoke of the primary forest 'hitting and eating' (*luw nokor*, lit. hit will-eat) these long-term secondary successions and 'making them rotten' (*kor ma sokor*, lit. rotten cause become). They pointed out locations where secondary woodland and cane grass abut the forest and explained how with humanbeings absent the montane forest would slowly advance down the valleysides; they cited the Ak valley as a place where this is occurring, for with the abandonment of gardens there (following the establishment of administrative centres elsewhere, prompting people to move to be closer to them), the forest is engulfing the *obael* woodland and *gaimb* grassland down to the edge of the Ak river. Eventually the forest comes back again.

CHAPTER 8

CONTENDING WITH FOREST AND FALLOW: DEMONS TO REGROWTH

It is difficult to gauge Wola attitudes to their region's various plant communities, other than indirectly. They are not familiar with discussing ecological issues openly, expressing their thoughts and feelings in other less accessible ways. Nonetheless these are evident to some extent in their practices, in how they attempt, or not, to manage natural vegetation, and they are also discernible in some of the supernatural beliefs to which they subscribe, in which are detectable various contradictory ideas not readily amenable to verbal expression. This chapter substantiates these assertions, arguing that it is possible to discern something about Wola attitudes to primary forest from the fear they express in its believed demon denizens, and more directly, to access their perspective on secondary successions through their approach to the management and manipulation of regrowth.

I. FOREST AND DEMONS

The Beech Montane Forest

The *pel* southern beech (*Nothofagus* spp.) epitomises the ecology of the forest (Figure 8.1). A prominent canopy tree in the highland forests of the Wola region, it reaches maximum heights of 30 metres or so, with branches giving a nearly level or domed canopy, which is rather open for tropical forest with up to 8% daylight penetration (Ash 1982). The lower trunk sometimes has buttresses. It is surprisingly shallow rooting for such a large tree, with roots going only 40 centimetres or so deep and spreading a few metres. It can be unstable, and in high winds my Wola friends always become very anxious in the forest, fearing tree falls. The tree grows stunted on shallow or poorly drained soils, and in exposed cloudy ridge top locations. Its upper altitudinal limit extends somewhat beyond that of cultivation. It is long-lived; tree growth rates, radiocarbon dates, and parenchyma ring counts indicate that it may live up to 500 years or so (Walker 1966:514). Contempoary beech forests occur as either nearly pure *Nothofagus* stands or as mixed genera stands; in the Wola region it is common for discrete patches of beech forest to occur surrounded by multi-genera forest (Robbins & Pullen 1965; Kalkman & Vink 1970; Walker 1966). According to Ash (1982), beeches overall comprise between 10% and 20% of the canopy trees in lower montane forest, which complies with quadrat survey findings in Wolaland. The local variations in *Nothofagus* distribution relate to topography and consequent complex interactions associated with differences in microclimate (cloudiness, humidity, windiness, etc.) and ground level processes (water drainage, solute movement, particle transport, etc.).

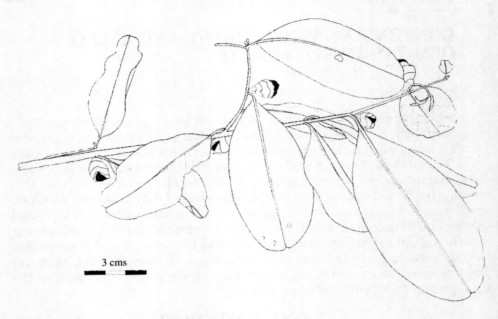

Figure 8.1 Southern beech (*Nothofagus*).

The montane Wola environment suits *Nothofagus* beech. The genus favours cloudy regions where precipitation is continuously high, and soils well drained. It particularly likes ridge top locations (Hynes 1974). It requires a reliable supply of water to compete successfully, but cannot tolerate overly wet, poorly aerated soils. Ectotrophic mycorrhizae, which play a part in phosphorus mobilization, have been reported for *Nothofagus* (Ash 1982), which is significant for a region where soil P fixation is high. Beech populations, gregarious and non-allelopathic, characteristically form extensive single genera, even one species stands, which is fairly unusual for tropical forests. A mixed stand usually comprises no more than two or three beech species. This relates to the tree's life cycle. It requires trees close together to ensure effective wind pollination of its flowers which blossom sporadically throughout the year, producing seeds which the wind in turn disperses only limited distances. Few seeds germinate into established seedlings, those doing so favouring mossy substrates (such as occur on fallen logs). Alternatively, trees may regenerate in close proximity to one another by producing suckers, sometimes sending these out horizontally on the soil surface for several metres, and producing along them several shoot and root systems; other suckers may originate from damaged saplings and even from underneath fallen logs.

Both seedlings and suckers require an open tree canopy if they are to complete successfully and grow into mature trees; under shady subcanopy conditions with less than 10% daylight penetration, such as prevail even under the relatively open *Nothofagus* canopy, they grow slowly and few survive to become saplings (Ash 1982). The proportion of *Nothofagus* seedlings and saplings to those of other trees that may grow into the canopy is, at 6.9%, almost one half of the ratio of mature beech trees to other canopy trees, suggesting that more beeches may nonetheless survive to become forest giants. The growth rate of young plants increases markedly with the creation of gaps in the canopy through senescence or tree fall. These occur regularly with dieback, sometimes of several trees and even entire patches of forest; beech forest often features local areas of regeneration (Kalkman & Vink 1970). Dieback is associated with a combination of unfavourable weather and pathogenic attack; pathogens include *Phytophthora* soil fungi and Carambycidae wood boring beetles (Ash 1982). High wind or minor earthquake damage to roots may increase host tree susceptibility, encouraging infection.

The beech is not a colonising genus, regeneration scarcely extending beyond the canopy of *Nothofagus* stands (Walker 1966; Kalkman & Vink 1970), and this is interpreted as evidence that today's stands are relics of once more widespread beech forests upon which other genera have encroached. Palynological evidence supports this view, indicating that *Nothofagus*, an ancient genus, migrated rapidly to higher elevations in Papua New Guinea as the climate warmed during the Pleistocene, establishing extensive beech forests which other genera then gradually invaded to give today's mixed montane forest (Walker 1970). The role of human activity in this is difficult to gauge currently, but pollen analysis suggests that it has varied over time.

Rainforest Attitudes

There is a popular image of non-industrial peoples like the Wola as more in tune with forest environments and more conservation conscious than us (Ellen 1986; 1993). A brief time in the beech forest with them soon undermines any belief in the universality of this attitude. It is disconcerting to a Westerner imbued with 'Green' environmental concerns about rain forest destruction and so on, to accompany them and see how they cut down saplings and herbs apparently indiscriminantly with swipes of their bush knives, hack at trees with axes as they pass by, and so on. When out hunting, men sometimes fell gigantic beech and other trees in the chase (if they are too large to climb easily or have an inaccessible hole thought to be a possible marsupial refuge) and tear large gaps in the canopy and understory, the trees breaking down nearby vegetation as they crash to the ground. Other times they fell several trees to allow nut-bearing pandans access to sunlight, again causing considerable damage. And they sometimes clear sizeable clearings to establish swiddens in which to cultivate crops, and their attitude towards resulting secondary regrowth is by and large disinterested, reafforestation is not an issue that apparently concerns them.

It is as if the Wola are again acting recklessly regarding their natural resources, here destroying rain forest. We may anticipate that ideas about and attitudes towards the forest will vary widely in New Guinea given its cultural diversity, and differing pressures on land resources between regions (Morauta *et al.* 1982). The forest is largely intact in some places, where the human population is too small to have disturbed it greatly (for example, in the the Muller Ranges — Sillitoe 1994). But in others, the population is large and deforestation over many generations extensive, giving rise to considerable areas of grassland (notably in the Central Highlands — Brown & Podolefsky 1976; Robbins 1960; Henty 1982). Yet even here considerable areas of montane forest remain. What seems remarkable is not that highlanders, like the Wola, have deforested the areas they have, but that people who have engaged in plant cultivation for nearly 10,000 years (Golson & Gardner 1990), coupled with an evident lack of any manifest conservation ethic, has not resulted in far more extensive destruction of forest. The Wola region, for example, has large areas of forest intact regardless of their apparent disregard for conservation; some 52% of their region is under primary forest and the percentage rises to over 80% on less heavily populated territories (see Chapter 7). Their belief in contrary forest demon spirits captures the inconsistency.

The attitude of the Wola to their resource rich rain forest is somewhat ambivalent. They enjoy the forest but are sometimes wary of it, they value it but are destroying parts of it piecemeal. They cannot conceive someday having no forest to visit. The forest, they say, can be a dangerous place for them, particularly when deep in it, far from settled locales with familiar homesteads and gardens. Although it offers them a range of valued resources, they approach it with respect and caution. It supplies a variety of widely used and esteemed raw materials, and is home to many animals which hunted provide a welcome meat supplement to an otherwise largely vegetable diet, but it sometimes exacts a heavy price for this produce. It is possible to lose your way in unfamiliar regions, which can be scarry, and even dangerous, in remote forest. And accidents are more likely here, shallow rooted trees and branches blow down in high winds and can hit passers-by, the unlucky may trip over on the frequently uneven and slippery terrain and injure themselves, and so on. And fatalities are known; one such poignant episode concerned my friend Mayka Kot, who was buried when a large tree he was felling towards Mount Kerewa caught in the canopy and swung backwards on lianas, ploughing him into the ground, such that relatives had to dig out his mangled corpse.

Forest Demons

It is difficult to encourage the Wola to talk emotively about the forest, to comment beyond saying that it is a large place and sometimes a dangerous one, that they value it as somewhere to hunt and a source of raw materials, and so on. But they catch its dangers and express something about their relationship with it, and explain some of the misfortunes that befall people there, in their fears of forest demons. These are a declaration about imponderable feelings, a comment on irreconcilable sentiments. The Wola signify something of their deeper attitudes and ambivalence to the forest in two arbori-

therianthropic spirit creatures called *saem* and *iybtit*, which inhabit different forested regions, one lurking at higher and the other at lower altitudes.

The **saem** forest demon inhabits the forest of higher altitude regions, to the north and north-west, in the directions of Mount Kerewa and the Doma Peaks according to those who live in the Was valley. They think that it is a half human and half vegetable creature, with bark and moss covered torso and

Plate 8.1 The montane beech forest home of forest demons.

Plate 8.2 A man camouflaged like a forest demon, rushing to and fro to keep spectators back at a dance.

limbs from which sprout leafy foliage.[1] It attacks strangers who wander onto its territory, usually at night. It cannibalises its victims, whose corpses are sometimes found after an attack, although more often they disappear. Others are found deranged and unwell, sometimes bruised and scratched but other-

---

[1] Strauss & Tischner (1962) relates that in the Hagen region they talk about a similar spirit man called Pöngönts, around whom there is a cult, whose back is like jagged, moss covered rock.

wise externally unharmed. They may be damaged internally, and demand ritual attention.

People say that the higher altitude *dalep* pandans (*Pandanus brosimos*), with their hard-shelled nuts, are planted by *saem* spirits, from single sprouting nuts called *dolimba*, though human-beings claim rights to trees when they find them and harvest their clusters of nuts (ownership of a pandan goes to the man who first finds a young screw-pine and clears away the vegetation around it so that it can grow without competition). And the alpine grasslands on the tops of high mountains like Hat Waenmaep (Mount Kerewa) are the *howma* ceremonial grounds of the *saem* demons, where they meet, after human-beings who stage large social events like wealth exchanges and dances on the grassy *howma* clearings of their settlements (Sillitoe 1979; Lederman 1986). Apropos this parallel, one man will sometimes dress up like a *saem* ogre at a dance, tying bushy foliage, fern fronds and clumps of moss on his body, together perhaps with a gourd mask, to frighten back spectators and give the dancers space to perform. Otherwise watchers crowd up to the performers and cramp them. The *saem* impersonator rushes to and fro, sometimes brandishing a rattle or shaking together bow and arrows in a threatening gesture, to keep the spectators back. Spectators appear to treat it more as harmless fun than as terrifying gesture (Plate 8.2).

When they are in the vicinity of higher altitude *saem* homelands, the Wola never shout or even speak loudly for fear of attracting the attention of the demons there. Some men told me that they converse in whispers and never refer to the *saem* by name, nor use certain other words like *hobaen* (a *Rapanea* tree), *wolaya* (a *Dimorphanthera* vine), or *nor* (eat — instead they say *konbiy*). While they could not explain the import of these taboos, everyone agreed that transgressing them would result in low cloud coming down and darkness descending to obscure the land. Those present would become very wet and cold, and would lose their way because the *saem* would take away their *konay* 'senses', such that they would be likely to walk over a bluff, into a river, down a pothole, or come to some other such grief.

The **iybtit**[2] forest demon inhabits the forest at lower altitudes, to the south and south-west, in the Lake Kutubu and Mubi river regions. It has a human-like form too, with a mixture of marsupial fur and birds' feathers covering its body. People disagree over the details of the appearance of forest demons, their grotesqueness depending on how vivid their imaginations are, but I have been told by some people that the *iybtit* has cassowary-like claws growing from its finger tips, and according to one man I spoke to it has four eyes around its head. They think that it lives in holes, in the ground and trees. One of its favourite homes is a hole in a *hael* fig tree, and another is the hollow that results after the strangling *wol* fig has killed its host (the rotting of the host tree's trunk leaving a cavity within the fused root network of the mature fig).

The *iybtit* ogre makes a "wuwuwuwu" noise in the forest, except when it attacks its victims when a bang is heard. Once, while in the forest in the

---

[2] This forest demon has several other names, among them *iybdaibol*, *iybmaelem* and *iybterayol*.

Sabim region of the Ak valley, we heard a sharp report like a gun going off, caused I think in retrospect by the snapping of a tree trunk as it hit the ground when blown over by a high wind. We were all somewhat overwrought by the strong wind, and my colleagues interpreted the noise as an *iybtit* attack. They were very concerned that we should have all heard it. The reason, it transpired, was that if you hear the bang of an *iybtit* attack then you are unscathed and it has missed you. But those who do not hear it are in mortal danger. They may appear alright for some two or three weeks after the attack, but then fall ill and die. Persons with this condition diagnosed require the performance of a curative ritual to save them. The bang heard is the twang of the bow carried by the *iybtit* monster. It fires a broad-blade *solomba* arrow (Sillitoe 1988), which has a hollow cane shaft. It fires the arrow into its victims and sucks their blood up through the hollow drinking-straw-like arrow. The name *iybtit* may derive from this, it meaning literally 'come-hole' or 'fluid-hole'. It shoots not only humans but also animals, and for target practice pandan nuts; you can tell when a demon has shot a ripening nut because it turns rotten, which is unusual.

Demon Attacks

Some people think that there are male and female *saem* and *iybtit* demons, and that they cohabit and have children. But they are not sure because no one has ever seen one and survived. It is the males that are dangerous and attack human-beings. Some men told me that they heard say that the Foi living at Lake Kutubu killed a forest demon in the past and ate it, being cannibals. The nearest anyone I know has come to seeing one and surviving is a man who lives at the settlement of Salaenda, only half a day's walk from Lake Kutubu, who was assaulted by an *iybtit* one dark night. It is unclear exactly what happened to him, he was confused and terrified in the attack, but people say that he was covered in cuts the next morning. He attributed the attack to felling a large tree with a hole in it when clearing a garden, in which an *iybtit* demon family resided.

Sometimes a *saem* or *iybtit* attack leads to madness. It is usually women who are afflicted in this way, and they may run amok attacking people and property. One name for insane behaviour is *iybtit aum^{e}boliy* (lit. forest-demon insane); and in banter someone may refer to another whose actions or thoughts they think are silly as an *iybtit*, 'twit' or 'idiot'. It is also believed that instead of inducing beserk behaviour forest demons may take away someone's *konay* or mental faculties, reducing them to the helpless baby-like state of *konay na wiy* (lit. thoughts not have), as happens at higher altitudes near the *saem* dance grounds. Individuals so afflicted will wander around the forest completely lost, roaming aimlessly about, they are *haeret hezay* (lit. path mad). When they fail to return home, their desperate relatives may go in search of them before they come to any harm. If they do not soon come to their senses when found their worried relatives may resort to a ritual cure.

In one case of *saem* induced confusion which I witnessed two women, Puwgael E'row's wife Saliyn and her co-wife's eldest daughter Kolkabiy, were out collecting edible fungi in the forest and failed to return home. They slept out in the open in the forest huddled together bemused at the foot of a large tree. Saliyn was a middle aged woman of good sense, she was a hard-working and reliable wife by reputation, no one thought her frivolous or silly. It struck me that the collection of fungi might be significant, and when I enquired people told me that on at least two other occasions women from the settlement of Haelaelinja had suffered similar paraphrenia while out collecting mushrooms in the forest.

The Wola identify a wide range of different fungi, which they refer to collectively as *sez* (see Chapter 9). Some are edible and they consume them when available. They collect them infrequently because most of the time there is only the occasional fungus to be found in the forest. But sometimes large flushes occur. According to local lore this happens on those infrequent occasions when heavy rain follows an extended dry period, such conditions stimulating fungi to produce fruit bodies. When a bumper sporangia crop occurs, women may go into the forest to collect the fungi, returning with string bags bulging with caps. The Wola recognise some forty-six named varieties of *sez* fungi as edible. By far the most common edible variety they collect being the *muwnaen* bracket fungus (*Grifola frondosa*), of which they identify several named sub-varieties.[3]

Some of the fungi that grow in the forest will have hallucinogenic properties, possibly even some of those cited as edible (such as the *goizmayja* mushroom (*Inocybe* sp.)).[4] When they are out collecting *sez* fungi in the forest all day, women may cook and eat some for a snack. It is possible that occasionally someone might consume an hallucinogenic fungus, either inadvertantly collecting an inedible type or an edible one that may have hallucinatory side effects for some individuals (Reay 1961). The Wola do not think that there is any connection between consuming fungi and mad behaviour. But if there is, this may partly account for why women more often than men are said to lose their *konay* senses in *saem* or *iybtit* demon attacks in the forest, going to gather fungi more often than men.

Ritual Interventions

Not all encounters with *saem* and *iybtit* forest spirits are fatal. Those which are not frequently require ritual intervention to avert death. Each demon has its own ritual. These are similar over some points, notably both require the

---

[3] They distinguish between these by their cap colour: *aegael* is yellowish, *haezort* is whitish, *mugumb* is reddish, and *sebhibiy* is brownish.
[4] I have no evidence of the Wola deliberately eating such fungi to achieve a 'high' feeling

Plate 8.3 The decoration worn during the *hwiybtowgow* curing rite.

slaughter of pigs to appease the attacking spirit, but they differ in detail. They also differ over short distances between different communities, to judge from people's comments, and probably over time too at the same place, the Wola adhering to nothing akin to standard canonical practice. But they are consistent in overall aim, meaning and expression (see Chapter 4).

Saem Rite: The ritual performed in the event of a *saem* demon attack is called *hwiybtowgow*.[5] It requires the preparation of a small cleared area, where relatives of the ailing victim build a small lean-to shelter called a *saem aenda* (lit. *saem* house), with an earth-oven hole excavated inside it. They also erect here a short length of sapling fence called a *komay*, faced with leaves from the *shiyp* (*Chisocheton ceramicum*) and *bat* (*Bubbia* spp.) trees poked behind spars with their pale underneaths outwards to make a light coloured wall.[6] It is common practice to hang some exchangeable wealth on the fence too, like pearlshells and cowrie shell necklaces. One side of the fence, where the shelter is, I was told is that of human-beings and the other side is that of the spirit force. The fence contains the demon and protects the

---

[5] A similar ritual may also be performed for sickness diagnosed as the result of ancestor spirit attack.
[6] The Wola employ screens of similar construction in other ritual contexts (see Chapter 4).

participants from its presence. Those preparing the site also place a hollowed out *hongok* tree fern stump (*Cyathea* spp.) adjacent to the leafy barrier. This wood is particularly suitable in such ritual contexts, one man pointed out, because it has robust qualities which they wish to see in the sick individual, being very hard and rotting slowly.

Two men, usually relatives of the ailing *saem*-assaulted individual, perform the ritual. They decorate themselves in a singular manner. They divide their bodies in two with a vertical line running between their eyes, down their noses and through their navels, and one half, including one arm and leg, they paint red and the other black. And they frequently sport black cassowary feather pompons or foliage wigs on their heads. When asked what they intended this decoration to achieve, men said that it made those performing the ritual look fearsome and this helped to scare off the attacking *saem* demon.

Relatives take the sick person to the prepared area, where they club and butcher a pig. Others heat up stones to cook the pork in the *saem aenda* shelter oven pit. The heart, lungs, liver and kidneys they remove separately. They prepare a small hot-stone, leaf-lined oven in the hollowed out tree fern stump, in which to cook the heart and lungs. The liver they wrap up in a separate parcel with hot stones to cook. And from the kidneys they carefully remove the layer of surrounding fat, for use at the end of the ritual. The two officiating at the ritual take a piece of *mondba* aroid leaf (*Alocasia* sp.) and wrap it in a *shiyp* meliad tree leaf. They hold this parcel in their mouths and rush to and fro on the spirit side of the fence, aggressively rattling bow and

Plate 8.4 The *komay* leaf wall erected for the *hwiybtowgow* curing rite.

arrows together and making a loud "shshshsh" noise. They repeat this over the sick person. The object of their performance is to frighten off the attacking *saem* demon. Their actions are likened to shooing away a pig; the Wola equivalent to "shoo, shoo, shoo-off" is "shshshsh" (if a pig breaches a garden fence for instance, it is chased out with cries of "shshshsh"). The *mondba* leaf ensures a vigorous shooing performance. The aroid leaf is caustic, even through another leaf, and the burning sensation that it causes provokes vehement "shshshsh-ing" from the performers, as they clear pungent saliva from their mouths through clenched teeth.

The liver and kidneys, which soon cook, the sick individual and two demon-scarers share. The heart and lungs, which are also ready before the pork in the *saem aenda* earth oven, are divided between all those present. When cooked, they also share out the pork among themselves, and prepare to leave immediately, packing leaf-wrapped parcels of meat into their net bags. It is important that they are all ready to depart in a hurry. One of them takes the kidney fat and kindles a fierce fire on which to burn it. This is for the attacking *saem* forest spirit, which also partakes of the blood spilled with the pig's slaughter and the smell of its singed bristles. The man blows on the fire so that the fat flares up with a greasy smell, and mutters "*saem njen nabay*" (lit. *saem* yours eat), inviting the demon to take it. Then he shouts out "weeeeeeee" and all those present rush off, leaving the *saem* behind to have its fill and return to the forest.

The strategy of the ritual is twofold. On the one hand, the relatives of the sick individual are trying to scare the attacking *saem* spirit off. And on the

Plate 8.5 Two men stand with aroid leaf parcels in their mouths as in the *hwiybtowgow* rite.

other, they are attempting to persuade it to relent and return to the forest by offering it the smell and flavour of slaughtered pig. It is even plausible that on some unconscious level they are trying to convince the forest demon that it is better to act reasonably, as they pressure each other through ceremonial exchange wealth manipulations. The valuables hanging on the *komay* fence suggest this interpretation, symbolising the wish of those performing the ritual to induce tolerable behaviour, as they regulate social behaviour among themselves by continually passing wealth around.

Iybtit Rite: The ritual performed to ward off an *iybtit* attack is staged away from any settlement at some high point in the adjacent forest. It is necessary that the sick person and some relatives go to a suitably elevated location overlooking the forest so that the *iybtit* might hear them calling to it, when they shout out inviting it to come and "eat" both the blood they will spill slaughtering a pig and the aroma of its singeing bristles. They summon the demon to the rite by yelling out the names of places where it could be lurking, enticing it to come and share the pig, as in the following call:

Oba	*haelaib.*	*Ebera*	*haelaib.*	*Momiya*	*haelaib.*	*Boriya*	*haelaib.*
Place name	anyone there.	Place name	anyone there.	Place name	anyone there.	Place name	anyone there.

Showmay	*iyba*	*muwk*	*iyb.*	*Weeeeeeeeeeee.....*
Pig	blood	take	come.	Cry to attract spirit's attention.

The victim's kin prepare a watertight bowl out of a hollowed out *hongok* tree fern stump, lining it with large leaves, like those of the banana, or with sheets of flexible bark. When they club the pig, they hold its bleeding snout over the bowl, filling it with blood. Some of them then butcher the animal, heat stones for an earth oven, and put the pork into a leaf-lined oven-pit to cook. Meanwhile someone who knows the required incantation takes the creature's kidneys and spears them onto a skewer-like sliver of wood. He prepares a torch of split *baerel* wood (*Cupaniopsis* sp.) and holding the kidneys over the blood-filled bowl he roasts them so that the grease melting from their fatty integument drips into it. He recites an incantation such as the following as he renders the fat surrounding the kidneys:[7]

Onyuw	*Koba*	*hael*	*iysha.*	*Itiy**	*woliya*.*	*Karoway*	*lor**	*sizay.*
Down there	place name	fig tree	trunk.	Tree	strangling fig tree.	Cut	say	cooling sizzle sound.

Koba	*lay*	*hael*	*ayow.*	*Karoway*	*lor**	*sizay.*	*Karoway*	*lor**	*sizay.*
Place name	emphasis marker	fig tree	emphasis marker.	Cut	say	cooling sizzle	Cut	say	cooling sizzle sound.

---

[7] Hoboga Makwes told me this version of the incantation. The asterisks mark Huli words, which give some added efficacy and mysticism to the incantation because they are foreign.

Koba	*hael*	*iysha.*	*Iysh*	*woliya**	*iysha.*	*Karoway*	*lor**	*sizay.*
Place name	fig tree	trunk.	Tree	strangling fig tree	trunk.	Cut	say	cooling sizzle sound.

Those present share out the roasted kidneys, no particular significance I was told attaches to their consumption. The use of *baerel* wood to roast them has no apparent symbolic import either, people use this wood because it burns fiercely with strong heat, barbecuing meat efficiently.

The sick individual assaulted by the *iybtit* demon drinks the blood and kidney fat mixture out of the bowl, using the hollow stalk of a *gaimb* cane grass infloresence (*Miscanthus floridulus*), called a *suwkunuwmb*, as a straw through which to suck it up. This counters the demon's blood-sucking attack through its hollow arrow, which has reduced the victim to his or her current sickly state. The ritual is akin to a blood transfusion, restoring lost blood to the unfortunate anaemic patient. The Wola think that the addition of kidney fat is necessary to ensure that the imbibed blood will "sit well" inside the person. Blood on its own they say would not remain fluid but congeal and cake inside sick people, making their condition even worse by blocking up their internal organs.

The incantation reiterates the rite's aim of reversing the victim's blood loss. When someone loses blood, the Wola say that the skin turns pale, and the body becomes hot and feverish. The incantation describes how the attacked individual's anaemic body will cool off when he or she sucks up the blood to replace that lost in the *iybtit* assault, like the sizzle of hot stones in an earth oven when doused with water. The reciter of the incantation also refers in the same phrase to the rite cutting through the demon's hollow arrow straw, so thwarting its sucking of the victim's blood. He admonishes the *iybtit* demon to return to its tree-hole home too, in the trunk of a *hael* fig or tangled roots of a *wol* strangler, and not to attack people. It is to go and leave the unfortunate victim alone.

After the patient has sucked up the fatty blood mix, the proceedings become more relaxed, as those present wait for the pork to cook. They have driven off the *iybtit* spirit and thwarted its leaching assault. They share out and eat the meat, which women and children may also consume but not, I was told, couples who have recently engaged in sexual intercourse (if they did, they would nullify the ritual by contaminating it through association with procreative activity, and the sick person would not recover). The kin of the victim anticipate an improvement in his or her condition within a few days.

The *iybtit* curative ritual contrasts with that performed to overcome a *saem* forest demon attack. It places less emphasis on impelling the spirit back to the forest, either by fear or persuasion, and more on treating the surmised cause of the victim's malaise, namely blood loss. While it expresses a wish that the attacking *iybtit* demon should go away, back to its arboreal abode, and the pig offering again suggests an exchange-informed inducement to comply and act temperately, an equally prominent theme is the attempt to treat the patient's forest-demon-diagnosed symptoms.

Forest Demon Sensibilities

There is little call today, in an anthropological context, to rehearse the well-worn argument about the intellectual soundness of beliefs and practices like those of Wola forest demonology. The forest can be a hazardous place. Accidents, sometimes fatal, occur there, and on occasion people fall sick after travelling there. But Wola beliefs, unlike others elsewhere (e.g. African witchcraft beliefs), are apparently not overly taken up with questions like why that individual, there and then? They are concerned more with how to deal with spirit attacks wherever and whenever they occur. The idea that sickness and misfortune are the result of the capricious acts of spirit beings is common in Wola cosmological beliefs. Ancestor spirits attack relatives and descendants in a similar arbitrary way, and the Wola practiced a range of ritual interventions to ward off their sickness-inducing, sometimes fatal, attacks. Social identity is not at issue with demons either, although it is with ancestor spirits, which only attack relatives and descendants. The Wola entertain no idea of taking redressive or preventative action against others over these supernatural attacks. They take the pragmatic view that all fall ill sometime or have accidents, apparently arbitrarily, and curative steps need to be taken when they occur. They are not associated with the malicious or wrongful activities of the living. When human-beings are involved, it concerns a different class of supernatural attack, namely sorcery, which demands ritual interventions of its own in revenge (Sillitoe 1987).

On another tack, Wola beliefs in these demons could be interpreted in a more psychological, less rationally-culturally-relative vein, as a manifestation of the universal human fear of the unknown, of strange places. We have all, at some time, experienced feelings of anxiety in foreign places, especially wild and remote ones; I certainly have in the rain forested mountains of Papua New Guinea and can sympathise with Wola demon expressed fears. This is a shared apprehension I maintain, not merely a post-modern transference of my experience to their words as we have walked through and talked about the forest and its spirits. But to interpret the capriciousness of forest demons functionally as a cultural expression of some ill-defined human fear of the unknown advances us little in understanding Wola demon spirit beliefs. It is what these beliefs may suggest about their attitudes to their natural environment, and their forest in particular, that are of more interest here.

The deep forest is not somewhere that people should go lightly. It houses dangerous spirits which may strike them down if they are reckless or thoughtless. This could be interpreted as promoting a regard for the forest, perhaps even a degree of environmental awareness, although it would certainly be going too far to construe it as a conservational outlook. While it would be utterly distorting to impose ethnocentrically on the Wola contemporary Western worries about the environment, the need we perceive for its protection, it is nonetheless possible to suggest that their beliefs intimate a parallel disquiet, albeit expressed in an entirely foreign way. While they are not an expression of any apprehended need for forest conservation, they nonetheless intimate that you should beware indiscriminantly damaging the forest, you may offend a demon. (If outside logging interests threatened the forests of their region however, and the Wola became aware of the implications of deforestation, these beliefs could conceivably reinforce any conservation urge they kindled. They might even expect the demons to protect their forest.)

But there is another side. The forest demon complex of the Wola is not exactly, by all events, some inchoate expression of a need for forest conservation. For a start, the forest is too vast for the Wola to conceive of its destruction, taking days to walk through in some directions before reaching human settlements. The implications are not that the Wola feel uneasy about cutting down the forest, when they are travelling, hunting or clearing new tracts for gardens. Indeed it is plausible to interpret their demon beliefs as paradoxically endorsing such action. While the 'nature versus culture' antinomy is inappropriate, the Wola having no analagous categories so far as I am aware (and the unconscious imputation of these ideas is untenable with post-modern criticism), it is plausible to suggest that by destroying the forest they are exerting some control over frontier areas, driving demons from their homeland by depriving them of anywhere to live. Where there is no rain forest, there are no demons. But the forest is to too vast to think of entirely depriving forest spirits of any forest in which to lurk. And the idea of destroying all their forest would anyway be horrendous to the Wola. The ambivalence they feel is captured in their demon fears. They enjoy visiting the forest to hunt, collect raw materials, and so on, but they fear its remoteness and danger. They value the forest highly but on occasion they destroy it, notably in establishing new swiddens. They want to embrace it, without it touching them.

The believed presence of demons reveals a tension. The Wola destroy the forest, steady population expansion obliging them to clear new land, but they are not indifferent in doing so. It is as if they are somehow aware that it is another race, the existence of undistrubed forest against their encroachments over time. Clearance has been gradual and restrained, not wholesale. When they destroy forest to establish new gardens, they usually do so on the fringes of the forest, eating slowly into it from already settled areas. It is convenient of course, adjacent to existing cultivations and homesteads. Human activity has already interfered with the forest here too, in collecting firewood, raw materials, and so on (Flenley 1969 notes the same for the Wabag region, where markedly disturbed forest locations all occur fairly close to human habitation). It is not where demons reside, they lurk in remoter and less disturbed forested regions. Over many generations, each nibbling away at the forest edges, destruction of large forested areas has occurred. But extensive forested tracts have survived, considering the substantial population density and the great period of time for which agriculture has been practised in the New Guinea Highlands. The Wola cultivate areas under secondary woodland or cane grass far more often than they farm virgin forest sites; for example, only 8% out of 293 gardens surveyed in the Was valley were on sites under primary forest before cultivation.

It is both supernaturally safer and physically easier to clear secondary regrowth. Until relatively recently the Wola had stone tools only, which would have further restrained forest clearance, although again seen over many centuries their ancestors cleared large areas in their main valleys. The demon image, ambivalent denizen of forested regions, is embedded in a wider set of natural and cultural circumstances. The natural resource base, population density and available technology cannot be overlooked in the configuration of enigmatic demon beliefs. The soil resources of this montane environment are critical, supporting an intensive cultivation regime, which on some sites amounts to a semi-continuous system (see Section V). Without these, and at the population densities involved in the Highlands, far more forest destruction would have occurred, under the consequent extensive shifting agricultural regime, possibly with some other sylvan beings lurking in the forest. The Wola are not the innate conservationists of the popular imagination, nor are they wanton destroyers of forest, as depicted by opponents of shifting cultivation (FAO 1957; FAO-SIDA 1974; Watters 1971). Their relationship with their forest is far more subtle and equivocal, more practical and lived, less pedantic and intellectual.

II. FALLOW VEGETATION

The preference that the Wola show for clearing secondary regrowth, sometimes short-term fallow, over primary forest when establishing new cultivations raises further issues pertinent to natural resource management, or its apparent mismanagement. It is widely accepted that the natural vegetation which regenerates following abandonment of a swidden site under a shifting cultivation regime is important to the restoration and management of fertility. During this, usually extended fallow, the soil rests and recuperates (Nye & Greenland 1960). This rotation of land contrasts with rotation of crops to manage soil fertility. The secondary regrowth pumps back up to the surface minerals leached down the profile while the soil was exposed under shallower rooting crop plants. It amasses nutrients both from the soil and atmosphere and holds them against loss, as described (see Chapter 2), incorporating them into a tight and efficient natural ecocycle, making them available when the site is next cultivated and firing releases them. The standing vegetation also offers an extended period of effective cover, protecting the soil from erosion losses, allowing the accumulation of new topsoil.

Secondary Regrowth Successions

The broad sequence in which plants recolonise cleared sites is well known to Wola gardeners. They take the change in the composition and size of plants comprising different communities over time as largely marking the different stages of succession which they name (Table 8.1[8]). They distinguish

---

[8] The data in this table come from 4300 m² of quadrat surveys; see *Singapore Journal of Tropical Geography* Vol. 16 for details.

Table 8.1
The vegetation characterising different stages of regrowth after garden abandonment

Plant families:	REGROWTH TYPE:					Plant families:	REGROWTH TYPE:				
	Gardens	Herbaceous	Coarse grass	Cane grass	Woodland		Gardens	Herbaceous	Coarse grass	Cane grass	Woodland
Crops:						Rubiaceae (2spp)	unc		unc	rre	rre
Acanthaceae	unc	rre	rre			Scrophulariaceae (2spp)	unc			rre	rre
Alliaceae	unc					Selaginellaceae	unc	unc	unc	rre	rre
Amaranthaceae	mde					Solanaceae (3spp)	unc	rre	rre		unc
Araceae	mde					Thelypteridaceae (4spp)	unc	rre	unc	unc	rre
Convolvulaceae	abn	abn	unc			Tiliaceae	cmn		mde		unc
Cruciferae (4spp)	unc					Umbelliferae (2spp)	unc	cmn	cmn	unc	
Cucurbitaceae (3spp)	unc					Urticaceae	cmn		cmn	rre	rre
Gramineae (3spp)	mde	rre	rre			Violaceae			cmn	rre	rre
Leguminosae (2spp)	unc					Woodsiaceae		cmn	cmn		
Malvaceae	rre					Zingiberaceae (5spp)	unc			rre	rre
Musaceae	rre				rre	**Trees & Woody Shrubs***					
Pandanaceae				mde	rre	Apocynaceae				unc	unc
Passifloraceae					rre	Aquifoliaceae (2spp)				unc	unc
Rubiaceae	rre					Araliaceae (5spp)			cmn	cmn	cmn
Grasses and Herbs:						Araliaceae	unc			unc	unc
Amaranthaceae (2spp)	rre					Asclepiadaceae				unc	unc
Anacardiaceae				rre		Cunoniaceae (3spp)				unc	mde
Apocynaceae				rre	rre	Cyatheaceae (4spp)				cmn	unc
Araceae	rre					Elaeocarpaceae (3spp)				mde	abn
Aspleniaceae				rre	rre	Euphorbiaceae (12spp)	abn			cmn	unc
Athyriaceae	unc			rre	unc	Fagaceae (2spp)					mde
Balsaminaceae	rre	unc		rre	rre	Gesneriaceae					rre
Boraginaceae	mde	cmn	unc			Icacinaceae					rre
Commelinaceae	unc	unc	unc			Indetermined					cmn
Compositae (10spp)	rre	abn	cmn	unc	unc	Lauraceae (4spp)	cmn	cmn	cmn	cmn	cmn
Cucurbitaceae (2spp)	cmn			rre	rre	Liliaceae	cmn	cmn	cmn	cmn	mde
Cyperaceae (2spp)	mde	unc		rre	rre						

Table 8.1
Continued

Plant families:	REGROWTH TYPE: Gardens	Herbaceous	Coarse grass	Cane grass	Woodland	Plant families:	REGROWTH TYPE: Gardens	Herbaceous	Coarse grass	Cane grass	Woodland
Dennstaedtiaceae (3spp)			rre	rre	rre	Loganiaceae (2spp)				cmn	mde
Dioscoreaceae				rre	rre	Malvaceae					rre
Ericaceae		rre		rre	rre	Melastomataceae (4spp)	cmn			abn	cmn
Gleicheniaceae				rre	rre	Meliaceae		cmn		rre	rre
Goodeniaceae			rre		rre	Monimiaceae					unc
Gramineae (19spp)	cmn	abn	abn	abn	mde	Moraceae (8spp)	cmn			abn	cmn
Haloragidaceae	rre			rre		Myrsinaceae (2spp)	abn	cmn		mde	cmn
Indetermined (4spp)	rre		unc	rre	rre	Myrtaceae (4spp)				cmn	unc
Juncaceae				rre		Ochnaceae				abn	mde
Labiatae (2spp)	unc	unc	unc	rre	unc	Pandanaceae	abn			unc	abn
Leguminosae (5spp)	rre	mde	mde	rre	unc	Piperaceae (2spp)				cmn	unc
Liliaceae				rre	rre	Pittosporaceae				unc	
Lindsaeaceae	rre			rre		Proteaceae				rre	
Loranthaceae				rre	rre	Rubiaceae (2spp)				unc	unc
Lycopodiaceae (2spp)				rre	rre	Rutaceae (2spp)				rre	mde
Marattiaceae				rre	rre	Sabiaceae				unc	
Oleaceae				rre	rre	Sapindaceae (4spp)	cmn			unc	mde
Oleandraceae				rre	rre	Saurauiaceae (3spp)	abn		cmn	cmn	abn
Orchidaceae (2spp)			unc	rre	rre	Symplocaceae				rre	rre
Polygalaceae			cmn	rre	rre	Theaceae				unc	mde
Polygonaceae	cmn	abn	cmn	rre	unc	Ulmaceae (2spp)	abn	cmn		rre	abn
Polypodiaceae (3spp)	cmn			rre	rre	Urticaceae (6spp)	abn		abn	rre	cmn
Rhamnaceae				rre	rre	Verbenaceae				cmn	mde
Rosaceae (2spp)	unc	cmn	unc	rre	rre	Vitaceae				unc	rre

abn = abundant, cmn = common, mde = mediate, unc = uncommon, rre = rare (for **Crops**, and **Grasses & Herbs**: abn = >100 plants 10 m^{-2}, cmn = 25–100 plants 10 m^{-2}, mde = 10–25 plants 10 m^{-2}, unc = 1–10 plants 10 m^{-2}, rre = <1 plant 10 m^{-2}, and for **Trees & Woody Shrubs**: abn = >0.5 plants m^{-2}, cmn = 0.1–0.5 plants 10 m^{-2}, mde = 0.05–0.1 plants 10 m^{-2}, unc = 0.01–0.05 plants 10 m^{-2}, rre = <0.01 plant 10 m^{-2}). Number of species per family in brackets (no entry = one species only)

the following principal stages in the regrowth sequence (see Chapter 7): cultivated sites (*em*), recently abandoned sites (*mokombai*), and long abandoned sites (*gaimb* or *obael*). They may sub-divide these into further named phases. The line between the different stages is not clear cut, sites pass gradually from one to another as their vegetation profiles change. The time they take to pass between them also varies, and not all sites follow the same sequence either, nor feature all of the named phases.

The first cultivated *em* stage the Wola divide into two phases called *waeniy* (lit. new) and *puw* (lit. go). They distinguish between these according to the progress of the sweet potato harvest: gardens in which tubers are not yet mature, or from which they are harvesting them for the first time, are *waeniy*, and gardens beyond this phase, from which they may be harvesting tubers for a second or subsequent time, are *puw*. Any newly planted *waeniy* garden will have some fast growing weed species germinating across it within weeks of planting, notably *Polygonum nepalense*, *Crassocephalum crepidioides*, *Bidens pilosus*, *Cynoglossum javanicum*, *Adenostemma lavenia*, *Viola arcuata*, *Isachne arfakensis* and *Impatiens* sp. (Henty & Pritchard 1973), together possibly with the occasional tiny tree seedling (e.g. *Trema orientalis*, *Cryptocarya laevigata*, *Dodonaea viscosa*, among others). These early weed colonisers remain through into the *puw* phase, and may have grown into sizeable plants, particularly if little weeding has occurred. Some coarse grasses are likely to have joined them too, which when fully developed will probably dominate the vegetation during the early phases of abandonment, including *Arthraxon hispidus*, *Paspalum conjugatum*, *Leersia hexandra*, *Ischaemum polystachyum*, *Setaria sphacelata* and possibly seedling *Miscanthus floridulus*, together with some other plants like *Erigeron sumatrensis* and *Rubus rosifolius*.

The size to which these plants grow and their numbers depend to some extent on the assiduousness with which people weed cultivations. A thorough weeding can result in a *puw* state garden having a *waeniy*-like colonising plant profile, and an absence of any weeding give a more recently planted site a longer established appearance. The extent to which people weed gardens varies widely between sites, and particularly according to crop cover. They are unlikely to spend considerable periods of time weeding sweet potato, especially when established, because its dense creeping foliage gives excellent ground cover against which few weed species are effectively able to compete, but when harvesting tubers women often pull up any weeds evident, so that many mounds enter the *puw* phase effectively clean weeded.

Passage from the *em* cultivated garden stage to the *mokombai* recently abandoned stage is commonly gradual, sometimes with women still harvesting from parts of a garden whilst having effectively abandoned others where sweet potato vines are in marked decline. The Wola sometimes call a site in the early phase of abandonment *taengbiyp*, after one of the grasses which is commonly prominent at this time (*A. hispidus*). This phase is similar in vegetative composition to the late *puw* one except that crop plants are less evident, or as the Wola express it 'there are not many sweet potato leaves' ('*hoday shor onduwp inj*'), and the herbs, grasses and seedlings are larger and more established across the site. The vegetation that marks the shift from *em* garden to *mokombai* abandoned site can vary considerably between locations; it foreshadows the vegetation community that is likely to dominate in years to come. The change from garden to abandoned site may also on occasion occur swiftly and have relatively little to do with natural regeneration, notably when pigs break into a cultivation and damage it seriously by rooting it over, prompting the gardening household to abandon it as *showmay mokombai ma sayor* (lit. pigs abandoned cause it-becomes), even though coarse grass and large herbaceous regrowth may not be far advanced.

The later *mokombai* phase is likely to be dominated by the prolific coarse grass *I. polystachyum*, and if so, is named *bol* after it. It is a relatively uniform succession, covering sites in a thick waist-high sward of grass. In its later stages, clumps of *Miscanthus floridulus* sword grass and robust tree saplings become evident. Until it reaches this later phase, a site may be recultivated fairly easily by pulling up the grasses, assuming that fence or other barriers enclosing the site are sound, and garden sites may be brought back into cultivation after two or more years under such grassy fallow. But once robust cane grass clumps and/or sizeable tree saplings have established themselves on a site it is unlikely that anyone will recultivate it for many years, after which time span a household will clear it of long-term fallow and enclose it as a new garden. Sometimes, on heavily disturbed sites where fertility is a problem (i.e. where topsoil is thin, repeated burning occurs, etc.), *Imperata* grass may come to dominate and the Wola call such a phase of succession *senz*, after this grass. Although not a widespread succession, it can be long-lived. It eventually gives way to other successions if interference ceases and as conditions improve, with tree saplings and such establishing themselves.

When left alone, sites finally pass under a longer term mature regrowth succession. If grassland surrounds them, cane grass with some trees and tree ferns will likely come to dominate the community, whereas trees like *Ficus* and *Saurauia* will initially come to dominate if adjacent to forest, and hardwoods like *Lithocarpus*, *Castanopsis* and so on later. The Wola distinguish two principal kinds of mature succession accordingly, *gaimb* cane grassland, and *obael* secondary woodland (see Chapter 7). The *gaimb* succession is named, following established convention, after the cane grass *M. floridulus* which predominates, as noted previously. The *obael* succession features more trees, the first soft-wooded species of which (e.g. *Trema*, *Cryptocarya*, *Dodonaea*, *Macaranga* etc.) eventually mature and die.

If left uncultivated, a site may pass through a series of successional changes, until a sufficient period of time has elapsed that all those who saw the garden or knew of its existence have passed away, when it becomes *em imb* (lit. garden not-cultivated) and its cover equates in the minds of the living with virgin vegetation that no one has ever cleared for the purposes of cultivation. Cane grassland can also become *em gaimb uw imb shay* (lit. garden canegrass is not-cultivated becomes), although it may eventually pass under a wooded succession if left undisturbed for long enough. It is not ridiculous for people to maintain that they have never cultivated some areas of cane grassland because, while the establishment of this succession currently depends on some site interference, this can happen in ways other than through garden cultivation (such as felling trees to allow pandans to flourish, or intensive collecting of firewood near settlements disturbing forest so much that cane grass establishes itself, or natural events like landslips, in which humans are not involved, occurring and affording cane grass a colonising opportunity).

Managing Secondary Regeneration

Regardless of their familiarity with, and detailed knowledge of, the foregoing regrowth sequence, the Wola are strangely casual about the secondary regrowth that colonises an abandoned garden site. They make little effort to manage it. They certainly entertain no ideas about rehabilitating forest regrowth to keep the demons content. They contrast with the Maring, who according to Rappaport (1972) refer to the seedlings that colonise a cultivated site as the 'mother of the garden' and tend them towards the end of the cultivation period. These small trees portend the secondary regrowth that will restore the site's ferility during the fallow period, and the Maring nuture them even though they compete with their crops and oblige them when robust saplings to abandon gardens before they have harvested all of the crops, thus Rappaport argues protecting the soil from serious depletion by making harvesting more laborious as returns decline. The Wola are unlikely to allow regenerating vegetation to inhibit the harvest of crops still yielding them adequate returns. They customarily weed gardens while harvesting from them, and only allow invading plants to grow to any size as sites near abandonment, and tall effective crop competitors like bananas and sugar cane remain, which can still be relatively easily harvested when ready.

It seems that the Wola are again acting in a perverse manner, not attempting to manage their natural resources to maximum effect but courting their degradation, off-handedly exploiting and abandoning land to chance spontaneous natural colonisation and regeneration. According to the Wola they have scarcely any control over this process. Whether it is cane grass or tree seedlings that colonise a site, let alone what species of grass or tree, is largely they say beyond their control. It is more a function of the surrounding vegetation and what seeds it supplies to a site, with trees more likely adjacent to forest and grasses are more probable to dominate in grassland.

The actions of gardeners may nonetheless influence to some extent the sequence of natural regeneration. The length of time for which they cultivate sites and the thoroughness with which they clear them of natural vegetation may effect the species composition of subsequent regeneration. If they coppice and pollard some trees, these may shoot again, ensuring that some of the site's original species recolonise it, as before clearance — so long as men do not kill them by ring-barking or repeated cutting back during the cultivation period. Trees that commonly send up shoots again include *Lithocarpus, Ficus, Artocarpus*, among others. They may short-circuit the gradual recolonisation succession sequences described above, and give a cover of mature forest trees far sooner than is normal on other sites.

Plate 8.6 A fence line planted with palm lilies (*Cordyline*), which may serve as boundary markers years later if the site is recleared.

The Wola also influence the species composition of some abandoned garden sites by what they plant on them during cultivation. They may cultivate some trees and shrubs which will remain when they abandon a garden, so influencing the long-term composition of secondary regrowth. Among those which yield edible and usable products are the *naep* casuarina (*Casuarina oligodon*), *aenk* pandans (*Pandanus julianettii*) and *aegop* cordylines (*Cordyline fruticosa*), together with the occasional *poiz* fig (*Ficus wassa*) and shuwat fig tree (*Ficus dammaropsis*), cultivated in gardens and around houseyards for their edible leaves (Sillitoe 1983). They commonly plant these trees around the edges of gardens, where they remain in erratic lines years after abandonment marking the course of long rotted fences once enclosing sites; they are often cited as important evidence in disputes over rights to cultivable land. Sometimes the Wola plant them within gardens too, commonly along internal boundaries.

Other trees and shrubs cultivated along garden fences and ditches, but rarely in gardens, include *gonkliyp* (*Graptophyllum pictum*), *shonon*

(*Acalypha* sp.) and *duwk* spurges (*Euphorbia* sp.). These are cultivated as ornamentals, particularly adjacent to houseyards, and also to mark fence lines, in the event of land disputes years later. They may also help in soil conservation. The Wola say that these plants have extensive root systems and when planted along the edge of trenches enclosing gardens they serve to bind the soil firmly together and reduce the chances of slumping with consequent erosive loss of soil. Gardeners do not always purposefully plant these trees: when they use their timber for fencing they may cutting-like take, the sprouting and rooting stakes growing into saplings.

According to Wola comments, they favour these trees because the ground beneath them remains relatively clear, lending a clean aspect to houseyards, which they value. But this is scarcely a desired trait from the perspective of soil fertility restoration, where a good litterfall and build-up is thought an important aspect of any efficient nutrient cycle. Again it is as if the Wola are acting unwisely regarding the management of soil fertility, seeking trees that apparently minimise organic matter returns. The exception is the *naep* casuarina, which they acknowledge affects a site's fertility. They point to the bed of needles that accumulates under this tree as beneficial to soil fertility. They say that these give the soil a soft and loose feel, the resulting low bulk density and porous structure they think is particularly beneficial to crops. When they plant crops into soil with a liberal needle content, in the vicinity of casuarinas, they obtain good returns: these trees improve adjacent soil fertility, as a friend expressed it '*naep shor obuw suw hemem menkoroga*', 'casuarina needles they change into the best *hemem* soil'.

Casuarina and Fertility Management

There is considerable interest in the potential of casuarina in soil fertility management (Diem & Dommergues 1990; Thiagalingam 1983). It may grow into a sizeable tree up to 30 metres high, an erect and graceful evergreen with an open feathery crown (its name alludes to the resemblance between its drooping and filamentous foliage and the cassowary's plumage). Once thought to be a very primitive tree, it is now considered highly evolved (Midgley *et al*. 1983). Its foliage comprises modified cladode branchlets, segmented and needle-like, with leaves reduced to whorls of tiny teeth-like scales. But of more interest regarding soil fertility are the woody nodules and filamentous mats present on the roots. These are readily visible just beneath the soil surface, and occur on roots to considerable depths, in large masses extending from the base of the tree's trunk out to the near the drip line. Moisture availability and soil aeration influence nodulation, hence they are densest on surface roots where oxygen is most available.

The perennial woody nodules result from infection of root hairs by the actinomycete *Frankia*. This bacteria-like microorganism, which has a symbiotic relationship with the casuarina, is able to fix nitrogen directly from the atmosphere, requiring oxygen for the process. The tree fixes roughly comparable amounts of nitrogen to legumes with their *Rhizobium* symbionts (BOSTID 1984). The nitrogen is added to the soil through the decay of dead rootlets and leafy litter. Analysis of some soil samples collected in the Was valley around casuarina trees compared with samples collected from adjacent open fallow land indicate that nitrogen contents under the former are some 1.24 times higher on average:[9]

Other research in the Papua New Guinea Highlands indicates that soil nitrogen contents are markedly higher under established casuarina trees than other vegetation, with 0.43 to 0.73 %N compared to 0.31 %N under young casuarina, and between 0.21 and 0.28 %N under other vegetation on the same

Table 8.2

Analysis of soil samples, collected in the Was valley under casuarina trees compared with samples collected from adjacent open fallow land

Site location	%N	
	Under Casuarina Tree	Adjacent Open Land
Honael (nr. Iybweb)	0.95%	0.83%
Honael (nr. Tomb)	1.02%	0.85%
Waenshowiyba (nr. Momiya)	0.91%	0.85%
Tombem (nr. Ganonkiyba)	1.01%	0.66%

soils; on average the %N is some 2.3 times higher (Thiagalingam & Famy 1981). The differences in nitrogen levels reflect to some extent differences in time under casuarina, besides soil variations. The evidence suggests that soil nitrogen builds up gradually under casuarina, increasing at between 0.015% and 0.018% a year, which represents an annual increase to 15 cm. depth of some 315 to 378 kg N ha^{-1} (Parfitt 1976).

The roots of casuarinas also form symbiotic relationships with mycorrhizal fungi, both ectomycorrhizae, where fungal hyphae penetrate between root cells (e.g. *Cenococcum*, *Pisolithus* and *Hymenogaster*), and endomycorrhizae, hyphae penetrating into cells (e.g *Glomus*). The intricate and ramifying hyphae network facilitates the uptake of minerals, notably phosphorus, and some trace elements. The presence of mycorrhizae also influences the nitrogen-fixing activity of *Frankia*, with marked reductions in fixation where they are absent in phosphorus-deficient soils. A range of further unidentified microorganisms interact with casuarina roots too, to produce dense mats of 'proteoid roots'. They increase the surface area of roots greatly and probably also help the tree to absorb phosphorus, among other nutrients (experiments show that they double tree growth in low phosphorus soils — BOSTID 1984);

---

[9] Soil samples, collected from either around the base of casuarina trees (under the canopy) or from adjacent land (away from the canopy), were combined and sub-samples taken for analysis using the Kjeldahl digestion method.

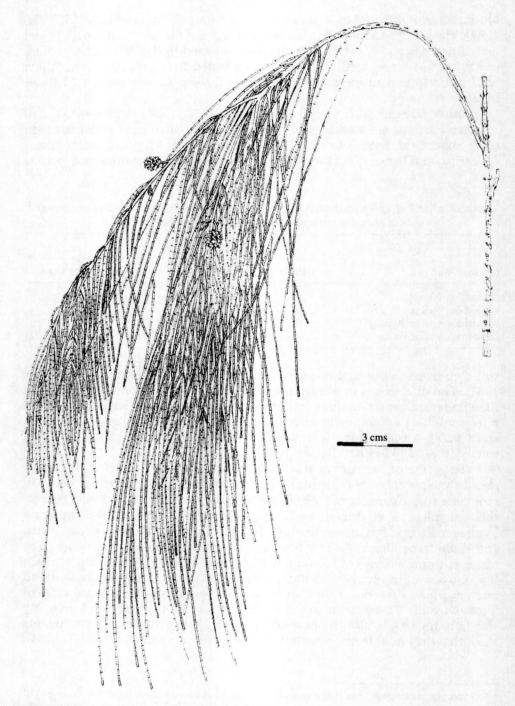

Figure 8.2 Casuarina (*Casuarina oligodon*).

Plate 8.7 A casuarina tree showing characteristic wispy cassowary-feather-like foliage (*Casuarina oligodon*).

this is a notable trait given the phosphate fixing capacities of the soils of the Wola region. These microbial symbiotic relationships allow the casuarina to grow vigorously on poor soils deficient in nitrogen and phosphorus. Its young seedlings are able to compete aggressively with weeds, even the notorious *Imperata* grass. The dioecious tree sets abundant seed and is easy to propagate. The Wola transplant naturally germinated seedlings to the locations where they want them. They call female trees that bear cones *way*, and male coneless ones *tuw* (these are the same terms they use for female and male animals); they say that they collect seedlings for transplanting from under *way* trees only.

In addition to improving soil fertility, casuarina improve soil structure, as the Wola observe, and help control erosion. Besides encouraging soil microflora, the persistent and inter-locked mantle of fallen foliage helps to rebuild soils and affords effective protection against run-off. The mat of mycorrhizal hyphae also reduces erosive processes by binding soil particles together. The casuarina has the attributes of the ideal secondary regrowth tree for managing soil fertility. It even has the advantage of not having to be felled when preparing a site for a garden because its feathery, open foliage allows sufficient light through to the crop below (perhaps with some judicious lopping of lower branches); it makes a useful shade tree. It has a range of other uses too. It supplies unequalled fuelwood with a high calorific value that burns to leave relatively little ash, valuable hard and durable timber for construction purposes and artifact manufacture, and gives good shelter, acting as a windbreak around houseyards.

Regardless of its manifest benefits, which the Wola readily acknowledge, they cultivate relatively few casuarinas. Men plant these trees and own them individually. In the majority of cases they possess so few that they do not think it ridiculous to be asked to count them (unlike the number of pandans they own, to which question they invariably respond 'very many'). During a survey of seventy-six men, only four individuals claimed ownership of twenty or more casuarina trees and had to estimate the number to which they had rights. The average number of trees owned per man was seven (ranging from 0 to 90, $\sigma = 13.3$). The comments of the Wola about casuarina planting tally with these statistics, suggesting that they plant them more for their ornamental effects than as a soil improvement measure. The occurrence of trees supports their remarks, they occur noticeably more frequently adjacent to houses and around *howma* ceremonial grounds than they do across garden sites. On a recent return trip I was struck by the increased number of casuarina evident, notably around settlements. These trees are the legacy of the Forestry Department's now defunct seedling distribution service from Nipa station. They also reflect the fact, as one man pointed out to me, that for over two decades the communities that I know well have not been routed in warfare, when the enemy does all the damage it can, chopping down casuarinas, burning houses, and so on.[10] The beneficial qualities of these trees notwithstanding, the evidence suggests that the Wola consider soil resources adequate on the whole, not regularly requiring regrowth management with casuarinas or any other species, to sustain their fertility to meet their cultivation needs. Again their off-hand attitude suggests short-sightedness regarding natural resources, but soil fertility and nutrient cycling under different vegetational communities vindicate the soundness of their fallow practices and intimate the derivation of their intriguing agricultural regime's sustainability.

---

[10] The significance of this point was brought home to me on a visit to Nipa, where post-election warfare had resulted in the felling of large numbers of casuarina that had previously graced this government station.

CHAPTER 9

INTO THE SOIL: NUTRIENT CYCLING AND DECOMPOSITION

The cycling of nutrients is integal to any vegetation community's ecology, but the proportions taken up, amounts stored in plants, rates of cycling, and so on vary widely. Differences between virgin and colonising successions are commonly taken, together with related changes in biomass, as yardsticks by which to assess the impact of human interference. The virgin vegetation covering the greater part of the potentially cultivable land of the Wola region is montane forest, and the cycling of minerals within it indicates fertility status before humans interfere with the vegetation. While mineral cycling under mature forest rarely features in the recuperation of cultivated sites, as it may in some agricultural systems, because few areas under cultivation are left abandoned sufficiently long for such forest to re-establish itself on sites, nevertheless it serves as a base line from which we can assess changes consequent upon cultivation. It is a starting point against which we can appraise the impact of cultivation on fertility and productivity, under the different regrowth successions that result, notably secondary woodland and grassland (both short coarse and tall cane grasses). A change in vegetational communities consequent upon human activity need not spell massive nutrient losses, degradation of land resources and diminished agricultural potential, although it may in the short run considerably reduce biomass and species diversity.

Decomposition is a critical step in nutrient cycling, whatever the vegetational community, breaking down complex organic molecules into inorganic ones capable of up-take again by newly growing plants. It involves countless organisms from the miniscule through to the large, both insects and other animals, and also the intriguing saprophyte kingdom that lives heterotrophically on decaying vegetable matter, colonising organic residues, deriving energy from decomposition and rendering nutrients available for plant up-take. This phase of the ecological cycle takes us finally into the soil, where a great deal of decomposition takes place. The biological world of the soil is truly remarkable, although mostly unseen and largely unknown to many, including the Wola. It is where a myriad of organisms that play a key role in facilitating the cycling of nutrient minerals within ecosystems are constantly active: without their activities nutrient cycling, and the wheel of life, would soon cease to turn.

I. NUTRIENT CYCLES

This discussion of mineral nutrient cycling starts with rainfall inputs and finishes with plant biomass, following the movement through the system of the major mineral nutrients of nitrogen, phosphorous, potassium, calcium and magnesium, under different vegetation successions from rainforest through to grassland — as summarised in Figures 9.1 to 9.4. The review of montane forest nutrition draws on the valuable work of Edwards and Grubb (1977, 1982) in the Bismarck Range, conducted in rainforest floristically, edaphically and climatically similar to that of the Wola region. Studies of mineral nutri-

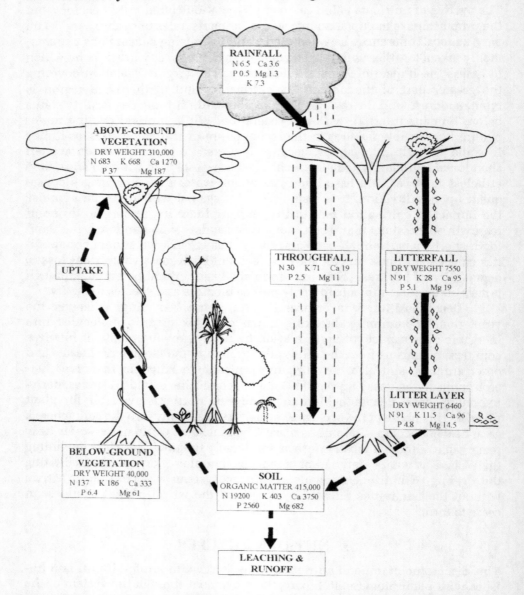

Figure 9.1 Mineral cycling in a New Guinea montane rainforest (values in boxes = kg ha^{-1}; solid arrows = annual mineral transfers of box values; broken arrows = unmeasured pathways) [after Edwards 1982].

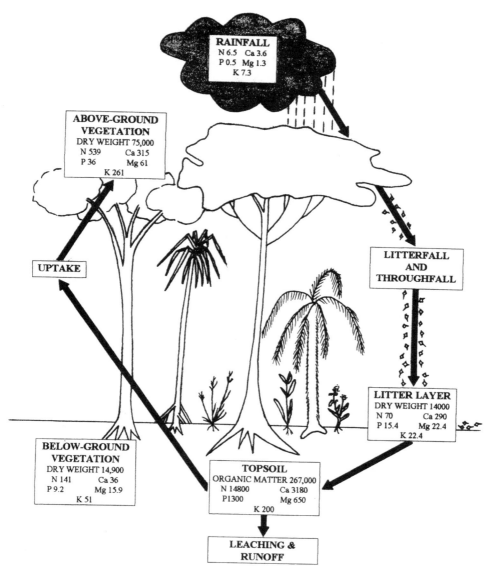

Figure 9.2 Mineral cycling under secondary woodland (values in boxes = kg ha^{-1}; arrows = mineral transfer pathways).

tion in long-lived forest conventionally focus on short-term throughfall and plant litter turnover, the cycling rates of which indicate the availability of minerals for, and their incorporation into, the slower growing, larger woody parts, thus shadowing the regeneration cycle of the trees that comprise a substantial part of any forest's biomass. The evidence suggests that while the production of woody parts is reduced in montane forests compared to lowland ones, the rates of cycling of minerals through foliage are similar, inviting

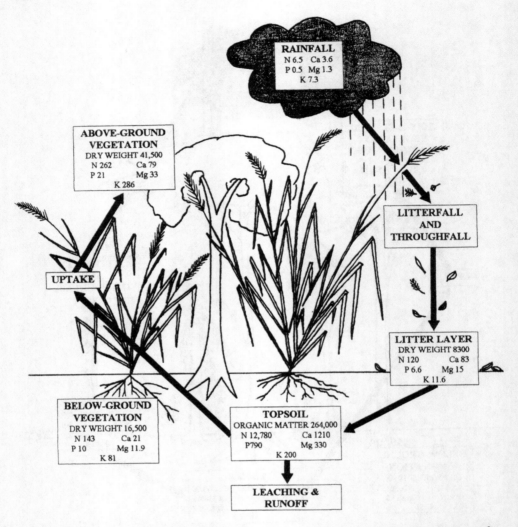

Figure 9.3 Mineral cycling under secondary cane grassland (values in boxes = kg ha^{-1}; arrows = mineral transfer pathways).

limited comparisons (Edwards 1982; Grubb 1977). The data on regrowth vegetation, both secondary forest and grasslands, come from measurements and estimates made in the Wola region for this study.[1] There is little comparative information on these vegetation communities, work elsewhere in the tropics on savanna vegetation, although superficially similar structurally, concerns quite different vegetation regimes.

---

[1] I am grateful to Dr. P. Wort of the ADAS laboratories for analyses of plant material. The biomass estimates were made on a 500 m^2 plot of *obael* secondary woodland; 500 m^2 of *gaimb* cane grassland and 5 m^2 of *bol* coarse grass, including the destructive harvesting and complete weighing of a representative sample of plants, roots and all.

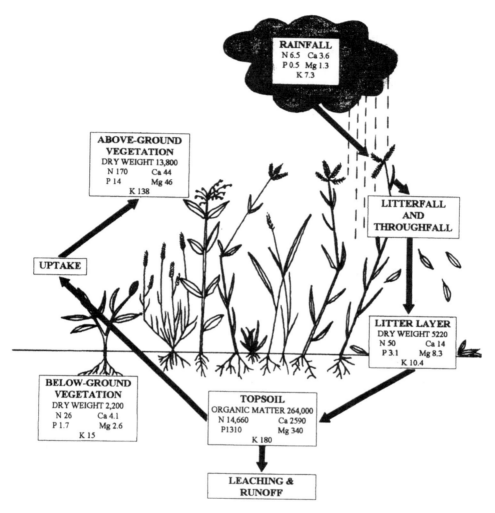

Figure 9.4 Mineral cycling under secondary coarse grasses (values in boxes = kg ha^{-1}; arrows = mineral transfer pathways).

Rainfall and Throughfall Inputs

The input of minerals from rainfall is fairly small in the New Guinea Highlands, which is expectable in a region remote from, and around 2000 metres above, the sea. The small amounts arriving may to some extent come from nutrient elements put into the atmosphere by the burning of vegetation elsewhere, particularly in heavily populated Highlands regions (Edwards 1982, after Ungemach 1969; also Pivello & Couthino 1992).

The rain has considerably higher concentrations of minerals once it has passed through the forest canopy, leaching them from vegetation on its passage. The proportion of rain that reaches the forest floor as throughfall is some 68% on average, which is similar to that reported for other tropical forests (Edwards 1982), although it varies significantly (ranging from 30% to 80% of rainfall), depending on the amount of rain falling and the architecture of the canopy. The proportion of rain reaching the ground under secondary successions varies widely: under secondary woodland it is a little higher than under primary forest at some 71% overall, although it too covers a considerable range (from 50% to 80% of rainfall), whereas the proportion falling directly on the soil under dense cane grass is markedly less at some 42%, again with a considerable spread (from 20% to 70% of rainfall). The longer and heavier the rain and the more open the canopy, the greater the throughfall in all vegetation communities.

The contribution of stemflow to rain reaching the forest floor is very small, at less that 1% of total throughfall; again it is proportional to rainfall. It is probably higher under secondary successions. The quantity of minerals leached from the forest canopy correlates strongly with amount of throughfall; concentrations in the rain fluctuate throughout the year, increasing as rainfall decreases. The interception and storage of rainfall by any canopy, which dictates the proportion passing through it as throughfall, depends on rain-fall intensity and evaporation rates. The daily interception of New Guinea montane forest is higher than that reported for other rain forests, at 4.5 to 7.0 mm per rainy-day minimum (Edwards 1982). The frequent daily sunny periods that characterise the Highlands help dry the canopy out (see Chapter 3), the heavy epiphyte population increases interception capacity, and the many small leaves without drip tips reduce run-off, all increasing the storage capacity of the forest.

In forest the principal throughfall source of phosphorus, potassium, calcium and magnesium is probably leaching from leaves, although the frass of phytophagous insects, which are estimated to graze some 10% of forest leaves annually (Edwards 1977), might be a significant source too. The decomposition of dead twigs and branches up in the tree canopy (considerable rotting occurring there before they fall as litter), may also in addition supply nitrogen, but little potassium. Differential leaching of nutrients occurs, potassium being leached relatively more, notably above phosphorus and nitrogen. The substantial nitrogen concentration of forest throughfall, which is not readily leached from leaves, is perhaps attributable to biological fixation of N_2 by epiphyllous blue-green algae and bacteria found on the leaves of smaller trees and shrubs in the lower canopy (the cations and phosphate in leaf-leachates probably promote their activity); the canopy's epiphytic soil might also contain some microbial nitrogen-fixing organisms (Edwards 1982).

The throughfall data for montane New Guinea forest suggest a lower rate of mineral turnover via this route than occurs at lower altitudes (Edwards 1982; Nye 1961; Bernhard-Reversat 1975). The leaching of elements from the canopy depends upon rainfall, temperature and the residence time of water on foliage. While rainfall is high in the Highlands, temperatures are <10°C lower than in lowland forests (see Chapter 3), and leaf area indices, to which rain residence time relate, are appreciably lower at ~5.5 m^2 m^{-2}, montane forest leaves being thicker and tougher textured on average (Edwards & Grubb 1977).

Litterfall

Throughfall is the major pathway of potassium cycling (the litter ratio quotient under montane forest is 2.5), whereas litterfall is the major pathway for other minerals (with quotients under forest of 0.5 for phosphorus, 0.3 for nitrogen, 0.2 for calcium and 0.6 for magnesium). This cycling occurs on an average annual forest litterfall of c. 7.6 t ha^{-1}, of which c. 6.4 t is non-woody (96% leaves) and the remainder woody material, all contributing to a litter layer of c. 6.5 t ha^{-1}. The comparable mass of litter layers under secondary successions varies considerably, from c.14.0 t ha^{-1} under *obael* secondary woodland through c. 8.3 t ha^{-1} under established *gaimb* cane grassland to c. 5.2 t ha^{-1} under mature *bol* coarse grasses (where none of these communities have been heavily rooted over by pigs). The spread of these litter statistics probably reflects differences in decomposer populations and activity under different vegetational successions, and tractability of material produced to decompostion and differences in relative nutrient contents, in addition to differences in the actual amounts of litter produced and plant turnover times. The turnover time of the standing leaf crop under virgin forest (estimated at 8.3 t ha^{-1}) is some fourteen months, with understorey trees and sapling leaves having somewhat longer lives by a couple of months (Edwards 1977).

Both leafy and woody litterfall values under-represent actual dry weight production, particularly under primary forest. The withdrawal of minerals from leaves before abcission and insect grazing damage result on average in a 20% weight loss, losses varying between tree species according to their leaves' anatomy and palatability to insects. In addition, the considerable decomposition of woody material before it falls as litter doubles dry weight losses to 40%. The litter production values cited are fairly high, allowing for these factors, and suggest a fertile region supporting well-developed vegetational successions. Litter production and mineral turnover are remarkably uniform whatever a site's topography and species composition, whether ridge, valley, or slope (Edwards 1977, 1982). This suggests that climate dominates litter production, with peaks during heavy rains. Sites differ in the amounts of litter that persist on the forest floor, with drier locations like ridge tops having greater quantities. The species contributing to litter also probably influence decomposition rates, which will accelerate as nutrient contents increase, particularly phosphorus and nitrogen. Any consequent local differences in nutrient mineralisation rates have no noticeable effects on vegetation growth.

The state of the litter on forest floors in the Highlands suggests vigorous decay, leaves carrying extensive fungal mycelia and rhizomorphs, and having frass and comminuted particle encrustations indicative of voracious soil organism activity. After forty-six weeks on the forest floor, in a litter bag disappearance experiment, Edwards (1977) found that leaves had lost 43% of their initial dry weight. The litterfall and standing litter layer data suggest a mean annual litter disappearance rate of c. 1.2, which is a low decay constant k compared to tropical forests at lower altitudes, where values range from c. 1.3 to c. 4.7 (Scott *et al.* 1992; Nye 1961; Morellato 1992; Jordan 1985).

Litter bag monitoring of fallen cane grass vegetation indicates a disappearance rate of c. 0.9, reflecting perhaps the comparative resistance of this secondary succession to decompositon (the rooting of pigs, which is common in cane grassland, may assist by breaking up the material). These low values probably result from lower average temperatures in montane New Guinea, reducing microbial activity. Nonetheless, they are high for a mountain environment, reflecting fertile soils and the production of relatively readily decayed litter under forest (Edwards 1977).

Regardless of generally slower turnover rates, except where lowland forests are on poorer soils, the annual throughfall and litter returns of phosphorus, potassium, calcium and magnesium in New Guinea's montane forests overlap with those reported for lowland forests (Nye & Greenland 1960; Scott et al. 1992; Jordan 1989; Dantas & Phillipson 1989). Nitrogen returns are consistently lower, which is unexpected with the substantial amounts arriving annually in throughfall, and suggests fixation in phyllosphere and soil. The slower rates of organic matter decay and nutrient mineralization compared to lower altitude tropical forests result in a slow-down of mineral nutrient cycling and, particularly regarding nitrogen and phosphorus, can be expected to have knock-on effects on plant growth, making sweeping arguments about Highland ecology, and the impact of human farming activities on it, founded on general lowland-rain-forest-derived-notions of tropical forest, rather questionable.

The calcium and magnesium concentrations in recently fallen leafy litter in primary forest are similar to those in healthy leaves on trees suggesting little leaching or withdrawal before abcission, relative to totals present. The nitrogen and phosphorus concentrations are respectively 15% and 31% lower, indicating significant translocation of these elements from senescent leaves. The concentration of potassium is considerably lower, at about one-half of healthy leaves, indicating rapid leaching of this ion both before fall and on the ground. The concentration of minerals in falling leaves also varies over time, with potassium fluctuating the most and phosphorus the least, depending on rainfall and hence leaching losses. Analysis of green and dead leaf samples from two common secondary woodland trees (*Cryptocarya laevigata* and *Trema orientalis*) suggest nitrogen declines of around 12%, phosphorus 40% and potassium 50% in leaves shed under this regrowth succession. Translocation and leaching before litter fall from cane grass leaves (*Miscanthus floridulus*) — which remain attached to the parent plant by a stout sheath long after they have died — is of a different order, with nitrogen 50% less, phosphorus 70% and potassium 90%; calcium concentrations relative to total dry matter increase and magnesium declines by some 40%.

The rapid decline in potassium continues on the floor of montane forest, where concentrations fall by over one-half again in the more decomposed ground litter layer compared to recently fallen leaves. The decline in other mineral nutrients in rotted forest litter is notably less. The up-shot is that potassium disappears at about twice the rate of the litter, magnesium only a little faster than the litter, and nitrogen, phosphorus and calcium a little more slowly, resulting in slightly higher concentrations as decomposition proceeds (Edwards 1982). Similar trends may be anticipated under other vegetational successions.

The nutrient status of well decomposed, mixed litter under different vegetational sucessions reveals some intriguing trends, which are particularly pertinent when considering human management of land resources (Table 9.1). The indications are that there are no dramatic differences between communities regarding nutrients stored in litter layers, when the differing litter masses that occur under different successions are taken into consideration. The forest litter is more acid than the others and secondary woodland less so. Regarding nutrient cations and nitrogen, the coarse grasses, which comprise many brief garden fallows, broadly compare favourably with primary forest, as do the other two secondary successions of woodland and caneland, although the latter evidences somewhat reduced phosphorus and potassium contents. Indeed the tendency for litter phosphorus and potassium levels to have declined under all secondary successions is noteworthy in an agronomic context. Even cropped areas, where some litter remains unincorporated after cultivation, compare with other vegetational communities with respect to litter nutrient status, except for declining potassium levels, one of the factors likely to be limiting under cultivation. Similar trends are evident in the nutrient data on more recently fallen, less decomposed litter; although coarse grass litter has the lowest NPK contents, these are not dramatically less than other vegetational successions. Rotting litter is one of the principal sources of plant-derived nutrients tapped by crops, and the indications are that wide discrepancies do not occur between different vegetational successions regarding this possible source of nutrients. Whatever plant communities gardeners clear in cultivation, the litter derived component of their nutrient supplies will not differ wildly, which is pertinent given Wola cultivation practices.

Table 9.1
The average nutrient status of comminuted litter layer samples from under different vegetation communities. (Nos. in brackets = standard deviations)

	pH	P (ppm)	K (me/100 g)	Ca (me/100 g)	Mg (me/100 g)	CEC (me/100 g)	C (%)	N (%)
Primary forest (n = 10)	4.5 (0.2)	42.9 (41.1)	1.5 (0.2)	19.9 (5.9)	4.0 (0.9)	41.5 (2.4)	32.8 (3.9)	2.0 (0.1)
Secondary forest (n = 3)	5.4 (0.3)	22.4 (19.6)	1.2 (0.4)	42.7 (9.3)	10.4 (1.1)	37.8 (5.6)	23.3 (4.4)	1.5 (0.2)
Cane grassland (n = 2)	4.9 (0.1)	17.1 (19.5)	0.9 (0.01)	19.2 (18.6)	3.9 (3.4)	32.2 (7.0)	27.4 (6.5)	1.6 (0.1)
Coarse grasses (n = 1)	5.3	34.3	1.9	28.7	9.2	37.4	31.4	1.9
Garden crops (n = 6)	5.0 (0.3)	27.3 (23.0)	0.8 (0.1)	27.1 (9.1)	4.6 (1.4)	38.0 (5.2)	30.8 (2.5)	1.7 (0.1)

Soils and Organic Matter

The fertility status of soils under different vegetation communities evidences a trend similar to litter. A comparison of the nutrients available in a sample of soils under different vegetation covers, reveals remarkably little consistent variation between successions (Table 9.2).[2] The acidity of topsoils is remarkably consistent across successions (subsoils varied considerably, but not significantly, between communities). The levels of phosphorus in topsoils, on the other hand, vary signficantly between different vegetational successions, evidencing a broadly declining trend as community biomass decreases (coarse grasses confound the trend in large part because of a high result for one site). This decline in phosphorus levels is notable regarding agricultural potential (subsoils showed no significant pattern of variation regarding phosphorus).

Table 9.2
Measures of topsoil chemical fertility compared with site vegetation cover
(values in brackets = standard deviations; n = 60 sites; degrees of freedom = 5, 54)

	VEGETATION COVER						STATISTICAL SIGNIFICANCE		
	Primary Forest	2y Forest	Cane & Trees	Cane Grass	Grasses	Crops	F	P	
pH	5.02 (0.66)	5.02 (0.35)	4.92 (0.73)	5.06 (0.40)	5.05 (0.30)	4.92 (0.42)	0.14	0.98	NS
Phosphorus (ppm)	17.1 (12.1)	8.9 (7.6)	7.2 (5.8)	5.3 (4.2)	8.8 (7.4)	3.7 (4.1)	3.88	0.005	S
Potassium (me/100 g)	1.04 (0.57)	0.52 (0.28)	0.72 (1.01)	0.52 (0.29)	0.47 (0.22)	0.40 (0.22)	1.82	0.13	NS
Calcium (me/100 g)	13.9 (11.2)	11.8 (7.3)	7.5 (7.5)	4.5 (3.4)	9.6 (8.8)	9.2 (5.4)	1.69	0.16	NS
Magnesium (me/100 g)	3.04 (2.53)	2.91 (2.14)	1.80 (1.37)	1.47 (0.96)	1.52 (1.12)	2.96 (2.35)	1.5	0.21	NS
CEC (me/100 g)	31.3 (5.2)	26.5 (9.3)	27.9 (7.2)	25.4 (3.4)	30.7 (6.4)	26.7 (11.6)	1.0	0.43	NS
Carbon (%)	25.9 (6.6)	16.7 (7.5)	13.6 (7.5)	16.4 (4.4)	16.5 (5.0)	11.0 (5.1)	7.37	<0.005	S
Nitrogen (%)	1.41 (0.62)	1.09 (0.51)	0.81 (0.46)	0.94 (0.18)	1.08 (0.36)	0.84 (0.39)	2.69	0.03	S

---

[2] Each vegetation succession comprises a sample of ten topsoils, the analysis results assessed statistically using the ANOVA procedure to assess whether soil fertility varies significantly according to site vegetation cover — site vegetation equated with different treatments, and soil properties with dependent variables assumed to vary with plant cover. (The level of significance accepted as indicating a significant relationship S is $P < 0.05$, N.S. = not significant). See Chapter 12 for further details.

The variations in the availabilities of the metallic elements in topsoils are not statistically significant, nor follow any obvious patterns, going randomly up-and-down between different vegetational communities and displaying considerable variability, as the standard deviations indicate (likewise subsoils). The possible exception is topsoil potasssium which tends to fall as plant community biomass declines, which is noteworthy because, as already noted, the supply of this major nutrient perhaps limits crop yields. Potassium is likely to occur with decreased availabilities under secondary successions. The cation exchange capacities (CEC) of the soils, important regarding nutrient ion retention, show a similar pattern to pH, topsoil CEC being remarkably consistent across vegetational communities (while subsoil CEC evidences more variability, albeit unpatterned and not significant). The levels of nutrient cations down a profile depend upon their leachabilities and the proportion of available bases taken-up and recycled by plants. There is a marked concentration of potassium, calcium and magnesium towards the top of soil profiles under forest, which Edwards & Grubb (1982) suggest probably results from roots pumping these bases up from depth, with them accumulating near the surface with litter and throughfall deposition. Nonetheless, there is a noticeable difference between topsoils and subsoils regarding the levels of these cations under all vegetation successions, not just under forest. This 'base-pumping' parallels that reported in lowland rain forests, although perhaps too much should not be made of it as a montane forest phenomenon, for these forests are notably different regarding soil-plant relationships, and have very shallow rooting trees on the whole; differences between them are particularly evident with respect to their distribution of organic matter between soil and plants.

The other statistically significant variations in topsoil properties under different vegetation successions concern carbon and nitrogen contents. The carbon percentage of topsoils expectably reveals a highly significant variation between vegetation communities, carbon levels showing a downward trend as community biomass declines. The more substantial the plants comprising a community, predictably the greater the organically derived soil carbon levels: locations under primary forest have the largest carbon contents and those under crops the smallest. The nitrogen percentage in topsoils parallels that of carbon, which is expectable, the greater part of the nitrogen in soil being stored in organic matter. Soils under primary forest tend to have the largest nitrogen contents and those under crops the smallest. (There was no significant nor patterned variation in subsoil percentage carbon or nitrogen). It would be erroneous to read too much regarding agricultural potential into these variations in organically derived soil properties under different vegetation successions, especially in the absence of any significantly patterned variation between communities regarding other soil nutrients. It should also be noted that regardless of variations between them, the carbon and nitrogen contents of all soils, whatever their vegetation cover, are substantial, and adequate to meet any cropping needs, which is again noteworthy with respect to local agricultural practices.

The high accumulation of organic matter is a striking feature of soils under rain forest in the mountains of New Guinea; it is substantial under other vegetational successions too (see Section V). The organic matter content of the soil exceeds that of plants by up to three or four times (Edwards & Grubb 1977). The relationship varies with topography. The instability of soils on steep slopes prevents accumulation of organic matter to large depths, holding down the soil:plant organic matter ratio across large areas of forest to about unity; the high ratios characterise more level locations, like ridge tops. The compost-like soil that accumulates in the crowns of large trees, largely comprising the remains of epiphytes, contributes a further ~1 t ha^{-1} of high organic matter substrate to the forest eco-system. Organic matter rich soils are common throughout the New Guinea Highlands, on a variety of parent materials (Haantjens & Rutherford 1967; Wood 1987; Bleeker 1983; Radcliffe 1986). The accumulation of large amounts of organic matter in them is attributed to low temperatures and wet conditions inhibiting mineralisation, as cited for rainy tropical mountains around the world; plus in this region the complexing of humus with amorphous allophanic clays of volcanic origin, and their hydrated iron oxide weathering products (Sowden *et al.* 1976; Wada 1977).

An appreciable proportion of the total supply of some minerals, like nitrogen and phosphorus, is unavailable to plants as a consequence of this high organic matter content locking them up (exacerbated further for phosphate with its allophanic mineral fixation). This build-up of organic matter intimates that mineralisation rates are low and that New Guinea montane vegetation, unlike lowland equivalents, must be well adapted to the relative nutrient shortages resulting from slow rates of release and supply (Edwards & Grubb 1982).[3] The amassing of nutrients in the soil due to organic matter accumulation marks New Guinea montane forest off from lowland rain forests, where the plant biomass contains the larger part of the mineral nutrient capital, holding upwards of 90% of the total available nutrient pool (Jordan 1985). This further compromises efforts to generalise about Highland forest ecology, the effects of human activity on it and resulting secondary successions, from the conventional viewpoint of what comprises a typical tropical rain forest, indicating some noteworthy differences in the cycling of nutrients.

Biomass and Nutrition

High soil fertility doubtless contributes to the substantial stature of lower montane rain forests in Papua New Guinea, with their high trees of sizeable girth. The estimated average total biomass of the country's montane forests is around 350 t ha^{-1} (Edwards & Grubb 1977), which accords with my limited observations in the Wola region. This is less than the biomass of the country's lowland rain forests (Paijmans 1970); except where southern beech dominates, when montane forest biomass may equal or exceed that of lowland

---

[3] Researchers in South America were among the first to report such soils and discuss the consequences for forest growth (Jenny *et al.* 1948; Popenoe 1959).

forests. The large biomass of the forest puts the high organic matter content of the soils in perspective, dramatically underlining the large amount stored there, greatly in excess of that in live plants. The biomass of other plant communities is considerably less, being expectably smaller the more recently cultivation or disturbance has occurred. The average total biomass of coarse grass and herb dominated communities, on sites aboandoned for over a year, is 16 t ha^{-1}. This increases to some 58 t ha^{-1} under established cane communities, which include some trees and tree ferns. Under secondary woodland successions the biomass increases again to about 90 t ha^{-1}, reflecting its smaller dimensions compared to primary forest and soft wooded trees.

The total above and below ground biomasses of the different communities, according to their distributions of different floristic life-forms, show some interesting trends (Table 9.3). Mature or near-mature trees dominate the forest. While climbers and epiphytes, pandans and tree-ferns, shrubs and herbs, and seedlings and saplings are conspicuous members of the forest community, they account for only between two to four per cent of its above-ground biomass. But they make a larger contribution to mineral cycling and organic matter production, relative to their biomass, because thinner plant axes have higher concentrations of all minerals. Plant tissue phosphorus and potassium is largely present in living protoplasts, and woody parts proportionally have considerably less living tissue than leaf and green twig cells. Furthermore, ground-dwelling semi-succulent herbs have particularly high concentrations of potassium, calcium and magnesium, possibly because these nutrients have more time to accumulate in their longer-lived understorey leaves. Magnesium concentrations are also higher in small tree and sapling leaves than those of larger trees, perhaps reflecting the higher chlorophyll levels, and hence higher chloroplast magnesium concentrations, of shade tolerant leaves.

Table 9.3
The contribution of different floristic life-forms to the biomass of different vegetation communities

	ABOVE GROUND BIOMASS							BELOW GROUND BIOMASS
	Trees (gbh > 30 cm)	Tree ferns	Shrubs & Saplings	Climbers & Scramblers	Epiphytes	Cane grass	Grasses Herbs & Seedlings	Roots
Primary forest[4]	84%		2%	1%	0.5%		0.5%	12%
Secondary forest	63%	1.5%	4%			12%	3.5%	16%
Cane grassland	2%	2%	3%			63%	2%	28%
Coarse grasses/ herbs							86.5%	13.5%

[4] After Edwards & Grubb 1977

Mature or near mature trees are also prominent in secondary forest regrowth, but cane grass clumps also make up a notable part of this community's biomass. Tree ferns, shrubs and saplings also comprise a larger proportion of the vegetation, as does the ground floor cover of grasses, saplings and herbaceous plants. Large trees are uncommon in cane grassland, where cane predominates, and absent from recent grassy regrowth (unless a pollarded tree left standing by a gardener reshoots). Tree ferns, shrubs and saplings are again more evident under cane, as they are under secondary woodland, but the ground floor layer of grasses and herbs comprises a smaller proportion of the cane community, due to shading by dense cane clumps and the frequently higher levels of pig rooting in caneland adjacent to homesteads. Another notable feature of cane successions is the large contribution of the below ground component proportionally to total biomass, cane grass having thick root mats. Recently abandoned or fallow grassland communities rarely feature any substantial plants. Regarding differences in nutrient capital stored in the plants of different vegetation communities, it is important not to equate these solely with biomass and assume that the less this is the smaller nutrient capital will be proportionally, because of the different concentrations of minerals in different life-forms and in the same plants at different stages in their life cycles.

A thoroughly convincing investigation of nutrient cycling is difficult to make because of the complexity of living plant communities — different life-forms, different species within each and different aged plants may exhibit significant variations in concentrations, turnover rates and distribution of mineral capital between leaves, twigs, stems, buds and so on. Also, the proportions of wood to bark to foliage, and so on, may vary greatly. Nevertheless, we may achieve a fair idea overall of the amounts of minerals involved and their distributions (Figures 9.1 to 9.4). Broadly speaking, in montane rain forest the mean concentrations of the various elements are similar for different life-forms, in woody axes of like girth and in leaves, except for those of epiphytes with low nitrogen concentrations (probably due to low leaching of nitrogen from host tree foliage, when other elements are in adequate supply). The mean concentrations of the various elements in different parts (stems, leaves, twigs etc.) of the various life forms that comprise secondary successions are also broadly similar, although potassium is higher in the stems and roots of grasses, while calcium is lower in their stems and leaves (green and dead), and nitrogen too is lower in their leaves. Up to 60% of the primary forest's above-ground nutrient capital is in tree trunks, and some 20% in branches. The percentages predictably fall under secondary successions to nearer 50% in trunks and branches under *obael* woodland and a mere 4% under *gaimb* cane grassland. Up to 70% of a nutrient's above-ground capital under cane grassland is in cane stems, and 40% in cane leaves, whereas under secondary woodland up to 6% is in cane stems and leaves respectively (Table 9.4).

The mean concentrations of nitrogen, phosphorus, potassium and magnesium vary between different parts of primary forest trees as follows: trunk wood < branches < bark < twigs < leaves; for calcium the order of bark through leaves is reversed (Grubb & Edwards 1982). The percentage of the capital in leaves decreases along the series phosphorus > nitrogen > magnesium > potassium > calcium, which is noteworthy given that phosphorus and nitrogen are the two nutrients identified as most likely to be limiting growth (Edwards 1982). The order is the same for the leaves of softer wooded secondary trees, except calcium is second highest. The percentages of nitrogen and phosphorus vary between different parts of secondary regrowth trees along the series: trunk < roots < twigs < leaves; for potassium, roots < trunk < twigs < leaves; for calcium, roots < twigs < trunk < leaves; and for magnesium twigs < trunk < roots < leaves. The series for secondary grasses has the following order for nitrogen, phosphorus, calcium and magnesium: stems < roots < leaves, and for potassium the order of stems and roots reverses. Regarding mineral capital, the percentage distribution of nitrogen and phosphorus are roughly comparable with one another, as are calcium and magnesium, but potassium differs from them all.

The greater part of the living forest's mineral capital is held in the wood of large trees. This has further implications for the use of lowland forest comparisons in montane contexts, to assess the relative impact of human cultivation activities on natural resources (Jordan 1989). While the burning of cut vegetation liberates some of the locked-up nutrients for crop uptake, it releases only a small fraction of those held in the forest stand. It is incorrect in the New Guinea Highlands to suppose that forest destruction rapidly releases a large proportion of the nutrient elements stored on a site and threatens to lose them. In excess of 90% remain in the large unburnt material and below-ground roots. Some suitable trees men split up and use for fence stakes. They eventually decay. Others they leave as trunks across sites, also to decay in due time, a slower process at the lower temperatures of higher altitudes than in lowland tropical rain forest. Others they ring-bark, pollard and leave standing like gaunt skeletons, sometimes using them later as a convenient source of firewood (removing elements from the site and releasing them by burning to the atmosphere, for the rain to bring back to earth and fertilise elsewhere). In all, the release of minerals locked-up in ligneous matter is a very slow process, allowing time for new vegetative successions, whether crops, grass or trees, to exploit them.

The restricted destruction of vegetation cut down when clearing gardens applies not only to primary forest but also to secondary woodland and even to cane grassland sites. When the Wola clear these successions they do not incinerate all of the material they cut down. They leave a proportion as ground litter to rot down and slowly release its store of minerals. They destroy a greater proportion of the vegetation cleared from secondary successions than from primary forested stands, the proportion destroyed increasing as biomass falls, with secondary woodland < cane grassland < coarse grass fallow. Again, gardeners commonly remove some material to rot elsewhere; for example they frequently throw the intractable rooty stumps of cane grass clumps outside garden fence lines to get them out of the way.

Table 9.4
The percentage distribution of nutrient elements between different fractions of above-ground vegetation under various successions

	Iyshabuw Primary Forest*					Obael Secondary Woodland					Gaimb Cane Grassland				
	N	P	K	Ca	Mg	N	P	K	Ca	Mg	N	P	K	Ca	Mg
Trees (gbh > 30 cm)															
Trunks & Branches	69.0%	59.2%	75.3%	79.1%	75.2%	46.3%	39.6%	47.8%	58.0%	47.3%	1.8%	1.3%	0.8%	4.4%	1.7%
Twigs	4.7%	7.6%	4.5%	5.1%	4.0%	10.2%	15.3%	8.3%	6.9%	10.8%	0.4%	0.5%	0.1%	0.5%	0.4%
Leaves	12.0%	14.4%	6.2%	5.9%	8.7%	18.2%	19.7%	13.6%	19.2%	14.4%	0.7%	0.7%	0.2%	1.5%	0.5%
Shrubs & Saplings (inc. trees gbh < 30 cm)															
Woody axes	3.0%	3.3%	3.8%	2.2%	3.1%	4.9%	6.4%	4.3%	4.1%	5.4%	6.5%	7.1%	2.6%	10.6%	6.4%
Leaves	1.1%	1.1%	0.6%	0.5%	1.2%	1.1%	1.2%	0.8%	1.2%	0.9%	1.4%	1.3%	0.5%	2.9%	1.0%
Tree Ferns															
Trunks						1.4%	1.7%	2.8%	2.6%	4.1%	2.5%	2.5%	2.2%	9.0%	6.5%
Fronds						0.5%	1.0%	0.5%	0.2%	0.4%	0.8%	1.5%	0.4%	0.6%	0.7%
Climbers & Scramblers															
Woody axes	2.5%	3.3%	2.9%	2.8%	1.6%										
Leaves	1.8%	2.1%	1.3%	1.2%	1.1%										
Cane Grasses															
Stems						6.3%	2.7%	6.6%	1.4%	2.8%	45.4%	47.8%	72.7%	32.0%	43.4%
Leaves						5.6%	6.2%	6.4%	2.8%	6.1%	40.4%	37.3%	20.5%	38.4%	39.3%
Epiphytes	2.2%	3.3%	1.7%	1.2%	1.8%										
Grasses, Seedlings & Herbs[†]	3.7%	4.8%	3.5%	2.0%	3.1%	5.5%	6.2%	8.9%	3.6%	7.8%	<0.1%	<0.1%	<0.1%	<0.1%	<0.1%

* Data source: Grubb & Edwards 1982.
[†] Grasses, seedlings and herbs comprise 100% of plant held N, P, K, Ca and Mg under *bol* coarse grasses

The limited destruction of woody material, where the bulk of primary forest mineral capital is locked up, has considerable environmental and agronomic implications. While sites under *obael* soft-wooded secondary forest, tall *gaimb* cane grass, or coarse *bol* and other short grasses, hold less mineral capital overall in their standing vegetation, lacking the large woody stores of *iyshabuw* montane rainforest, they may nonetheless supply, in the short-term, when cleared and cultivated, comparable amounts of nutrients for crop uptake, notably of the three major nutrients of nitrogen. potassium and phosphorus which most commonly limit agricultural production (Table 9.5). In other words, the nutrient inputs derived from cutting and burning cane grass and soft-wooded trees, even from the short grasses and herbs of brief fallow intervals, may compare with the minerals released with the limited destruction and burning consequent upon the clearance of montane forest vegetation, which amounts largely to the incineration of foliage and smaller woody stuff. This corresponds in some regards with Wola assessments of differences in the fertility of soils under different vegetational communities as more often a physical matter than a nutritional one; when commenting on the cultivation potential of land under cane grass, for instance, they are more likely to refer to the dense mat of roots binding and compacting the soil firmly so that other roots have trouble penetrating it, than they are to make comments related to nutritional status.

It is not only the large amounts of nutrients stored in slowly destroyed woody material that averts a massive release and loss of minerals from sites under primary forest following their clearance, but also the large proportion of the total held in the soil's organic fraction. The slow release of minerals from organic matter is important both to the conservation of nutrients in New Guinea's montane forests and to their cycling through the eco-system. The cycle is rapid to compensate. The quantities of minerals returned yearly in litter and throughfall, expressed as a percentage of the quantities in the

Table 9.5
The estimated release of nutrients following the clearance of different vegetation successions for cultivation (kg ha^{-1})

	iyshabuw primary forest	*obael* secondary woodland	*gaimb* cane grassland	*bol* coarse grassland
Estimated biomass of fraction of above-ground vegetation destroyed in cultivation	100,000	40,000	40,000	15,000
Nutrients involved:				
N	220	287	253	170
K	215	139	276	138
P	12	19	20	14
Ca	410	168	76	44
Mg	60	33	32	46

standing forest vegetation, indicate that compared to biomass stored mineral capital, short-term nutrient cycling is fairly rapid:

Nitrogen	Phosphorus	Potassium	Calcium	Magnesium
18%	21%	15%	9%	16%

It is also probable that the forest only locks away in new wood a small percentage of the minerals that it takes up annually.

The short-term cycling rates and proportions of nutrient cations in plants in Papua New Guinea's montane forests are within the lowland forest range. It is unlikely that calcium or magnesium are ever deficient in these montane forests, the exchangeable quantities of these cations in soils being far in excess of those cycled annually. The amount of exchangeable potassium held in soil relative to that cycled in a year is not so large, but the rapid and tight cycling of this cation probably ensures its adequate supply. (The low concentration of potassium in stream water, compared to the considerable calcium and magnesium leaching losses detected, is further evidence of the efficiency of the potassium cycle — Turvey 1974; Edwards 1982).

It is nitrogen and phosphorus nutrient supplies that most likely limit forest growth. The short-term cycling rates of these nutrients are high relative to the amounts held in plants, the smallness of the biomass share indicating that long-term cycles are slower than in tropical lowland forests. The withdrawal of appreciable amounts of nitrogen and phosphorus from leaves before abcission might be further evidence of shortages (Edwards & Grubb 1982). And while the total quantities of nitrogen and phosphorus in the soil are vastly larger than those in plants, only small proportions of these reserves are in readily plant-available forms. This is attributable to soil organic matter complexing of phyllosphere fixed nitrogen and the fierce fixation of phosphate by soils containing volcanic ash. It is difficult to assess the extent to which shortages of these nutrients limit forest growth. The tiny proportion of nitrogen cycled for example, and its low mean concentration in vegetation compared to lower altitude forests, might have a functional explanation (i.e. it may result from biological factors unrelated to nitrogen availability, such as montane vegetation having thicker cell walls) or it might be evolutionary (i.e. an adaptation to low nitrogen supply). Whatever, total soil nitrogen levels may have built up through continued accumulation to such a large degree that although only a very small percentage is available annually, this is sufficient to ensure a non-limiting nitrogen supply. The follow-on regarding crop cultivation and secondary successions is that while nitrogen supplies may remain for a long time more or less adequate, phosphorus is likely to be limiting, and potassium too with disturbance of the forest's tight cycling regime. The agronomic implications are intriguing and inform the next Section's efforts to 'unearth' the foundations of the not-so-shifting Wola agricultural regime's sustainability.

II. DECOMPOSITION

Decomposition is a crucial part of any nutrient cycle, whatever the vegetation community. It releases the mineral capital locked up in dead materials for cycling again, making it available for incorporation once more into complex organic molecules. The chain of decay and decomposition involves many complex interactions between a wide range of organisms, both inside and outside the soil, going from tiny microbes upwards. These are responsible for the breakdown of both vegetative and animal material remains, their incorporation as organic matter in the soil's humus fraction, and finally through mineralisation the release of their nutrient elements. They comprise a complex technical field of research not readily accessible to non-experts. Even the preliminary step of comprehending the classification schemes used to order the vast array of organisms involved is not easy. Different approaches employ divergent criteria with a barrage of associated technical terms.[5] A large part of this complex technical terminology applies to soil microbiology, where current research on decomposition concentrates and, while crucial to decay processes and mineral cycling in the New Guinea Highlands as anywhere, it is largely beyond the point in an ethno-environmental discussion of decomposition. The traditional soil lore of the Wola is, so far as I am aware, largely oblivious to the existence and activities of microorganisms. Their knowledge is restricted to what the naked eye can see: the invertebrate mesofauna and vertebrate macrofauna, together with some saprophytes, notably those that occasionally produce prominent fruiting bodies.

Fungi and Microflora

The Wola classify fungi into a single named *sem* 'family' or primary taxon called *sez* (see Chapter 7). They recognise over fifty named different types of fungi within it, equivalent to the genus/species level of Western mycology; sometimes they distinguish varieties of these, notably of the common, consumable fungi.[6] In any Wola list of names, the more common fungi occur first,

---

[5] One may, for example, classify organisms according to their size (e.g. microbiota, mesobiota and macrobiota), variations in their nutritional strategies (e.g. chemoheterotrophs, chemoautotrophs, photohetereotrophes and photoautotrophs), their functional ecology (e.g. epigeic, anecic or endogeic species), their different strategies for liberating energy from organic substrates (e.g. obligate aerobes, facultative anaerobes and obligate anaerobes), or one may use purely technical laboratory-based discriminations (e.g. gram staining for differentiating between bacteria). Regarding the Protista group of organisms (bacteria, algae, protozoa and fungi) the boundary between animals and plants even becomes confusingly blurred.

[6] For assistance with identifications of organisms discussed in this chapter I am grateful to Steve Frankel for advice and Kew Herbarium for arranging identifications of fungi; the staff at the British Museum (Natural History), Papua New Guinea Department of Primary Industry (Entomology Section), the Royal Museum of Scotland (Edinburgh), and National Museum of Victoria (Melbourne), Val Standen of Durham University Zoology Department, John Murphy of Hampton and Bill Dolling of Hull for assistance in identifying invertebrate specimens; and Jim Menzies of the University of Papua New Guinea and Brian Coates of Brisbane for vertebrate identifications.

particularly those eaten like the *muwnaen* varieties, which come top. Fungi occur more in some regions than others. According to the Wola a wide variety occur where *pay* chestnut oaks (*Castanopsis acuminatissima*) are abundant, and people who live at such places have a considerable knowledge of many different fungi. Those living in the Was valley, for instance, where there are relatively few chestnut oaks, point to the Nembi valley where there are many more and say that those living there are aware of fungi unknown to them.[7]

According to the Wola, the ideal conditions for *sez* fungi growth are a period of fine and dry weather followed by a wet spell. The fine interval they say causes soil and rotten wood to dry out and crack, and when these crevices fill with rainwater they supply the favoured environment of fungi, affording hyphae the opportunity to sprout forth fruiting bodies, which grow rapidly. They recognise five stages to the fruiting cycle of fungi:

---

1. *iysh lay* (lit. tree hit) = the small protuberances evident as fungi start to emerge.

    ↓

2. *hiyn bay* = the fruiting bodies are small and immature.

    ↓

3. *bat say* (lit. wide become) or *baka biyay* = the fruiting bodies are large, tender and collectable if edible.

    ↓

4. *kat hay* = the fruiting bodies are mature and tough, although still edible.

    ↓

5. *shor domay* or *shor pung biy* (lit. leaf rotten does) = the fruiting bodies are old and rotting, sometimes infested by *sez-hiym* insect maggot larvae (edible types are leathery, likely to induce vomiting if eaten).

Several kinds of *sez* fungi are edible and collected, particularly when flushes occur after a dry period followed by a wet spell. The Wola always cook them before eating, by baking them in hot embers, steaming them in earth ovens, simmering them in bamboo internode tubes or today boiling in pots. If they were to eat them raw they say that they would vomit and have stomach pains.

---

[7] When we collected fungi for identification, I noticed that those helping me frequently turned for assistance to women married into the Was valley from the Nembi region.

When a fungi flush occurs, as opposed to the occasional fruiting body occurring here and there in forest or regrowth, the different species appear in a rough sort of sequence which, for the commoner fungi, approximates to *hert* (*Russula* sp.) ripening early, followed by *aelgit* (*Lentinus araucariae*), *naen* (*Pleutrotus djamor*) and *mongowshuwt* (*Polyporus arcularius*), and then the abundant bracket fungus *muwnaen* (*Grifola frondosa*) which comprises a goodly part of any harvest. The Wola acknowledge that some fungi favour particular locations. The large, stout *hulba* (*Boletus erythropus*) and *haeriypaend* (*Pholiota* sp.) fungi, for example, occur around the base of *pay* chestnut trees (*Castanopsis acuminatissima*), whereas the *dimbul* (*Strobilmyces velutipes*) and *elkondiyt* (*Russula* sp.) fungi favour both chestnuts and *kaeriyl* oaks (*Lithocarpus rufovillosus*), and the *bortngaelngael* fungus (*Pycnoporus sanguineus*) favours the dead wood of *bort* silkwoods (*Cryptocarya laevigata*) in sunny locations, although it is common also on *aenk* pandans (*Pandanus julianettii*), *saemow* mimosa (*Albizia fulva*) and *waen* woolly cedars (*Trema orientalis*). The names given to some fungi signal something about them, like the names given to some other plants (see Chapter 7), for instance their preference for certain trees, such as *bortngaelngael* (*bort* silkwood), *kaeriylpak* (*kaeriyl* oak), and *waenhael* (*waen* woolly cedar). Other names reflect something about the appearance, smell or taste of the fungus; *mongowshuwt* for instance, refers to the navel and alludes to the dimple in the centre of the cap, *shiyortombor* (*Lentinus umbrinus*) means 'cassowary-stomach', and *hael* simply 'ear', after the likeness of some bracket fungi to that organ. And the name *bordorwiy* (*Pholiota austrospumosa*), meaning literally 'dry-burn-is', celebrates the occurrence of flushes of this fungus after the burning off of cane grass, and *nuwpiriy* (*Collybia* sp.), literally 'old-bag', alludes to the resemblance between the tatty gills of this fungus and a holey old string bag.

Diseases and Pests

Fungal parasites, together with bacteria and viruses, are responsible for much plant disease, and reductions in crop yields (Pearson 1982). The Wola do not associate diseases with *sez* fungi. They recognise certain gross categories of pathogen (compare Table 9.6 which lists some of the common disease causing pathogens reported for the Southern Highlands Province [after Waller 1984]). They call stem rots on living plants *maend bay*, including those which cause entire adult plants to topple over, and also any damping-off-like diseases responsible for the wilting and death of young plants. These blights of living plants contrast with the rotting of dead vegetation, the general term for which is *pung biy* or *domay*, particularly when it occurs with an offensive smell. The rotting of a senescent standing plant, usually with drying and browning of foliage is *gaip say* (lit. dry becomes).

The greyish-white powdery mildews and bacterial moulds of various colours the Wola refer to as *huwguwmb daebiliy*. Leaf spot necroses, galls and rusts are *kond lay* (lit. scars hit), after their resemblance to human decorative body scarification, and leaves deformed with folds and crinkles due insect

Table 9.6
Common crop pathogens and diseases of the Southern Highlands Province
[after Waller 1984]

Crop	Common Name	Genus & Species
Sweet potato	scab	*Elsinoe batatas*
	dieback & stem rot	*Phomopsis ipomoeae*
	tuber rots	*Ceratocystis fimbricata*
		Fusarium solani
		Cylindrocarpon destructans
Taro	root rot	*Pythium sp.*
		Phytophthora sp.
	leaf spot	*Phyllosticta colocasiae-esculentae*
Ginger	white leaf spot	*Phyllosticta zingiberi*
Sugar cane	ring spot	*Leptosphaeria sacchari*
	leaf stripe	*Ramulispora sacchari*
Highland pitpit	tar spot	*Phyllachora setariicola*
	leaf blight	*Cochliobolus setariae*
Maize	leaf blight	*Septosphaeria turcica*
	rust	*Puccinia sorghi*
Beans	anthracnose	*Colletotrichum spp.*
	web blight	*Sclerotium sp.*
	angular leaf spot	*Phaeoisariopsis griseola*
	white leaf spot	*Ramularia phaseoli*
Winged bean	leaf spot	*Cercospora kikuchii*
Cucurbits	downy mildew	*Pseudoperonospora cubensis*
	leaf blight	*Didymella bryoniae*
	anthracnose	*Colletotrichum lagenarium*
Crucifer spinach	leaf spot	*Cercospora nasturtii*
Amaranth spinach	leaf spot	*Leptosphaerulina trifolii*
Hibiscus spinach	leaf spot	*Cercospora malayensis*
Cabbage	black rot	*Xanthomonas campestris*
	black ring spot	*Mycosphaerella sp.*
Bananas	sigatoka	*Mycosphaerella musicola*
	freckle	*Guignardia musae*
	anthracnose	*Colletotrichum musae*
Pandan screw-pine	leaf spot	*Leptosphaeria ? obtusispora*
Cordyline	anthracnose	*Colletotrichum sp.*

attack and disease are *hunduwm menay* (lit. folds carry), like the folds women put into their sedge skirts (Sillitoe 1988). The yellowing of leaves prior to abscission, both due to disease and senescence, is *shor baebray*, and a leaf fretted by the grazing of insects is *deret biy*. The Wola also recognise and name some tuber and fruit diseases, like *dimbuw* for any watery softening rot of tubers (also called *sinjbiliy* in taro). A similar condition in fruits is *tag day*, which also refers to over ripe and squashy, though possibly still edible, fruits. A totally rotten fruit or tuber is *maend bay*, like any rot infected plant. Discoloured browny rot marks in sweet potato flesh are *kol* or *haeriy* and similar blemishes in taro corms are *hezaembuwl*. Black scab on sweet potato tubers is *hokay haerok* (lit. sweet-potato soot) and on taro corms is *ma hok*.

When they discuss the above conditions the Wola may talk about them as sicknesses of plants, and when asked why they occur to some plants and not to others they may draw an explicit parallel with human sickness, pointing out that the reason why some people fall sick while others remain healthy is not clear to them. Just as their ancestors spirits and forest demons act capriciously (see Chapter 8), so do whatever forces are responsible for plant disease. They are not aware what these forces are and seem to lack any systematic explanation of plant disease, although they know that if a plant is injured, for instance through insect attack, then infection and disease are more likely to occur, likewise if infected material is introduced to healthy plants.

Pathogenic microflora cause some disease but serious outbreaks seldom occur. The majority of local crop pathogens are widely distributed across the region, but several major ones of recently introduced crops are currently absent from the Southern Highlands (Shaw 1984). While a range of pathogens occur on traditional food crops (Table 9.6), they are infrequently severe, and are rarely cited locally as a reason for abandoning gardens, even those under semi-continuous cultivation for lengthly periods, where their build-up might be anticipated to be a problem. Nonethless, the long-term cultivation of some sites undoubtedly encourages the build-up of soil borne diseases, somewhat limiting their productivity, though within locally acceptable bounds. And the vegetative propagation of many crops (Sillitoe 1983) results in the accumulation and transmission of pathogens, notably viruses, in planting stock.

Parasitic microflora doubtless act, with soil and other biotic factors, to constrain productivity; poor soil conditions, weed competition and insect pests will interact with pathogens to hold down crop yields. The scattered location of gardens, and the intercropping of diverse crops nevertheless prevents airborne disease build-up to epidemic proportions. And over the generations, the Wola have selected resistant strains of traditional crops according to their performance under natural selective pressures, choosing those that can tolerate local pathogens and pests, thus ensuring yield stability, albeit at lower levels than achievable with selectively bred, higher yielding but more vulnerable cultivars given elaborate protection (Thurston 1991). Crop protection measures are otherwise few.

Damage by animal pests increases the vulnerability of plants to attack by certain pathogens, opening them up to infection. Pests can also directly limit crop productivity, sometimes seriously. Small rats (*Rattus* spp.) are sometimes a particular nuisance,[8] and gardeners take their threat seriously. They believe that if someone who has recently eaten meat looks at a taro garden it will suffer serious depredation, and some men go to considerable lengths to protect their crop, erecting large screens of leafy cane grass around their gardens to hide them from view. Other protection measures they take, called *honez tol deraey* (lit. rats lead outside), involve scaring the animals off. They may spread prickly fronds of *maen* hoop-pine (*Araucaria cunninghamii*) or stinging *niysh* nettle (*Laportea decumana*) around a garden ravaged by rats, the pricks and stings they receive to their noses they say make them leave the crop alone.

---

[8] Sillitoe 1983:223 illustrates how extensively rats can damage a garden, reducing crop yields in one case to less than 13% of the average.

Some birds can also be pests and seriously damage crops, like the *naway* Sulphur-crested Cockatoo (*Cacatua galerita*), *waliyn* Musschenbroek's Lorikeet (*Neopsittacus musschenbroekii*), and *hongol* Plain-breasted Little Parrot (*Psittacella madaraszi*), and one tactic to scare them is to suspend stout pieces of *paym* (*Pandanus antaresensis*) pandan leaf by vine, streamer-like from sticks driven in around a garden where they blow around in the wind, the fluttering movement people say scares birds which think that humans are moving around the garden. The *tagem* fruit bat (*Dobsonia moluccensis*) and various members of the parrot family also feed on ripening bananas and to prevent their depredations men commonly wrap the fruiting stalk up (strong polythene bags are popular for this today, where they previously used stout leaves).

Insects can also reduce yields markedly (Gagné 1982; Smith & Thistleton 1982). The *twenj* chafer beetle (*Papuana* sp.) for example bores into sweet potato tubers and taro corms, causing considerable local damage.[9] The *bombok* mole-cricket (*Gryllotalpa* sp.) and *mol* cricket (*Gymnogryllus angustus*) also damage sweet potato tubers. The *bombort* beetle larvae (*Olethrinus tyrannus; Arrhenotus* sp.) sometimes infest and consume sugar cane stems, and also the branches and aerial roots of pandans. The principal pandan insect pest is the *kobasond* caterpillar which eats into nut clusters, doing serious damage. Another less serious pandan pest is the *aenklet* bush-cricket (Mecopodinae), which eats crown leaves and nut spathes. The *eraygay* grasshopper (*Oxya japonica*) is a voracious sap-sucker which feeds on the leaves of many crops, sometimes opening them up to disease infection. The small *ed* caterpillar (*Pieris* sp.) attacks crucifers, amaranths and brassicas, women sometimes spreading *henk* tree-fern (*Cyathea* spp.) fronds over crops to reduce their depredations (according to them it is the bitter taste of the *henk-huwga* or spores on the underside of the fern fronds which deters the caterpillars). Another notorious pest of these crops is the larger *shomb* caterpillar (Pyralidae) which not only eats the leaves but also nibbles through the stems of plants, causing them to fall over, resulting in substantial crop losses where their numbers are high. The *ogimb-honaelok* wasp (*Heterodontonyx bicolor*) is a pest of ripening bananas, it 'spears' young green fruits, after which they become infested with *hiym* or maggots and rot, a small black area on the banana skin is evidence of their egg-laying, called *diyr waerak bombay* (lit. banana slime fill-up).

Nematodes too are reported a potentially serious crop pest throughout the Southern Highlands region (Bridge & Page 1982), causing root galls and necrosis, cracked and constricted tubers, and leaf chlorosis. The Wola refer to these symptoms of nematode attack by the same names they give to the similar disease symptoms listed above; cracks in tubers they call *kas* and constricted tubers *kunguwp*. Again traditional agronomic practices are effective at controlling these and other insect pests. Grassy fallow or cane and tree

---

[9] The Wola also call tropical ulcers *twenj*, which likewise bore into the flesh of human-beings via infected wounds.

regrowth give rise to plant communities that feature few crop pests. And long-term selection of tolerant cultivars keeps any infestations to tolerable levels, though the price of their more stable yields may be lower returns than optimally obtainable with selected less tolerant cultivars — so long as these remain uninfested (Bridge & Page 1982:43 report that some Southern Highlands cultivars are particularly poor hosts to local nematode races).

Animal Decomposers

Animals also play a significant part in the decomposition stage of any ecosystem cycle, and are particularly active in the soil and litter layer. It is probable that the activities of many animals, human-beings included, affect these terrestrial aspects of ecosystems in one way or another. The problem is drawing a line between those considered significant and others. It is customary to include in the soil fauna proper only those creatures which are permanently part of the soil ecosystem, together with those that regularly spend some active phase of their life-cycle in the soil or surface litter.

Some macrofauna noticeably affect soil and litter. Megapodes locally have a marked effect on the litter layer, building large mounds of heat-generating, decaying vegetation to incubate their eggs in. Various rats burrow into litter and soil, and build nests there, as do the bandicoot, silky cuscus and echidna. These animals also root around on the floor for food, the echidna turning over considerable areas with its long bony snout, in search of earthworms according to the Wola. The pig consumes earthworms and other soil organisms too; the Wola referring to earthworms as *pil* in the context of pig feeding. Locally around homesteads pigs can turn soil over to considerable depths in their search for earthworms, having a marked effect on soil structure and susceptibility to erosion (Humphreys 1984).

While some invertebrates spend little time on the ground, others reside there for goodly parts of their life cycles, sometimes apparently unknown to the Wola who have a hazy knowledge of how some grubs relate to their imagos. Some creatures are prominent in organic matter recycling elsewhere in the ecosystem other than in the soil; for example, some of the worms found in the soil-like debris that accumulates in the forks and crevices of trees supporting epiphytic and other plants play an infrequently noticed part, as do the larvae that inhabit rotting wood, reducing ligneous material and facilitating access by other decomposers. Those invertebrates that have little or no direct effect on the decay chain have a role in nutrient cycling elsewhere in the wider ecosystem, for example destroying foliage and fruits on plants. And many others make irregular and casual visits to litter or soil and have marked localised effects on the ecosystem (e.g. some insects, arachnids and myriapods).

The Wola classify invertebrates and reptiles into a single *sem* 'family' or primary taxon called *elelbiy* (lit. round-round-do, after the way in which many insects ceaselessly scuttle and fly about, 'round and round'); it is taxonomically equivalent to their upper level botanical classes like *iysh* and *ya* (see Chapter 7). Below this level occur the specific names they give to particular organisms.[10] Several of these cover sundry genera and species; they parallel English folk classes like butterfly and spider, the Wola equivalents of which are *hedbabuwk* and, for most arachnids, *aegolaend* (lit. walk-around-house, after the spinning of the web). On the other hand, many creatures they do not name at all, for example various beetles and larvae; they are all simply *elelbiy*. The Wola evince relatively little interest generally in insects. In comparison to the natural variety of invertebrate life, their culture has not elaborated greatly on its classification, which may be taken perhaps as further evidence of the commonsense observation that people tend to classify extensively only natural phenomena which feature prominently in their lives as food, pests, symbols or whatever (Hunn 1982; Brown 1984), which applies to few insects.

The Wola are fully aware that some of their names for insects cover several different organisms. They point out, for example, that *bombort* grubs are not all the same but vary from one tree wood to another. The hard woods like *pel* beech (*Nothofagus* spp.) and *kaeriyl* oak (*Lithocarpus* spp.) have large grubs that are firm and chewy to eat, whereas the softer wooded trees like *waen* woolly cedar (*Trema orientalis*), *hog* cola (*Sterculia* sp.) and *bort* silkwood (*Cryptocarya laevigata*) have smaller and more tender grubs. They sometimes distinguish between grubs, not according to their future adult form, but the name of the tree in which they occur, added as a prefix; for example *aenk-bombort* are grubs found in *aenk* pandan pith (*Pandanus julianettii*).

In contrast to this lumping together of different creatures under one name, the Wola occasionally give some organisms two separate names, one for the larval stage and the other for the adult, just as in English folk biology; for example, a *kaengap* grub which pupated in my house hatched into a *himakay* beetle (*Lamprima adolphinae*). They are fully cognizant of invertebrate metamorphosis, using different terms for the different stages observed in various creatures. The sequence for foliage-eating larvae is:

1. *mind* = caterpillar larval stage

    ↓

2. *saybol* = comatose larval stage (not all species they say go through this larval phase)

    ↓

3. *keshond* = chrysalis stage, which may also feature in some species a *kopenaend* = cocoon around the pupa

    ↓

4. *hedbabuwk* = the adult moth or butterfly imago.

---

[10] For a detailed catalogue of the insects and reptiles identified by the Wola, see *Science in New Guinea* Vol. 21.

The sequence for dead wood consuming grubs is:

1. *bombort* = the legless grub stage, a creature with powerful wood chewing mandibles

   ↓

2. *hogomb* = the quiescent pupal instar stage when the larva swells with *hobor* fat

   ↓

3. *wetwet* = the plump holometabola bursts its skin, having metamorhosed into a legged grub, which at this stage is very soft, white and immobile

   ↓

4. *bombort-injiy*[11] = the adult legged beetle stage, that develops from the pupated instar

The sequence for insects laying their eggs in carrion, some fruits, or fungi is:

1. *hiym* = the maggot larvae stage

   ↓

2. *keshond* = the chrysalis stage

   ↓

3. *pungun* fly or *aeretjimbiyziy* cranefly = the adult insect stage or *ogimb* wasp etc.

---

[11] The suffix *injiy* means literally 'mother', for example the general term for pigs is *showmay*, whereas a sow that has borne piglets is a *showmay-injiy*.

And the metamorphosis sequence for psychids is:

1. *saembuwt* = the small grub stage
   ↓
2. *saembuwt-injiy* = the adult grub stage, the black & yellow grub bearing, snail-like, a spiky silky cocoon
   ↓
3. *keshond* = the pupa stage, when the grub metamorphoses through a chrysalis inside its silken bag[12]
   ↓
4. *hedbabuwk* = the emergent adult moth or butterfly stage.[13]

Enquiries into the hatching of insect pupae aroused the interest of my Wola friends who said that they did not know beforehand what creatures would emerge. It transpired that they frequently are unable to specify what the adult appearance of a larva will be, for example the size and colouring of a particular caterpillar's imago. This is understandable in part, in a culture that uses names of wide reference for both life stages — for instance, most caterpillars are called *mind* and moths/butterflies are *hedbabuwk* — for this reduces the scope for cultural accumulation of observations over generations and the detailed transmission of such knowledge. It seems that there is an element of chance to the Wola mind what kind of butterfly or moth will emerge from any pupa. They maintain that for the larger ones it depends on what kind of pig's *wezow* (lit. shadow-cum-life-force) has entered and animates the creature, believing that the shadow-cum-life-force of slaughtered pigs inhabit lepidoptera; a dark moth or butterfly, for example, would previously have been a black-bristled pig, a buff one a brown pig, a large one a big tusker, and so on (those who saw a large *Coscinocera anteus* moth hatched in my house immediately identified it as a prized brown-bristled tusker). Another name for large butterflies and moths reflects these ideas, it is *showmay wezow* (lit. pig shadow-cum-life-force). These short-lived and sometimes beautiful pigs' shadows are a little awesome, although people are not afraid of them, unlike earthworms.

Soil Fauna

According to the Wola earthworms are not harmless creatures that merely croak, as the Kalam of the Western Highlands believe (Bulmer 1968), they are

---

[12] Only males pupate, females die as grubs.
[13] Informants denied that *saembuwt* psychids metamorphose into *saembuwt* stagbeetles, underlining that terminological congruence between larval and adult stages is irrelevant.

harmful animals that bite. And their bite, while not felt like that of a dog, causes aching pains. They describe the bite as *heb nay*, meaning a brushing together experience; this contrasts with an *aezay* bite which spears the flesh. It is not unusual to see a woman nursing a painful and sometimes swollen knee, which may even be infected with a *wiyp* boil, and attributing it to a bite received from a worm while kneeling in a garden heaping up sweet potato mounds, work that women commonly engage in. They are adamant that it is brushing encounters with earthworms, which they call collectively *gogay* (*Amynthas* spp., *Metapheretima* spp. among others), that cause their discomfort, and a European holding a large worm in his hand for many minutes with no discomfiture nor pain does nothing to disprove their belief — this merely proves what they already know, that Europeans have power to protect themselves from ailments that afflict them (like malaria) and leads to demands for some of the appropriate medicine to alleviate their worm-induced suffering (the attention that Western medicine gives to intestinal worms, distributing medicine to purge them, reinforces this attitude).

Since learning about Wola fear of earthworms I have periodically racked my brains for some 'explanation' of it, symbolic or otherwise, but have largely drawn a blank. They point to their strange appearance: not only no legs, hair or feathers, but no eyes, nose or even mouth, and yet they 'bite', but mouthless without being felt, and they apparently eat nothing but earth. The similarity with snakes might be taken as a possible clue to their earthworm fears, for the Wola believe that their ancestors' spirits, which can cause them sickness and death, manifest themselves as snakes. Yet they kill, cook and eat large pythons (*Liasis* spp.), but believe that if a worm merely crawls over some food the individual eating it will fall ill. The physical similarity between earthworms and some human parasitic worms might be thought another possible reason for their vermi-phobia, these intestinal parasites causing sickness and even death sometimes according to the Wola (worms crawling out of the orifices of corpses on occasion during the customary three to four day mourning period, and sometimes even out of the mouths and noses of the living, as well as being passed in faeces), but those I have spoken to deny that their fear of earthworms is because they resemble these horrible intestinal worms. Nor is it that living in the earth, a dark subterranean existence, necessarily makes earthworms fearsome, for there are several other soil dwelling animals that are not feared, some of which are edible; they include the *bombok* mole-cricket (*Gryllotalpa* sp.), the *mol* cricket (*Gymnogryllus angustus*), *kaezuwmb* cricket (*Gryllus bimaculatus*), *kaengap* stagbeetle (*Lamprima adolphinae*), among others.

In the context of an agro-environmental study, it is noteworthy that the Wola do not consider the soil's *gogay* worm population significant to its productivity,[14] unlike us. We hold it as common knowledge that worms are evidence of a good soil and promote fertility (Lee 1983). They are thought to contribute to decomposition processes by comminuting and mixing plant

---

[14] The neighbouring Huli think the same (Rose & Wood 1980).

litter and soil in the gut, and also by affecting the decomposing activity of microflora in the soil cast. By consuming organic matter, together with some mineral fraction, they assist breakdown, producing an argillaceous humus.[15] They also loosen the soil with their burrows, so improving its structure and promoting aeration and drainage (Humphreys 1984:33–37 considers earthworms the most dominant of soil moving fauna in the Highlands, their activities promoting soil conservation).

But the Wola do not believe that earthworms improve fertility, and they do not associate an abundance of them with a good and productive soil. And they may have a point when they insist that earthworms have nothing to do with soil productivity. An investigation of sweet potato tuber yields and earthworm populations revealed something of a negative trend (Figure 9.5), with yields tending to decline as the mass of worms increased, though not significantly (a similar trend was evidenced when yields were plotted against overall worm numbers).[16] Evidence from elsewhere corroborates these findings. In the neighbouring Tari basin significant negative correlations have been reported between the size of earthworm populations and sweet potato yields (Rose and Wood 1980); and the highest rates of earthworm casting in Simbu occurred on soils (Hydrandepts) with the lowest sweet potato yields (Humphreys 1984).

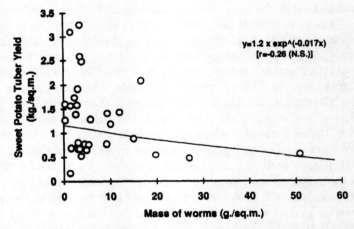

Figure 9.5 The mass of earthworms in the soil compared with sweet potato tuber yield.

---

[15] A population of 20 worms m^{-2} passing a staggering 50 kg m^{-2} y^{-1} through their bodies under sparsely wooded savanna on the Ivory Coast (Lavelle 1974).

[16] The method used to collect these data was to excavate completely single sweet potato mounds in different gardens (previously measured to calculate the areas involved). We weighed all of the sweet potato tubers dug up, and then sieved the entire mound to recover the earthworms in it for weighing and counting. The number of mounds in the sample = 31, selected at random by casting stones in those gardens where women cultivators agreed to allow me to excavate a mound. I thank Mayka Sal, Mayka Haebay and Wenja Neleb for overcoming their aversion to earthworms sufficiently to assist me in this to them eccentric task.

The implication is not that earthworms reduce yields, but that soil conditions which favour high earthworm populations are likely to be less favourable to sweet potato production. According to men with whom I have discussed this, the number of earthworms is likely to be higher when they clear a new garden than later because the soil they say has not dried out. Yields of sweet potato are likely to increase with subsequent cultivations of the site because exposure to the sun dries the soil out, also lowering the worm population. The implicaton is that the presence of earthworms in large numbers indicates a poor sweet potato soil because these creatures favour wetter locations not well suited to the crop, wetter soils not physiologically favouring tuberisation (see Chapter 4). It is also possible that worms favour places where plants grow less well, the plentiful rotting vegetation ensuring them a good supply of the food. In addition to soil moisture content, worm populations also correlate with soil acidity (Tuneera-Bhadauria et al. 1991), and it is possible that pH conditions which favour worms reduce the availability of certain nutrients, notably phosphorus, which could be particularly problematic in phosphate fixing soils, taking levels below even sweet potato's tolerance (see Chapter 12).

According to the Wola the presence of recently arrived foreign earthworms is evidence of particularly poor soils. They call these worms *senis gogay*, after the Pidgin word *senisim* for 'change' because when they enter an area the indigenous species vacate it and these new arrivals replace or 'change places' with them. They maintain that these earthworms, which started to arrive in any numbers along tracks in the Was valley in the early 1980s, signal hard and dry and 'strong' soils which sweet potato tubers cannot easily penetrate. They are fortunately restricted to a few locations currently, favouring flatter areas and the edges of paths. The positive side to their presence is that pigs relish them and very quickly grow big if fed on them, with a thick layer of subcutaneous fat, although in the process they turn the soil over heavily, possibly reducing its agricultural potential further. It is easy to identify areas that these recently arrived foreign worms have invaded because only these worms throw up casts, which the Wola call *gogay iy* (lit. worm excreta). Wormcasts may be relatively more nutrient-rich (as Humphreys' 1984:38 chemical analyses of Simbu casts shows) but the implication is not that earthworms necessarily increase nutrient supply rates, improving soil fertility. The casts of earthworms, comprising fine argillaceous material heaped on the soil surface, may also influence microrelief and susceptibility to erosion (see Chapter 6), their mucus binding soil particles and affecting aggregation (Lee 1967; Humphreys 1984:82). And their burrowing as mentioned, may also improve infiltration rates, although again the Wola maintain that to their knowledge neither casts nor burrows have any apparent effect on soil relief or loss.

It is difficult to make a reliable scientific estimate of soil fauna populations responsible for a deal of decomposer activity, although it is possible to give some indication for the Wola area. During other work on soils reported in this book, I noted soil animals seen, and also excavated a metre cube pit specifically to recover and count organisms (Table 9.7). These data compare

Table 9.7

Soil fauna population densities: organisms observed during (1) the digging of 85 soil profile pits (to ~50 cm depth, no systematic searching or sieving attempted); (2) the sieving of soil from 31 sweet potato mounds (to ~50 cm depth, total area excavated 107.43 m²); and (3) the excavation of a one cubic metre pit (numbers of organisms markedly decreasing at 23 cm, the A and AB horizons boundary, and no organisms evident from 28 cm, the AB and B horizons boundary)

ORGANISMS		(1) SOIL PROFILE PITS		(2) SWEET POTATO MOUNDS		(3) 1 M³ PIT
		Number Counted	No. Pits Seen In	Number Counted	Number Per sq. m.	Number Counted
None seen			16			
Earthworms	all sizes & fams.	90	41			
(Megascolecidae)	large			165	1.54	13
	medium			706	6.57	16
	small			532	4.95	71
(*Metapheretima annulata*)	medium					20
	small					3
(Glossocolecidae)	large			30	0.28	
	medium			675	6.28	
	small			71	0.66	
Larvae: beetle (Lucanidae)		9	6	109	1.02	13
(Elateridae)						5
moth (Hepialidae)		1	1	16	0.15	1
cranefly (Tipulidae)						2
cicada (Cicadoidea nymph) indet.		13	7	6	0.06	1
Pupae: beetle (Lucanidae)				1	0.01	~44
cicada (Cicadidae)		1	1	5	0.05	
weevil (indet.)		3	3	34	0.32	2
Mole-crickets (Gryllotalpidae)		1	1	2	0.02	
Crickets (Gryllidae: *Gymnogryllus* sp.))				1	0.01	
(Gryllidae: *Gryllus* sp.)						

Table 9.7
Continued

ORGANISMS	(1) SOIL PROFILE PITS		(2) SWEET POTATO MOUNDS		(3) 1 M³ PIT
	Number Counted	No. Pits Seen In	Number Counted	Number Per sq. m.	Number Counted
Beetles (Scarabaeidae)			3	0.03	
(Pythidae)			39	0.36	
(Carabidae)			5	0.05	7
(Elateridae)			8	0.08	
(indet.)	6	6			11
Bugs (Gelastocoridae)			15	0.14	3
(Pentatomidae)					1
Earwigs (Dermaptera)					14
Ants (Formicidae) with nests		27	not counted		~120
Centipedes (Chilopoda) large	7	7	21	0.20	11
small		6	47	0.44	
Millipedes (Diplopoda)			5	0.05	1
Spiders (Lycosidae) (Barychelidae)	2	2	1	0.01	
Rat (Muridae) burrow		1			1

with those from elsewhere. The density of earthworms reported under montane rainforest in the Bismarck Ranges of the Eastern Highlands region, for instance, ranged from 68 to 120 worms per cubic metre of soil; although numerous, the worms were confined largely to the top 25 centimetres (Edwards & Grubb 1982). Earthworms are reported to be highly abundant throughout the adjacent Simbu region, at 35 to 83 grams fresh weight per metre square, and cicada nymphs moderately abundant at 1.2 grams (Humphreys 1984). And earthworm cast observations there suggest annual casting rates of around 200 to 700 grams per metre square, which are low for the tropics, although this is expectable if the worms responsible for casting are fairly new arrivals as they are in the Wola region. In the neighbouring Tari basin, on wet peaty soils particularly favourable to earthworms, populations range from 93 to 302 worms per square metre, yielding 47 to 128 grams fresh weight (Rose & Wood 1980). And nematode populations in roots range from low to very high in the Tari region, at 12 to 290 mature females per gram (Bridge & Page 1982).

SECTION V

THE SOIL

Suw hemem ebayda wiy ngo tomb hombuwnja bokwa ebay ora popokor. Hokay diy diyr diy wol diy ma diy komb taguwt diy hombuwnja ebay aend sokor. Suw hundbiy na wiy, iysh pila na hae, suw iyba biriy ebay, obuw suw kolkol bay ebayda. Suw taebowgiy kor ora, obuw suw hundbiy waes araytol onduwp ora, suw kor ora. Suw pondiyp tay kor ora. Suw hul buriyda nay osisor bokwa diy ebay na popokor.

Where the most fertile soil is everything I plant will grow well. Sweet potato, bananas, sugar cane, taro and greens, everything will flourish. It has no clay subsoil [near the surface], no tree roots, it has plenty of 'grease' and it becomes favourably friable. Infertile soil is very bad, it has clayey subsoil [mixed with thin topsoil] and very many stones, it is very poor soil. Soil supporting ground orchids is particularly bad. The peds are strong and any plants that I plant will not grow well.

Waebis Homalow

CHAPTER 10

ETHNOPEDOLOGY: THE SOILS

The soils of the New Guinea highlands, like those in any region of the world, result from complex and ongoing processes. The foregoing environmental factors of climate, geology, topography and vegetation all contribute to them, interacting over time to produce today's soils. The time that these soil forming processes have been active is relatively short in pedological terms. Regarding the region's soils this is noteworthy: geologically and topographically youthful, it consequently has young soils.

Inceptisols dominate the Wola region. These are soils with a cambic horizon but few diagnostic features, an absence of extreme weathering and no accumulation of clay, nor Fe or Al oxides in horizons. They are the embryonic soils of humid regions, thought to develop quickly, due largely to alteration of parent materials. Soils of other orders cover small areas in comparison and are relatively insignificant. There are some Entisols, the youngest soils, that have little profile development being very recent soils. And there are a few Ultisols, the region's oldest soils, which have developed where there have been undisturbed conditions for considerable periods such that soil forming processes have proceeded sufficiently to produce these mature tropical soils. They are moist soils that develop under a tropical climate, featuring an argillic horizon and low base saturation (<35%). There are also a few Mollisols, but only young members of this order, represented by permanently immature, truncated soils that occur predominently on steep slopes where soil movement ensures arrested development. They are dark soils with a soft, crumbly, mollic epipedon dominated by divalent cations, having a high base saturation. Finally, there are localised areas of Histosols, the organic or bog soils with >20% organic matter, developed in watersaturated environments. The total is five of the ten USDA (1975) orders represented in the region, one predominating and the others of minor significance.

This chapter reviews these soils in cultural context. Like several of the environmental sciences reviewed in this book, the study of ethnopedology is largely overlooked.[1] This is a serious shortcoming because 'common or garden' soil is central to human subsistence, whether people are hunters, pastoralists or farmers. Unremarked soil is truly remarkable for all life. The awareness that land users have of their soils, their classification of them and understanding of

---

[1] The findings reported here reflect an emergent concern with others' perceptions of their soil environments, for while interest in ethnopedology is small compared to that in zoological and botanical ethnoscience, it has attracted some attention recently (see Ollier *et al.* 1971; Dvorak 1988; Landsberg and Gillieson 1980; Behrens 1989; Furbee 1989; Guillet 1989; and Philips-Howard & Kidd 1991; Kerven *et al.* 1995; Guillet *et al.* 1995).

pedological processes, and their management of them and husbandry strategies will be fundamental to their subsistence regime. This study starts with the classification of the soils found in the Wola region, before moving onto issues relating to cultivation and management. The classification of soil is notoriously difficult and a compromise is sought. No attempt is made to pretend an unalloyed emic indigenous view because my Western scientific knowledge inevitably informs any understanding that I have of soil. No amount of symbolic jiggery-pokery nor ambiguous structured double-talk can circumvent this epistemological conundrum, to pretend otherwise is dishonest. This chapter attempts to blend a soil science perspective sympathetically with indigenous ideas. Regarding the cross-cultural hybridisation of know-ledge, an intermingling of apperceptions may prove fruitful in advancing our overall understanding of phenomena. It is certainly nearer to the anthropological achievement when considering technical subjects than assumptions about translating one culture into categories intelligible to another.

I. SOIL CLASSIFICATION

The first pedologists to study the soils of the Wola region were members of a CSIRO team on a reconnaisance survey to assess land potential and use by the lands systems approach, and they came to the conclusion that they are primarily zonal soils (Haantjens and Rutherford 1967). After the classic Russian work on soil classification in the second half of the nineteenth century, they considered climate the dominant soil-forming factor, arguing that the rapid chemical weathering promoted by the region's warm and wet climate was the major factor responsible for its soils. More recently, other soil scientists have stressed parent material and degree of its weathering as the dominant environmental factor influencing soil properties and formation (Radcliffe 1986; Chartres & Pain 1984; Wood 1984). They are of the opinion that any classification of the soils of the region should take geology as the principal factor in defining soil classes.

In reality it is difficult and distorting to select one feature of the environment as dominant over the others in determining a region's soils because all will play some part. If any one of them were different, then the soils would not be what they are — a level landscape, for instance, would be less dynamic and unstable regarding soil formation, it might be less well-drained and only water loving vegetation possible and so on. It is more realistic, and less prejudicial, to ascribe equal weight to all of the environmental factors. But this presents us with the complex and bewildering mosaic of soil in the field, undifferentiated according to any factor selected as dominant around which to structure a classification, such as variation in climate, parent material or whatever. This is nearer reality than some contrived classification, for soil is a highly variable and complicated medium, essentially unsuited to the kind of hierarchical pigeon-hole classification demanded by Western science. Nevertheless we need to classify to start to understand the world we see. This conundrum has been a perennial problem in soil science: soil is not readily amenable to straightforward classification and yet demands some kind of ordering that we might further our understanding of it, and furthermore, any ordering presupposes some prior knowledge, in selecting those features taken as dominant by which to structure it, which may in turn condition, even determine, subsequent understanding achieved using the classification!

Problems with Soil Classification

Why is soil so difficult to classify? It has no sharp boundaries to begin with, no discrete readily isolated units (like individual plant or animal organisms) on which to base species classes. It comprises a continuous mantle across the Earth, featuring no natural sharply defined discontinuities. Soils grade imperceptibly one into another, comprising a continuum both laterally and vertically in space, whereas classification demands breaks or divisions, boundaries around discrete classes. We impose these artificially and somewhat arbitrarily on soils, lacking natural discontinuities for the purpose (e.g. the pedon and polypedon concepts of the USDA 1975, and the various mapping units of soil surveys), so distorting pedological reality at the out-set.

Furthermore, soil not only varies continuously over space but it also varies over time. It is a dynamic not a static medium and we need a flexible ordering scheme that can keep up as time passes by, can accomodate temporal variations, whereas the impetus of any classification is to fix phenomena in rigidly defined classes. If we are to conceive of a processual classification, moving soils conveyor-belt-like along as they develop and change, we shall need a sophisticated knowledge of pedological processes to predict the developmental path any soil will follow under prevailing environmental conditions. Our knowledge of soils is inadequate for this at present. Indeed we find ourselves forced to adopt a "cart-first" strategy because to achieve this level of knowledge we have initially to classify soils. We have first to order the teeming variety of soil to have any hope of understanding it, in some measure so determining beforehand what it is that we think we come to know about the processes involved in its ongoing formation.

Added to these problems of spatial and temporal variation, no single genetic nor other factor dominates and controls variations in soils (like reproduction defines classes in botany and zoology, and mineralogy in geology). The concepts of evolution which dominate in other natural sciences and underpin their classifications are less readily applied in soil science, where not only no dominant factor dictates variation but also where there are no common ancestors (similar soils may come from quite different parent soils, or different soils come from the same parent). Soils result from the interaction of several interrelated natural processes, which condition and affect one another. None are dominant and can be used indisputably to structure a classification, as illustrated by disagreements over environmental factors dominating soil development in the New Guinea Highlands.

We are left with the problem of deciding how to order and weight these different criteria in classifying soil. If we decide to use traditional hierarchical classification schemes, like those that have until recently dominated in the natural sciences, we are forced to select some features over others and use these to structure our classes, to determine where any soil will fall. One consequence is the separation sometimes of similar soils into widely different classes because they happen to vary on the arbitrary criteria used at higher levels for class definition. The classificatory schemes most suited, given these problems, to coping with the multivariate nature of soil classification are multi-coordinate ones facilitated today by computers. But they are still in their infancy, and have the major drawback of generating classifications that are difficult to visualise, comprehend and use, which in some senses negates the objective of classification to facilitate observation and understanding by showing relationships and organising knowledge in such a way that it can be readily passed on, understood and developed upon. Mind boggling classifications in tens of dimensions mitigate against these aims and obviate cross-cultural comparisons.

Approaches to Soil Classification

In the face of these somewhat intractable problems, two broad approaches to soil classification have emerged, each having its strong and weak points regarding their accomodation. We may gloss these two approaches as natural classification systems versus artificial classification systems (after Young 1976) each representing a phase in the history of scientific soil classification and attempts to accomodate and come to terms with these problems.[2]

The **natural** systems approach came first, originating with the classic Russian work of Dokuchaev and Sibertsiev last century and continuing with, among others, Glinka, Gerasimov and Neustruev in the USSR, Robinson, Kubiena and Avery in Europe and Marbut, Baldwin and Thorpe in the USA. It promoted many classificatory schemes and dominated soil classification up to the 1960s. It takes soils as natural products of their environments, assuming that their evolution depends on the physical conditions that pertain in the regions where they occur. Natural systems tend to select one of these environmental factors (e.g. climate, parent material or whatever) as dominant over the others and structure their classifications around it. They place emphasis on the genesis of soils. Their aim is to further understanding of soil origin and development, of the processes that give rise to particular soils. But they pre-empt this understanding somewhat by selecting the environmental factors presumed to condition genesis on subjective rather than objectively assessed grounds. It is these criteria that subsequently stand out in the classification and appear central to soil development.

---

[2] In a recent critical review of soil classification and mapping in Papua New Guinea, Humphreys (1990) refers to these approaches as nodal and artificial systems of classification.

We have here soil classification to further certain theoretical points of view regarding environmental and pedological processes. The environmental forces supposed prominent in soil genesis, by which classified, and the processes in which they feature, are poorly understood. Natural classifications and taxa definitions are consequently founded on conjectural hypotheses, and we have endless scope for academic debate and disagreement, researchers arguing over the emphasis to give to different criteria. One up-shot is a confused nomenclature, different writers giving different meanings to the same technical terms (e.g. podsol, laterite and so on), and similar meanings to different terms. The emphasis placed on presumed environmentally determined evolutionary processes in constructing taxa, rather that actual observed properties of soils makes this terminological imbroglio worse because the classes, relating more to environmental processes that substantive soils, are poorly defined and woolly regarding soils in the field. The schemes are descriptive of broad environmental and soil conditions, not analytical of soils themselves. The boundaries between taxa are unclear, indeed some are residual and even indefinable according to the environmental criteria taken as central to their definition.

The **artificial** systems approach to soil classification developed in response to the vague and unscientific taxa definitions of natural systems. It is epitomised by the monumental USDA (1975) system which has become ascendant over the past couple of decades, although there have been others. Analytical, not descriptive in approach, they focus on the close definition of different soil types according to observed properties. The intention is to facilitate the unambiguous identification and naming of different soils through the precise delimitation of soil taxa. They reflect the trend in modern science to quantify and define phenomena precisely, on the grounds that rigorous definition of terms and categories is fundamental to constructive scientific debate. Artificial systems are quite different to natural typological classifications founded on inferred genetic processes, aiming to classify soils according to current properties, regardless of processes of possible origin or future development. An artificial system is more akin to an elaborately keyed dictionary that facilitates precise and unambiguous definition, than an excogitative classificatory schema. Legalistic in form, closely defining the criteria used to place soils in different taxa, and often involving the measurement of parameters, class ascription is made exact — so long as the sometimes complex procedures for placement are followed correctly.

Artificial systems of classification are laudible attempts to introduce precisely defined terms and limits, often featuring quantification, to define taxa boundaries by observable and objectively assessable soil properties, rather than assumed and subjectively assessed soil processes. But considerable problems attend them. In Papua New Guinea contexts the USDA system has been criticised, for being too U.S. orientated, regardless of intentions that it should be a worldwide system, and covering tropical soils inadequately; for placing too much emphasis on arable land at the expense of steep or stony terrain, common in highland New Guinea; for requiring several expensive laboratory tests which serve no useful purpose beyond classifying soil; for placing considerable emphasis on argillation, which is sometimes difficult to identify positively without thin sections; for having soil taxa that rarely correspond to mappable units; and for having a complicated taxonomic nomenclature likely to confuse non-specialists and hinder local extension work (Humphreys 1984, 1990; Radcliffe 1986).

The new USDA nomenclature, thought necessary to avoid old confused though familiar pedological terms, is daunting and off-putting — although once mastered it conveys a considerable amount of information about a named soil. And the precision required for the definition of taxa, particularly their laboratory-based quantitative aspects, makes the system difficult to use; and the measurement of some of these variables (e.g. pedoclimate, water content) may be somewhat arbitrary, even illusory, as they are closely linked to fluctuating external environmental conditions (Duchafour 1982). Furthermore, it is unsatisfying, if not unsatisfactory, to classify soils by their properties alone, we wish to know how these properties came to be and how they may change further. Indeed we need to know, for soil is a dynamic medium; awareness of these processes is necessary in any comprehensive classification as they underlie changes that occur, soil properties being the product of on-going interactions and processes. Any attempt to classify soil by properties alone not only side-steps the issue that we improperly understand pedological processes, but literally applied is also distorting because it ignores an entire dimension along which soils vary, namely time. Any soil exists against time. The concept of the genesis of soil properties remains a feature of artificial analytic systems for this reason, albeit somewhat implicitly, unsystematically applied in the sequential arrangement of classes.

Natural descriptive systems, on the other hand, grounded in the environmentally dictated processes presumed to underwrite genesis and development, recognise that soils are dynamic entities that evolve and change. It might be argued further that their woolly taxa are a truer reflection of pedological reality, soils in nature being variable and vague entities, difficult to draw boundaries around, blending one into another. Nevertheless natural systems are ultimately unscientific. But they relate soils better to their locations and landscapes, being tailored to the local environs, whereas artificial systems fail to make sense of landscape relationships, of what is observed and experienced in the field; defined analytically and rigidly according to soil properties they fail to give readily mapped units, required by the soil surveyor. Nearly all classifications, both natural and artificial, are hierarchical schemes and have associated problems, separating similar soils that just happen to differ on a feature used to define taxa. They may fail as a result to show some significant relationships, highlighting the criteria used for classification and obscuring others.

Neither approach to soil classification is without its shortcomings. But any study, like this one, has to use some classificatory scheme to order and make sense of the data. In this event, how best to classify the soils of the Wola region for discussion? Regarding the classification of soils in Papua New Guinea as a whole, the historical trends related above have occurred over a short time span with natural descriptive schemes devised after the last war, largely by Australian CSIRO staff, giving way in the last ten years or so to the artificial analytical scheme of the USDA, which is now the principal classification used in soil investigations in Papua New Guinea (e.g. by the Department of Primary Industry's Land Utilisation Section, and the CSIRO's Division of Water and Land Resources). Another relevant scheme is the FAO one.

I cannot pretend to be able to select between these various schemes, with their positive and negative points, their irreconcilability relating ultimately to the intractibility of soils to classification at all. The one favoured depends to some extent on one's concerns: soil genesis and pedological processes or contemporary land use and agricultural potential. I shall use neither exclusively to order the following account, and shall cite major taxa names for the soils described from schemes following both approaches. I have decided to use a mixed scheme combining features of both natural and artificial approaches, each having something to recommend it, a rapprochement exploiting their combined strong points against their weak ones. And I give a prominent place to the local classification, producing an overtly hybridised cross-cultural scheme, in contrast to the usual anthropological covert hybridisation. The natural process element draws on previous soil classifications of the region (Rutherford & Haantjens 1965 and Radcliffe 1986), re-ordering their classes to some extent and putting no emphasis on any environmental feature. The properties element derives from the soil classification scheme of the local Wola people, supplemented in part by the USDA scheme.[3]

Ethnopedology: The Local Classification Scheme

It may at first seem extraordinary that the non-scientific Wola scheme is associated with the avowedly scientific USDA one. But it too uses objective properties and concerns practical agricultural issues relating to the definition of good, bad or indifferent soil. It is similar in this regard to our culture's pre-scientific classification of soils which, founded on practical experience, involved land use, notably ease of cultivation and crop yields on different

---

[3] After considerable thought I have chosen to ignore the recent USDA recommendation that volcanic ash soils be raised to a separate Andisol order, deciding instead to stay with their original Andept sub-order ranking, because moving these soils up the hierarchy to order level separates them from other similar Inceptisol order soils with which they effectively form a continuum across the Wola region; to distinguish them as a separate order would take them further away from the local scheme with its flexible adaption to a continuum concept.

kinds of soil.[4] The genesis of soil was not at issue. The USDA scheme has a similar focus, its central consideration being the accurate definition of soil taxa as a prelude to assessing their agricultural potential. The Wola could appreciate this objective, whereas the origins of soil and pedological processes are of little apparent interest to them. They are understandably concerned, as subsistence gardeners, in knowing and conveying something about the agricultural usefulness of land.

Local subsistence activities also affect the soil, agriculture intervening in natural pedological processes. Giving prominence to the local classification scheme may help us better to appreciate what happens to the soil when people cultivate it, for it will give us some idea of their experiences, those whose tradition is founded on practical use of the land. The local scheme also necessarily involves only a small region, known to the inhabitants in great detail. It has no universal pretensions, which give rise to enormous, even currently insurmountable, problems for scientific classifications because of the great variability in soils worldwide. It is an advantage to have only a limited range of soils to consider, in a region where there are few marked environmental variations. The incorporation of local classification schemes into soil surveys should also facilitate agricultural extension work and attempts to develop on local cultivation practices, introduce cash crops and so on. Indigenous people are more likely to be receptive to suggestions couched in terms of categories they know, modifying and extending on them, than they are foreign ones that appear to assault their understanding and expectations.

Furthermore, consideration of non-scientific classificatory schemes, like that of the Wola, so far as we can apprehend them, may tell us something about the nature of classification and the assumptions that underpin our notions of it. They inform us of different cultural trajectories which may take our idea of classification in different directions. This is particularly pertinent regarding the classification of soil, there being no wholly satisfactory way of classifying it, and new insights may be gained from novel approaches. The Wola notion of classification for example, is essentially fluid and flexible (see Chapter 7), which ideally suits it to the classification of a continuously variable medium like soil, unlike the rigidly bounded classes our classifications try to impose on it. The Wola name different kinds of soil according to observed properties (such as colour, texture, moisture, stoniness and so on), and they can combine and modify these endlessly to build up descriptive classes, referring to "some of this and some of that" and so on (for example *hundbiy sha araytol onduwp* as opposed to *hundbiy araytol* or *hundbiy tongom momonuw araytol haeruw*,

---

[4] At the dawn of the scientific era there were both Jethro Tull's classification of soil according to texture of tilth because he thought plants absorbed small quantities of the fine fraction (advocating the stirring of earth in cultivation to make this easier), and County Agricultural Surveys which classified soils into broad classes relating to surface texture and ease of cultivation, a classification coming direct from our still extant folk classification of soils into heavy and light depending on their clay and sand contents.

which broadly translate "very stony bright-brownish-clay" as opposed to "stony bright-brown-clay" or "stony bright-brown with gleyed-clay" etc.).

An elastic classificatory scheme like that of the Wola, which can describe a soil as a mixture of this and that property, being neither exclusively but lying on the ill-defined boundary between them, highlights how contrived is the division of soil into bounded classes. This flexibility is partly the product perhaps of the soils of the Kerewa-Giluwe region, which comprise a continuum due to the volcanic ash falls that have influenced them to varying extents. The effect of these ash falls has been to introduce gradual variation and not abrupt changes across the region. The impact of ash on an area depends on the amount involved, which varies according to such factors as distance from volcanic source, weather at the time of eruptions (winds spreading ash more widely and thinly), slope form (if steep, ash is more likely to have been swept off to some extent, if a depression, more likely to accumulate etc.), and so forth. The continuous nature of the resulting soil mantle makes it difficult for the profile orientated soil surveyor to draw classificatory lines, as made clear in a comment by CSIRO soil scientists on the distinguishing features of the dominant soil group they identified in the region: "in practice the features overlap to a certain extent and distinction in the field can be difficult in places" (Rutherford and Haantjens 1965:88).

The Wola name each distinct horizon revealed in a soil exposure as a different kind of soil, but they have no notion that certain horizon sequences comprise named profiles (this is common throughout the Papua New Guinea highlands — see Brookfield and Brown 1963:35; Wood 1984; Radcliffe 1986:80; Ollier *et al.* 1971; Landsberg and Gillieson 1980). While each horizon comprises a different soil type and has a name, there is no name for the profile as a whole and hence no attempt to classify soils by different profiles. This is a radical point of difference with Western soil science classifications. According to the Wola scheme soil horizons can occur in any order, and whatever this is, it merits no particular remark (which is not so eccentric in a region where frequent landslips can put subsoil horizons above epipedons — see Chapter 6). They may consequently recognise countless different horizon combinations and accomodate any number of variations; for example, when naming the soils exposed in 372 profile pits, people came up with a total of 52 different combinations (Table 10.1), one or two of which were considerably more common than others, as is usual in any region. Nonetheless, the range of different kinds of soils, of modal types as discussed below, is finite.

A further point of difference between Wola and Western soil classifications is that the former has no hierarchy of nesting classes organised according to certain governing principles (such as environmental factors, nutrient chemistry, profile morphology, soil age or whatever), which have proved inherently unsatisfactory in the scientific ordering of soils and stumbling blocks to the excogitation of some relationships between them. The Wola have a word for all soils collectively, which is *suw*, and may prefix the names given to particular kinds of soil with it (e.g. *suw hundbiy*, "ground bright-brown-clay"). It is a word of broad connotation and may refer to anything from a

Table 10.1

A selection of horizon sequences named and identified by the Wola (n = 372 profiles)

Profile Sequences (Read vertically)

Wola Horizon Terms	1	2	3	4	5	6	7	8	9	10	11	12	13	14	15	16	17	18	19	20	21	22	23	24	25
shor paenpaen	1																								
waip	2	1	1	1	1	1																			
dowhuwniy				1																					
pombray		2	2	2	2	2					1														
pa pombray						1																			
pombray araydol			2				1																		
pombray tiyptiyp							1	1																	
iybmuw pombray							1	1																	
pombraysha									2																
hundbiysha										2			2			2									
hundbiysha araydol												2	2												
payhonez	3																								
tiyptiypsha																							2		
hundbiy	4	3	3	3															2	2	2	3			
hundbiy araydol		3	3							2											3	3			
hundbiy iybmuw					4																				
omb																									
tiyptiyp				2	3																				
kolbatindiy						3																			
iybmuw									3																
iyb dor tilai								2																	
tongom										3	3														
pa tongom																									
tongom tiyptiyp														2	3										
iyb uw damiy																									
haenhok																							2	3	
haen																									2
Frequency	1	4	2	1	1	1	4	1	2	1	1	6	1	5	3	1	7	1	8	5	11	149	4	1	38
Percent	0.3	1.0	0.5	0.3	0.3	0.3	1.0	0.3	0.5	0.3	0.3	1.6	0.3	1.3	0.8	0.3	1.9	0.3	2.2	1.3	3.0	40	1.0	0.3	10

Table 10.1
Continued

Wola Horizon Terms	Profile Sequences (Read vertically)
shor paenpaen	1
waip	1,3 1 1 1
dowhuwniy	
pombray	1,3 1 1 1 1 1 1 1 1 1 1 1 1 2
pa pombray	
pombray araydol	1 1
pombray tiyptiyp	1
iybmuw pombray	1
pombraysha	1 1 1
hundbiysha	2 2 2 2 1 1 1
hundbiysha araydol	2 2 1 1
payhonez	3
tiyptiypsha	2
hundbiy	4 2,3 2,3 2 2 3 2
hundbiy araydol	2 3 2 3 2
hundbiy iybmuw	2
omb	
tiyptiyp	2 3 2 2
kolbatinddiy	3
iybmuw	2
iyb dor tilai	2
tongom	3
pa tongom	2,3 2 2,3 2
tongom tiyptiyp	
iyb uw damiy	
haenhok	2
haen	
Frequency	1 18 2 1 33 16 3 1 2 1 1 1 3 1 1 1 5 1 7 1 2 1 1 5 2 1 1
Percent	0.3 4.7 0.5 0.3 9.0 4.3 0.8 0.3 0.5 0.3 0.3 0.3 0.8 0.3 0.3 0.3 1.3 0.3 1.9 0.3 0.5 0.3 0.3 1.3 0.5 0.3 0.3

[The numbers in the main part of the table refer to the order in which the horizons occurred — e.g. 1 = horizon 1, 2 = horizon 2 etc. — and each vertical column of nos. = a profile].

handful of earth through to an entire geographical region (see Chapter 5) — a commonly heard phrase is *na suw*, which means "my ground" (i.e. my place, where I live) and is often followed by the name of a territorial location, such as *na suw Haelaelinja*, "my place Haelaelinja". Nearly the entire Kerewa-Giluwe region is mantled in *suw*, in a dark brown to black topsoil called *suw pombray* (lit: ground or soil black), which according to the Wola is essentially the same everywhere and does not vary in any consistent manner with changes in the underlying subsoil. While the topsoil comprises a single named taxon in their classificatory scheme, the Wola make a note of the way it varies in different places, notably in depth, *iyba* 'grease' content, strength (i.e. friability), stoniness and water content, all important considerations for cultivation, and may accordingly qualify the term *suw pombray* (for example, *suw pombray buriy iyba na bidiy*, "strong, black, 'greaseless' soil"). Here their concerns are in some way more akin to a land capability assessment than a classification of soil. Below this topsoil mantle occur a variety of subsoils between which the Wola make distinctions. The framework of their classification thus comprises:

all earth/ground (suw)
↓
relatively uniform topsoil + various subsoils

Using the Wola scheme to structure the following account of the soils of their region results in an F.A.O.-like non-hierarchical scheme, in which each unit is monocategorical and roughly equivalent. While the following accounts comply with the soil-property-oriented approach of the Wola and USDA schemes, differentiating between soils by observed properties, making few assumptions about processes, it would be unsatisfactory and limiting to omit what we know of soil genesis and process. The format followed is unashamedly a hotch-potch as a consequence, further making it somewhat akin to the international compromise F.A.O. scheme, the result being an intermixed system that falls between the natural and artificial, trying to relate and reconcile one to another.

II. SOILS OF THE WOLA REGION

The schema used, incorporating local discriminations and intimating the soil's continuous character, distinguishes five major kinds of soil in the Kerewa-Giluwe region: shallow soils, clayey soils, sandy and alluvial soils, gley soils, and peaty soils (Table 10.2).[5] The clayey soils dominate the region. They comprise those that derive from weathered sedimentary rock materials

---

[5] Technical profile descriptions of the soils described in this chapter may be found on the internet at the following address (URL):http://www.dur.ac.uk/~danøps/soil.html.

Table 10.2

The soils of the Wola region according to various classifications, including the indigenous scheme.

			WOLA (subsoil classes)		USDA		CSIRO	AFTSEMU	FAO
			MAJOR GROUP	MINOR GROUPS	SUB-ORDER	GREAT GROUP			
SHALLOW SOILS		skeletal arrested	haenbora haen hok		Orthent Rendoll	Troporthent			Lithosols Rendzinas
CLAYEY SOILS	SEDIMENTARY ROCK DERIVED SOILS	immature			Tropept	Humitropept	Shallow dark clay soils	Sedimentary soils	Cambisols
[Continuum of soils, affected to varying extents by ash falls]		mature	hundbiy	paphonez kas tongom	Humult	Tropohumult Dystrandept	Deep dark clay soils		Acrisols
	MIXED SEDIMENTARY & VOLCANIC ASH SOILS				Andepts		Humic brown clay	Mixed ash-sedimentary soil	
	VOLCANIC ASH DERIVED SOILS	low montane	tyrptyp	kolbatindiy		Hydrandept		Brown ash soil	Andosols
		high montane				Hydrandept	Humic olive ash soil	Olive ash soil	
ALLUVIAL & SANDY SOILS	ALLUVIUM	ash derived	iyb dor tilai		Fluvent	Hydrandept Dystrandept Tropofluvent	Old alluvial soils	Alluvial ash soil	Fluvisol
[Continuum of age & material mixes of deposits]		recent					Recent alluvial soils	Alluvial & stony colluvial soils	
	SANDY SOILS		iyb muw		Psamment	Troposamment			Arenosol
GLEY SOILS	VOLCANIC ASH SOILS	wet locales	pa tongom		Aquepts	Andaquept	Gleyed plastic heavy clay soils		Gleysols
[Continuum of soils, varyingly affected by modal class p.m.]	SEDIMENTARY SOILS					Tropaquept			
	ALLUVIAL SOILS				Aquent	Fluvaquent			
PEATY SOILS	TOPODEPRESSIONS	low montane	iyb iw damiy		Saprist Hemist	Troposaprist Tropohemist Cryohemist	Peaty soils	Organic soils	Histosols
		high montane	waip						
	OTHER				Folist Fibrist	Tropofolist Tropofibrist			

through to those soils derived exclusively from volcanic ash or tephra, and include the entire spectrum of soils between, that are composed of varying mixtures of both volcanic ash and sedimentary material, giving a continuous distribution. The alluvial soils comprise a continuum too, both of materials and in age. The older ones consist of redeposited volcanic ash and the recent ones eroded bedrock and redeposited clayey soil; those soils in between in age may comprise varying mixtures of both. The sandy soils are very localised in extent, occurring largely where occasional sandstone beds outcrop at the surface. The shallow soils too are very small in extent. Any of the above soils, either an undisturbed one in the volcanic-ash-sedimentary-rock series or one redeposited as alluvium, may be subject to wet conditions and become a gley soil, again presenting a continuous soil spectrum. And if the wet conditions are particularly severe and prolonged, peaty soils of high organic matter may develop. In summary, the Kerewa-Giluwe region is mantled with soils derived from sedimentary parent materials, variably affected by volcanic ash (from dominated by it, to no evident effects), with some alluvial redeposition, and some soils affected by high water contents leading to changes in their morphology.

Shallow Soils

***Haenbora* skeletal soil** (Troporthent; Lithosol): Skeletal soils occur where masses of limestone outcrop precipitously at the surface, too steep or unstable for any depth of regolith to build up. They commonly occur adjacent to sheer rockfaces, where the incline decreases enough from the vertical, or ledges occur, or there are sharp-edged ridges, sufficient for the accumulation of some weathered material.

The deposit that accumulates initially is commonly a soft whitish putty-like surface covering, which sometimes subsequently develops a brownish hue when some organic matter decomposition has occurred. It is typically a thin veneer, collecting in crevices and other irregularities on very steep limestone exposures. The Wola call this regolith *haen paenj* (lit. rock sap — they think of it as emanating from the rock rather like sap deposits leak from, and accumulate on, the trunks of some trees, like hoop-pines). It supports a sparse vegetation, ranging from small rock plants, notably mosses, to stunted ferns, short grasses and small herbs. This skeletal soil-like veneer is uncommon, restricted to limited locations of rocky outcrops. It is of no agricultural value.

***Haen hok* arrested soil** (Rendoll; Rendzina): On less precipitous, though nonetheless steep and unstable slopes, slightly deeper rendzina soils may develop. They occur predominantly on limestone, and sometimes very calcareous mudstones, being associated with very steep rocky limestone outcrop slopes, steep colluvial slopes and stony outwash slopes below limestone escarpments. The terrain results in their arrested development and they occur only as shallow soils, resting directly on bedrock.

Table 10.3
The nutrient status of *haen hok* rendzina soil
(n = 1 topsoil & subsoil)

	pH	P (ppm)	K (me/100 g)	Ca (me/100 g)	Mg (me/100 g)	Na (me/100 g)	CEC (me/100 g)	C (%)	N (%)	C:N ratio
topsoil	5.9	31.2	1.14	19.3	1.83	0.8	27.5	23.81	1.39	17
subsoil	6.3	56.3	0.18	5.12	0.52	0.78	15.8	6.8	0.69	10

The *haen hok* Rendolls have a black to very dark brown A_1 *suw pombray* horizon, 10 to 50 cm. thick, sometimes turning dark greyish brown with depth. They have a strong fine crumb or granular structure, becoming perhaps medium subangular blocky with depth, and are of friable to firm consistency. Weathered pieces of limestone and chert fragments commonly occur throughout the profile, increasing with depth. And some of these soils have small amounts of volcanic ash mixed in them. The A horizon usually sits directly on a weathered limestone surface of crumbly moist rock, called *haen hok*. It is quite distinct from hard consolidated *hat haen* limestone (see Chapter 5) or soft *haen paenj* regolith deposits.

These relatively youthful soils are formed as calcium carbonate ($CaCO_3$) is gradually dissolved by rain borne carbonic acid (H_2CO_3), and the products either retained on the exchange complex or leached down through the profile. The process results in a calcium rich clay residue that favours the rapid mineralisation of organic matter, which gives these soils their dark colour and well developed granular structure. The pH of these soils predictably increases gradually with depth, with leaching of some released cations and proximity to calcium rich parent rock (Table 10.3). They feature a neutral to weakly acid reaction. They are of moderate fertility with high base saturation and have one of the highest available phosphorus levels for the region (Rutherford and Haantjens 1965:94,97; Bleeker 1983:113). But this fertility, together with the promising physical properties of these soils, are frequently difficult to exploit agriculturally, even for the Wola who regularly cultivate precipitous sites, because of the very steep and rugged nature of the terrain on which they occur.

Clayey Soils

***Hundbiy* clay soils** (Tropept → Humult; Cambisols → Acrisols): When a heavy brown clay derived from sedimentary rock underlies the dark *suw pombray* A topsoil horizon we enter the *hundbiy* clay class of soils. They are the dominant soils of the Wola region, occupying the greater part of the area under sedimentary rocks, between the flanking volcanic zones (see Chapter 5). The distinguishing feature of these soils is their heavy clay subsoil, which is plastic and firm, and frequently imperfectly to poorly drained. They occur principally on steep to moderately sloping terrain away from volcanoes, and also on some gentle colluvial slopes and in depressions such as doline floors and slump alcoves. A variety of sedimentary parent rocks, both consolidated and unconsolidated, underlie them, notably limestones, and on occasion fine textured sediments like mudstones and shales. They are subsoils from which most, if not all volcanic ash has apparently been stripped by erosion.

The *hundbiy* clay soils have well developed black to very dark grey-brown A_1 *suw pombray* horizons, usually in the range 10 to 50 cm. thick. These topsoils are friable and have fine blocky to granular crumb structure. In undisturbed areas, a black to dark brown leaf litter horizon and root mat may cover them, as it may any soil. The Wola call the litter layer *waip* under forest and *kolomon* under cane grassland. The A_1 horizon overlies a yellowish brown or bright brown B subsoil called *suw hundbiy*, the boundary between them being characteristically clear and abrupt (although they may sometimes grade more gently into one another where organic compounds have moved downwards from the dark topsoil to give an intermediate brown, less clayey horizon, which the Wola identify and call *hundbiy sha* [literally brownish, rather than brown]). This B horizon has a weakly developed blocky structure, is friable to firm in consistency, and fine textured, usually sticky and plastic. Chert fragments, called *araytol* (lit: chert-dirt), are quite common in these soils, both in topsoil and subsoil, sometimes increasing markedly in size and number with depth (see Chapter 5).

The Wola distinguish three further types of clay subsoil, primarily according to differences in colour, which they call *payhonez*, *tongom* and *kas*. The *payhonez* type is a variety of *suw hundbiy sha* AB transition horizon. It is a greyish yellow brown, friable silty clay of fine to medium strong angular blocky structure. It is common under *pay* chestnuts (*Castanopsis acuminatissima*), hence the name of the subsoil, and also under brakes of *saezuwp* fern (*Dicranopteris linearis*). It might result from the deposition of organic compounds leached from the high organic matter content *pombray* A horizon into the *hundbiy* clay B horizon, possibly evidencing some translocation effect induced by chestnut oak and fern vegetation, their litter releasing some compounds that promote the leaching of organic complexes when in solution.[6] The occurrence of many varieties of fungi under chestnut oaks also suggests a somewhat different decompositional environment, the abundant mycelia could play some part in, or reflect the occurrence of, some such postulated chemical processes. According to the Wola, the *payhonez* soil develops into a good garden tilth, producing a medium crumb structure when exposed under cultivation, broken up and dried out somewhat by the sun.

The other two types of clay subsoil *tongom* and *kas* both occur below a normal brown *hundbiy* horizon according to the Wola, often a metre or more down. They are of limited occurrence. *Tongom* is a white clay, consisting mainly of kaolin, with lesser quantities of mica and illite and some quartz. It

---

[6] In a process somewhat analagous to the leaching induced by polyphenols released under conifers in temperate regions, which promote chelation, humus translocation and podzolisation.

is probably evidence of a relic gley soil on a site long ago when drainage was impeded for some reason (or alternatively a diatomaceous earth). *Kas* (also called *omb*) is a bright reddish brown to strikingly bright orange clay which is iron rich, consisting mainly of hepidocroite with some geothite and a trace of hematite. It is the clay the Wola fired to produce red ochre paint (Sillitoe 1988). An associated feature is *omb hul* (lit: *kas*-clay bones); these are small (up to one or two centimetres across), Fe rich laterite-like concretions found dotted here and there in pockets and sometimes discontinuous strata in *hundbiy* clay subsoils.

The movement of iron relates both to the age of the soil and the extent of leaching permitted in any location by the permeability of underlying strata and so on. These redder soils, the presence of concretions and so on are features that further support the impression of a continuum of soils mantling the Wola region. Differences in leaching rates and soil age between locations result in a sequence of soil types: Tropepts or Cambisols through to Humults or Acrisols respectively, according to the USDA and FAO classification schemes. The B horizons of all these soils either merge gradually into weathered broken-up rock or, if on limestone, may rest abruptly on the more or less fresh bedrock, and they are normally more than one metre thick. They have developed through processes of deposition, and physical and chemical weathering on these parent materials.

On the limestones that predominate across the Wola region, the aforementioned acid rain induced dissolution of carbonate is the foremost weathering process, Ca^{2+} and HCO_3^- ions leached from the system leaving an insoluble residue. The limestone however is fairly pure. After grinding-up and dissolving some limestone in acid, mimicking the weathering process, only 4.8% by weight remained. The residue was pale grey, not iron-rich brown in colour like today's clay, and under polarised microscopic investigation was predominantly coarse, sharp-edged chert grains (>90%, with some quartz, tourmaline and zircon).[7] The colour and dominance of sand sized particles, with very little fine clayey material, suggests that the weathering product of limestone has contributed relatively little to today's *hundbiy* clay subsoil. It possibly originated as marine or estuarine sediment, deposited before uplift occurred to produce today's mountains. Where chert occurs in the subsoil, it probably comes from limestone weathering, subsequently mixed up with the clay.[8] It is possible that the clay is also to some extent the weathering product of volcanic ash,[9] although volcanic derived clay minerals are not prominent. The clay consists mainly of gibbsite with some chloratised mica and montmorillonite, and small amounts of siderite which account for the brown colour. A variety of other clay minerals have been reported for

---

[7] Limestone sample from exposure at Sezkombor (Map 5.2). I am grateful to P. Burnham of Wye College for advice on this weathering assessment experiment.
[8] Similar to the clay with flints deposits in S.E. England (P. Burnham pers. comm.).
[9] Volcanic ash weathering over 50,000 years to produce one metre of kaolinitic clay (Mohr *et al.* 1972).

these soils by others, including kaolinite and metahalloysite, and occasionally illite and quartz (Bleeker 1983:93–95, 152–154). The presence of metahalloysite, gibbsite and kaolinite are possible evidence of volcanic ash contamination, the higher their contents the greater the influence of volcanic derived materials on the soil, moving them indisputably towards soils of mixed sedimentary and volcanic ash origin.

These soils are undoubtedly zonal in character, the moderate temperatures and high rainfall of the region markedly influencing their development. The relatively low temperatures, for the tropics, retard the breakdown of organic matter, contributing to the build-up of thick, dark topsoils. Some of this accumulated organic matter is in turn moved downwards by the heavy rainfall and deposited in the B horizon, resulting in some subsoils with relatively high organic matter contents. This downward movement and deposition of organic matter, together with iron and aluminium oxides and clay, is probably due to several processes: the straightforward leaching, under the high rainfall conditions, of both organic compounds and clay (partially destroyed by weathering in the acidic, unsaturated and organic rich topsoil); the movement of clay downwards as a negatively charged humus protected sol; the mobilisation of iron and aluminium oxide ions as chelates (complex ring structures of organic origin featuring several bonds between metal and complexing agent molecule); the independent translocation of humus in colloidal form; and the movement of compounds induced by volcanic ash contamination and the reactions associated with it. The clay content may increase with depth as a consequence of this movement, and where strong leaching features or the soil is of considerable age, such that leaching has occurred for a very long time, then argillation may be evident (being diagnostic for the USDA Ultisol order — viz Humults). In addition there may be a variable build-up of sesquioxides, giving a continuum related to age and severity of leaching conditions, its modal end-points reflected in the USDA and FAO soil type sequences: Tropept/Cambisol → Humult/Acrisol.

The severe leaching consequent upon the high rainfall not only gives clayey textured sesquioxide enriched soils but also removes bases and gives soils acid to strongly acid in reaction, with a mean about pH 5.0 (Table 10.4).

Table 10.4

The average nutrient status of *hundbiy* clay soils
(n = 107 topsoils, 82 subsoils; nos. in brackets = σ)

	pH	P (ppm)	K (me/100 g)	Ca (me/100 g)	Mg (me/100 g)	Na (me/100 g)	CEC (me/100 g)	C (%)	N (%)	C:N ratio
topsoil	5.16	10.56	0.66	13.35	2.65	0.75	28.19	16.69	1.01	17.6
	(0.58)	(11.5)	(0.77)	(15.25)	(2.44)	(0.60)	(7.93)	(7.67)	(0.41)	(10.12)
subsoil	5.29	3.74	0.19	3.69	0.56	0.70	16.27	4.17	0.33	14.6
	(0.61)	(8.83)	(0.39)	(5.20)	(0.86)	(0.45)	(6.21)	(2.07)	(0.18)	(6.87)

Nevertheless, while base saturation is low, soil fertility is generally moderate. The mean percentage nitrogen value is high in the topsoil at 1.0%, as is the mean organic carbon content at 16.7% (according to Bleeker (1983:93), these soils are more often in the region of 0.66% and 8.5% respectively). The high values may be partly due to volcanic ash in the topsoil, the allophane present, together with the relatively cool temperatures, inhibiting organic matter breakdown. The relatively high organic matter contents of these soils make a significant contribution to their fertility. The inhibition of organic matter breakdown results in a fair topsoil CEC, which decreases noticeably in the subsoil with depth. Exchangeable potassium shows a similar trend, having moderate values in the topsoil and low ones in the subsoil. Available phosphorus levels are also moderate to low, suggesting marked fixation, notably by organic matter. The trend is for older soils predictably to have somewhat reduced chemical fertility, as do those soils subject to intenser leaching due to permeable sub-strata, both of which fall towards the Humult end of the *hundbiy* clay soil continuum.

The Wola make wide use of these soils, cultivating them extensively. Where not under current cultivation, they commonly support large areas of cane grass secondary regrowth. They also support extensive areas of primary and secondary forest vegetation. They are suitable for cash crop development where they occur on moderately sloping accessible land, although they occur largely in less suitable rugged mountainous terrain on medium to steep slopes. Their modest fertility could prove difficult to maintain under continuous commercial cultivation, as evidenced by the traditional agricultural practice of green manuring sweet potato mounds in gardens cultivated repeatedly.

Tiyptiyp volcanic ash soils (Andepts; Andosols): In locations where substantial amounts of volcanically derived ash have accumulated soils of the *tiyptiyp* volcanic ash class occur. They are the second most common soil of the Wola region, predominating on its vulcanised eastern and western margins. They occur not only on the dissected slopes and ash plains of volcanoes, but also on mountain slopes and across valley floors some distances away from where substantial sized volcanic debris has been deposited, finer volcanic ash having been spread widely in the atmosphere across the region (see Chapter 5). They also occur on alluvially redistributed volcanic ash deposits, as old ash-derived alluvial soils located on fan surfaces and river terraces. All of these soils are formed on andesitic volcanic ash, deposited mainly during the Pleistocene, and their consequent considerable age (about 50,000 years), compared to recent coastal volcanic soils, has resulted in them weathering into mature soils characterised by deep, dark topsoils of high organic matter content overlying brown, gritty textured clayey subsoils with moisture contents continuously around field capacity in this wet region (hence the majority of these soils fall into the USDA Hydrandept great group; in those relatively few locations where drainage is better due to permeable strata, and leaching more severe, they fall into the low base Dystrandepts great group).

The *tiyptiyp* volcanic ash soils are usually deep, their depths depending to some extent on the thickness of the ash mantle. They have black to brownish-black friable A_1 *suw pombray* horizons characterised by a high organic matter content. These topsoils are fine to medium textured and have a granular or crumb structure. They have a low bulk density (<0.85 g cm^{-3}). They are usually in the range 20 to 65 centimetres thick, but can be thinner or thicker depending on terrain. They overlie a clay B horizon with a characteristically sandy feel to it, which is diagnostic to the Wola in identifying a soil as *tiyptiyp* ash, who refer to the sandy texture as *popo hae* (lit: roughness/grittiness stands).[10] It is not as slippery underfoot to walk on as *hundbiy* clay as a consequence, neither is it as sticky, coming away cleanly from a digging stick or spade when dug whereas *hundbiy* clay sticks tenaciously to them, having the property *paerai mbay* (lit: stick finish). Nor does *tiyptiyp* soil, according to the Wola, become *mondow*, which is soft sticky liquid mud, but is always well-drained, high porosity being another diagnostic feature of these subsoils. Nevertheless, regardless of their rapid permeability, these soils are continuously moist due to the region's wet conditions, and as a consequence they have been described as structureless. But when dried artificially they develop a strong subangular blocky structure. Another related diagnostic feature is that drying of these subsoils results in irreversible dehydration of clays into silt and sand sized aggregates. A thixothropic consistence or smeary property is a further characteristic, the soil going from a plastic solid to a liquid under pressure and then resuming its original structure when released. They are usually of friable to firm consistency. The depth of the B horizon can vary from about 30 centimetres to over 150 centimetres thick.

The boundary between the A1 and B horizons is clear and usually abrupt, as in many *hundbiy* clay soils. The B horizon is commonly brown to yellowish-brown in colour and merges into a thick olive brown to yellowish-brown C horizon, which is frequently speckled with dark minerals, and is massive and porous. But the nature of subsoil horizons can vary considerably depending on the extent of weathering. The B horizon may show a clear gradation from a firm highly weathered brown or reddish-brown subsurface layer to a somewhat less weathered yellowish-brown subsoil. But where weathering processes are retarded, at higher altitudes with their lower temperatures, the volcanic ash parent material is only partially weathered and forms an olive smeary 'structureless' compact sandy clay B horizon, speckled with dark minerals, faintly to distinctly mottled, more akin to the C horizon in profiles where weathering is more advanced. Soil scientists have consequently thought it necessary to distinguish a taxa of high altitude volcanic ash soils separate from lower altitude ones (below 2300 metres), called humic olive ash soils (Rutherford & Haantjens 1965) or olive ash soils (Radcliffe

---

[10] The same phrase they use to describe the glasspaper-like surface of scabrous *Ficus quercetorum* leaves which they use to smooth the surface of wooden artifacts (Sillitoe 1988).

1986). Further variation occurs in B horizons due to the presence of solid ash, cinders and pyroclastic material in variable amounts giving them a range of stone contents. The Wola refer to these hard bruise-like coloured concretions as *tiyptiyp*, from which derives the name of the soil *suw tiyptiyp* (lit: earth/ground volcanic-ash-concretions) abbreviated itself to *tiyptiyp*. Some soils contain considerable amounts of these hard variously blue, white and red speckled materials and are very stony, others have few to none in the upper parts of their subsoils. They vary in hardness from friable and breakable under the foot, called *tiyptiyp kolkol*, to rock hard and unbreakable (see Chapter 5). The other rock that sometimes occurs as stones in these soils is *huwbiyp* basalt.

The Wola also distinguish a minor volcanic ash subsoil group which they call *kolbatindiy*. It is a red medium textured clay in which fragments of volcanic ash, tuff and so on may occur. It is akin to the *kas* and *tongom* subsoils of *hundbiy* clay soils in that it occurs deep in the profile (below 50 centimetres.) overlain by *tiyptiyp* subsoil: the two volcanic subsoils often occur alternately together giving a series of bluish and reddish-pink Shanklin-sand-like layers. The clay content of *tiyptiyp* soils is also variable. The Wola refer to those that are clayey as *suw hundbiy tiyptiyp*, usually abbreviated to just *tiyptiyp*, or qualified variations of this name depending on the dominance of clay, for example *suw hundbiy sha tiyptiyp* (lit: earth *hundbiy*-clayish *tiyptiyp*-volcanic-ash) if the clay content is low, *suw hundbiy tiyptiyp sha* (lit: earth-*hundbiy*-clay *tiyptiyp*-volcanic-ashy) for the reverse, and so on. It should be borne in mind that in many locations *tiyptiyp* volcanic ash subsoils and *hundbiy* clay subsoils may occur overlying one another, even mixed up together, so justifying the interchangeable lego-like nomenclature of the Wola, which can be put together and modified to suit any combination of observed soil properties, whatever their origins (whether a bright-brown clay derives from weathered ash or sedimentary deposition or both is not relevant to people if a mineralogical assay is needed to distinguish between them).

The early CSIRO soil survey thought that the situation justified the identification of a transitional soil family (called Ivivar) where thin ash layers overlie sedimentary rock derived clay (Rutherford and Haantjens 1965:88–89) and the AFTSEMU survey of Upper Mendi went on to distinguish a mixed ash-sedimentary soil series (Radcliffe 1986:63). This raises again the issue of the continuous nature of the soils that mantle the region, with numerous intergrades existing between one modal class and another. It should be remembered that while the accounts given here tend to describe the clayey soils as if they are derived entirely from ancient marine sediments and the weathering breakdown of either sedimentary parent rocks or volcanic ash deposits, in reality each is often variably influenced by the other. Soils derived largely from ancient deposits and sedimentary rock weathering, for example, may be affected to varying extents by volcanic ash inputs (such as small recent ash falls, which account for the coarser texture of their topsoils, compared to their subsoils, and effect a degree of rejuvination). Or older ash layers may be too thin to obscure the influence of underlying rock or sedimentary derived materials, or tephra may have become intermixed with these through soil movements on slopes, and so on.

The continuous variability of soils is promoted further by the differential rates at which the soil processes that feature in genesis proceed, as illustrated above for volcanic ash soils by the brown and olive subsoils. The main processes at work are the neosynthesis of amorphous clay minerals, high humus accumulation, and the geochemical leaching of mobile products of weathering (notably bases), together with desilication. The soils that develop on volcanic ash depend initially on the nature of the extruded parent material. This can vary in composition but generally it comprisises noncrystalline small glass fragments, pieces of easily weathered feldspars and ferromagnesian minerals, together with varying amounts of quartz. It is andesitic, moderately basic in composition. The ease with which this material breaks down is significant, volcanic ashes being among the most rapidly weathered of materials, as disordered structures with no regular crystals. They are disordered due to quenching when explosively transferred from magma to ash shower, randomly solidifying in the chaotic and rapidly changing conditions of the volcanic eruption. They have many integral points of weakness, susceptible to weathering processes, which are promoted further by the gross open and vesicular physical structure of ash deposits, presenting numerous spaces for weathering, and consisting of many small pieces of debris giving many surfaces for attack (unlike massive bedded rock). Variations in soil mineralogical, physical and chemical properties can be attributed to differences in time since deposition and to differences in degree of weathering, the principal controlling factors of which are temperature and site drainage — as demonstrated by the more weathered lower altitude brown volcanic soils compared to the less weathered higher altitude olive coloured ones. The depth of ash is also relevant, being a function of both the violence of any eruption and volume of matter expelled, and distance from the volcano, the coarser material tending to fall in thick layers near it, the finer material in thinner layers further away, each subject to different weathering rates. Regardless of these variations, we can generalise and say that volcanic ash not only imparts distinctive properties to soils, but also that they follow broadly similar pathways of formation over a range of climatic and topographic conditions.

The generally accepted theory of volcanic soil genesis postulates a series of steps (Mohr *et al.* 1972). These start with the breakdown of the disorderly structured parent materials into small fragments under the rapid weathering conditions that prevail, and from these develop extremely disordered clay minerals. The amorphous non-crystalline structures of volcanic ash minerals promote the development of the disorganised and amorphous clay minerals characteristic of volcanic soils, as variably composed building blocks that break-up in random sequences, producing heterogenous fragments that lack sufficient ordered chain units to form sheet structures. Considerable quantities of both silicon and aluminium also go into solution as ions, aided by the high alkalinity induced by hydrolysis at the mineral surface, away from which there is a resulting steep fall in pH, leading to the co-precipitation of silicon and aluminium in a short distance. The consequent speed of reaction, combined with the random occurrence of the fragments, results in a considerably disordered precipitate product called allophane, which is a randomly structured 1:1 aluminosilicate which, as an amorphous gel, is difficult to characterise but approximates to $Al_2O_3.2SiO_2H_2$. Its open and porous structure retains much water, and what occurs next depends on how wet conditions are, for so long as water remains present no shrinkage can occur nor further cross-linking with consequent mineralogical transformations. Under continuously wet conditions allophane consequently persists and accumulates, which is the situation at moist cloud-enveloped higher altitudes in the Kerewa-Giluwe region where the olive ash soils occur, their dominant clay mineral being allophane (Radcliffe 1986:55).

The existence of two forms of allophane, called prosaically A and B, somewhat complicates the mineralogical series. Allophane A is the normal type in which both silicon and aluminium occur in the same lattice, it is found away from the soil surface. When this normal type of allophane loses water, probably through alternating wet and dry periods, shrinkage occurs and increasing cross-linking results, which leads gradually to more structural order and transformation into the next mineral in the series called metahalloysite, which is a common 1:1 layer clay mineral forming tubular sheets (lacking mica fragments to act as templates for flat ones). It is structurally similar to kaolinite except for an H_2O layer between successive layers — kaolinite being $Si_4Al_4O_{10}(OH)_8$ and halloysite $Si_4Al_4O_{10}(OH)_8.4H_2O$. This similarity points to the next step in the mineralogical series: further water is shed with the formation of kaolinite clay. When kaolinite comprises a considerable part of the clay fraction, the soil will have ceased to be one we should classify as a volcanic ash, being well on the way to becoming a senile tropical soil of the Ultisol order.

In allophane B, the silicon and aluminium occur in discrete lattices in the surface horizon. Organic matter forms a complex with the aluminium lattices preventing their precipitation with silicon to give normal allophane A, so inhibiting the development of metahalloysite. The formation of these stable humus-allophane complexes also results in the accumulation of considerable amounts of organic matter, due to the enormous surface areas involved affording protection against decomposition and the inhibition of microbial decomposition by allophane or its derivatives. The formation of these characteristic black, high organic matter topsoils is also promoted by the permanently wet climate and relatively cool temperatures of the Wola region, conditions that greatly enhance the accumulation of plant waste which cannot be broken down rapidly by soil organisms. It has considerable agronomic significance regarding nutrient availability.

Analyses of the mineralogy of volcanic ash derived clays from the highlands region of Papua New Guinea (see Bleeker 1983:74–87; Bleeker & Healy

1980), although they give somewhat conflicting information,[11] suggest that amorphous minerals, mostly allophane and gibbsite, together with halloysite, are the dominant minerals in the majority of soils of this region, putting them largely in the allophane stage of the mineralogical series, with some halloysite, and even kaolinite formation. They remain comparatively young soils relative to the mineralogical development cycle, their evolution inhibited by the wet and cool conditions.

In addition to the precipitation of clay minerals following weathering, the characteristics of which determine in some measure the charge and nutrient holding properties of any soil, there is also the concomitant release and movement of nutrient elements, the supply of which determine in considerable part the soil's fertility. When released from the parent material these are subject to leaching, and given the porous nature of volcanic ash soils and the region's heavy rainfall, losses of nitrates, calcium, magnesium and various micronutrients can be anticipated. The soils are of low base status due to this severe leaching and are consequently generally acid, in the pH range 4.5 to 5.5 (Table 10.5). But aluminium saturation is low. Rapid weathering ensures a supply of nutrients for plants in rich, relatively unweathered ash soils, it is only in the later stages of development that deficiencies show up. They have high CEC for the tropics (usually >25 meq 100 g^{-1}), their high organic matter contents being largely responsible for this, given the importance of pH dependent surface charge in these soils. The high organic matter contents are also responsible for their high nitrogen values. The exchangeable potassium values are moderate to high too. But available phosphorus contents are generally low, these soils having a high phosphate fixing capacity due to their organic matter contents and the amorphous oxides that characterise their allophanic clays.

The chemical fertility of these soils is less than might be thought, volcanic soils generally conjouring up images of very fertile land. It is highest in

Table 10.5

The average nutrient status of *tiyptiyp* ashy clay soils
(n = 12 topsoils & subsoils; nos. in brackets = σ)

	pH	P (ppm)	K (me/100 g)	Ca (me/100 g)	Mg (me/100 g)	Na (me/100 g)	CEC (me/100 g)	C (%)	N (%)	C:N ratio
topsoil	4.99	9.88	0.86	18.2	4.01	1.10	34.63	16.99	0.97	19.7
	(0.66)	(10.83)	(0.73)	(20.71)	(3.26)	(0.54)	(14.49)	(9.0)	(0.46)	(5.85)
subsoil	5.2	1.67	0.40	3.95	0.90	0.88	16.92	3.49	0.23	12
	(0.75)	(3.0)	(0.90)	(8.87)	(1.79)	(0.76)	(6.50)	(2.61)	(0.15)	(5.9)

---

[11] The confusion arises due to the generally diffuse and difficult to detect patterns obtained using X-ray techniques, resulting from the amorphous nature of allophane, the presence of mineral complexes, and intimate bonding between clay and organic matter.

the organic rich topsoil, with plant roots tapping some nutrients from the rapidly weathering, mineral rich subsoil reserve. Although the fertility of *tiyptiyp* volcanic ash soils is relatively speaking low, with phosphorus, and occasionally potassium too, as major limiting nutrients, land use is the same as on *hundbiy* sedimentary clay soils, with which they are broadly similar in terms of nutrient status, although perhaps with fixation of phosphorus more severe, while available levels of potassium are marginally higher (Radcliffe 1986:122–23). The Wola cultivate them equivalently, and they support a wide range of natural vegetation from primary forest to grassland. Regarding their nutrient deficiencies, the suggestion that the traditional agricultural practice of topsoil mounding leads to pronounced drying and wetting cycles that could promote the liberation of phosphorus and nitrogen is of interest, particularly for those contemplating cash agricultural developments in the region, for the fertility shortcomings of these soils, even where accessible and on moderate terrain, will be a hindrance to market crop promotion.

Alluvial and Sandy Soils

***Iyb dor tilai* alluvial soils** (Fluvent; Fluvisol): The old alluvial soils derived from redeposited volcanic ash, identified by both the CSIRO and AFTSEMU (see Table 10.2), are distinguished by topographic location (occurring on river terraces, fans etc.) rather than soil properties. The Wola class them as *tiyptiyp* volcanic soils, which they otherwise resemble, having organic rich black topsoils overlying deep, slightly stratified, ash-derived, brown clay subsoils. They distinguish them from recent alluvial soils, which they call *iyb dor tilai* (lit: water flood covers). These alluvial soils occur on river flood-margins and terraces, and in depressions subject to periodic flooding.

The *iyb dor tilai* Fluvents are quite variable, although similar overall in having little profile development. These young soils, deposited by river flooding, comprise stratified alluvial layers with marked texture variations. The Wola distinguish between strata, calling them *iyb muw* if they are gravelly deposits, and where they are finer, qualifying this term according to colour, *iyb muw pombray/hundbiy* (lit: water taken black/bright-brown). Both alluvial and sandy soils feature *iyb muw* material, the coarse sandy texture which it imparts to them giving them some similarity for the Wola.

The alluvial soils have a brown to brownish-black A_1 *suw pombray* horizon, with weakly developed medium granular structure, of friable consistency and sandy silt loam texture. The topsoil is of variable thickness, 30 centimetres or more deep, sometimes thinner. It overlies, and merges into, a structureless C horizon, which is brown to yellowish-brown in colour, becoming mottled greyish-yellow with depth if the watertable is fairly near the surface and gleying induced, as is common in locations adjacent to watercourses, putting the subsoil into the *tongom* class (see later). Stratification causes organic carbon content to decrease irregularly with depth (a defining characteristic of the Fluvent suborder). These soils may have various plant remains to depth (e.g. pieces of saturated wood, cane grass stem etc.) The subsoil is of variable texture, from sandy silt to clayey, in the latter event the soil falling into the *hundbiy* clay-like category of *suw hundbiysha*, rather than the *iyb muw* one. And it is of firm consistency, and may contain a variety of rounded waterborne stones, some large.

The nature of the deposits depends on the energy of the nearby watercourse and its transportational capabilities, which are considerable for the turbulent rivers of the mountainous Wola region. The local people are aware that these soils result from waterborne deposition, for on occasion some of them may be inundated, leaving behind a fresh layer of sediment. Pedogenesis is again limited and centres largely on the development of the A1 horizon through incorporation of decomposed organic matter, and the process of 'soil ripening' where draining and evaporation of excess water, aided by plant evapotranspiration, dries and consolidates the soil with the establishment of a regulated moisture regime suitable to a range of plants. The improved conditions result in the oxidzation of ferrous iron compounds to ferric forms, which give the soil its brown hues. The dominant clay minerals are montmorillonite with illite or kaolinite (Bleeker 1983:47).

The alluvial *iyb dor tilai* soils are only moderately fertile. Soil reaction is mildly acid at about pH 5.2 to 5.5 and base saturation is high (Table 10.6). They again have a fair CEC for the tropics. Available phosphorus and exchangeable potassium values are on the low side, probably because these are largely deficient/unavailable in the parent materials from which the alluvium derives. These soils, featuring regular in-washing of mineral material, are also lower in organic matter than the previous soils and have reduced nitrogen contents. But land use is restricted on them anyway because they are commonly sited in locations liable to periodic inundation, any crops cultivated there would be liable to washing away in floods or burial under a layer of sediment. Nevertheless in some places people recognise that it is a fertile soil worth cultivation regardless of the risks, and some families establish a few gardens on alluvial soils (notably of wet-loving taro). Some alluvial soils are cultivable with fair security if gardeners dig ditches to direct accumulated rainfall runoff or stream flow if banks are burst. These soils otherwise support a range of natural vegetation.

Table 10.6
The average nutrient status of *iyb dor tilai* alluvial soils
(n = 4 topsoils & subsoils; nos. in brackets = σ)

	pH	P (ppm)	K (me/100 g)	Ca (me/100 g)	Mg (me/100 g)	Na (me/100 g)	CEC (me/100 g)	C (%)	N (%)	C:N ratio
topsoil	5.25	12.93	0.46	20.75	2.01	1.07	22.65	6.24	0.49	16
	(0.13)	(18.42)	(0.23)	(19.34)	(0.90)	(0.16)	7.74	(4.74)	(0.24)	(9.5)
subsoil	5.48	12.23	0.17	13.05	1.12	1.10	14.83	1.68	0.20	8.5
	(0.15)	(17.07)	(0.03)	(8.09)	(0.76)	(0.62)	(1.71)	(0.51)	(0.04)	(1.3)

Iyb muw sandy soils (Psamment; Arenosol): The Wola distinguish Psamments initially on textural grounds, *iyb muw* being their word, as already noted, for any coarse feeling sediment, from medium sand to gravel, regardless of its mineralogy, colour, origin, and so on (whether occurring in stream bed or soil deposit). The sandy *iyb muw* soils are however well drained, which marks them off from texturally similar alluvial soils. These arenaceous soils occur in the few locations where Tertiary sandstones outcrop at the surface unmasked by tephra deposits (see Map 5.4). They are, together with alluvial soils, the least common soils in the Kerewa-Giluwe region, after the rendzinas and peats.

The *iyb muw* Psamments have a brown A_1 *suw pombray* horizon of moderately developed medium granular structure and friable consistency. The sandy loam topsoil is thin at 10 centimetres or so thick. It overlies, and merges into, undifferentiated and structureless sand, which is greyish-olive to orange in colour, and becomes progressively compacted and massive with depth, containing pieces of soft sandstone, until the hard consolidated sandstone bedrock called *haen naenk* is reached at fairly shallow depth (see Chapter 5). The water holding capacity of these porous soils is low. And they have been subjected to little in the way of pedogenesis beyond the formation of the thin A_1 horizon with the addition of organic matter from dead plant material, and possibly the decalcification of some calcium carbonate contained in any shell fragments. The mineral fraction is dominated by relatively inert quartz grains.

The sandy *iyb muw* soils are of low fertility. They are strongly acid, at around pH 4.7 and low in bases due to the leaching to which their thin topsoils are subject, overlying highly permeable sands (Table 10.7). They have a low CEC compared to other soils in the region, given their low clay content, and small mineral reserves too because of it. Phosphorus, potassium and nitrogen contents are all low. Fertility is largely dependent on the relatively low organic matter content of these soil's thin surface horizons. Land use is predictably restricted on them. Local farmers avoid sandy *iyb muw* soils by and large because it is known to have low fertility, they point to its thin *suw pombray* topsoil layer, which produces little of the *iyba* 'grease' essential to plant growth, and its compact subsoil, which is largely impenetrable to roots. It supports mostly cane grassland and patches of woodland.

Table 10.7

The average nutrient status of *iyb muw* sandy soils
(n = 3 topsoils & subsoils; nos. in brackets = σ)

	pH	P (ppm)	K (me/100 g)	Ca (me/100 g)	Mg (me/100 g)	Na (me/100 g)	CEC (me/100 g)	C (%)	N (%)	C:N ratio
topsoil	4.7	7.27	0.39	7.9	1.93	0.43	20.33	7.04	0.46	12.3
	(0.1)	(4.67)	(0.19)	(3.34)	(0.63)	(0.30)	(4.47)	(6.05)	(0.42)	(8.0)
subsoil	5.0	2.4	0.22	1.93	1.24	0.31	12.73	2.38	0.17	10
	(0.56)	(2.12)	(0.19)	(2.10)	(1.13)	(0.17)	(8.55)	(2.60)	(0.13)	(7.6)

Gley Soils

***Pa tongom* gley soils** (Aquent, Aquept; Gleysols): Any of the above soils may show gley features in wet locations, except for the *iyb muw* Psamments. These are more common on the heavier, less well drained *hundbiy* clay soils than they are on the more permeable *tiyptiyp* volcanic ash-derived clay ones, gleyed versions of which merge into the wet olive ash soil types of higher altitudes. The Wola call any gleyed soil *pa tongom*. It is the third most common soil type in their region. The distinguishing features of these soils are their high water contents and gleyed, grey coloured, frequently reddish mottled, subsoils. Poor drainage conditions, which in turn relate to topography and parent material, largely govern the formation of these soils. They may occur anywhere that the watertable is at, or near the surface, notably at spring sites, in poorly drained depressions and adjacent to watercourses, and even on ridges and slopes on slowly permeable fine grained parent materials like mudstones and siltstones.

The *pa tongom* gley soils have black to dark brown A_1 horizons, called *pa pombray* usually 10 to 50 centimetres thick and silty clay in texture, which are high in organic matter, and sometimes evidence a network of fine reddish brown mottles. They are wet and sticky, but when drier are seen to be weakly medium to coarse subangular blocky in structure, and friable. They may merge into a B horizon that is prominently brownish mottled grey, or merge directly into a C horizon that is greenish or yellowish grey. The grey gleyed heavy clay subsoil may also evidence bright reddish brown mottles. It is structureless, very plastic and very sticky. And very slowly permeable, it is waterlogged.

The genesis of these soils relates to the waterlogged conditions under which they occur. The anaerobic bacteria that function in these low oxygen conditions respire by using, in a preferred sequential order, a series of electron acceptors other than oxygen to oxidise energy yielding organic substrates. Two of these acceptors are iron and manganese, compounds of which, reduced to their Fe^{2+} ferrous and Mn^{2+} manganous states, are responsible for the characteristic grey and bluish hues of gleyed horizons. If conditions are part oxidising and part reducing, such that reoxidisation of iron occurs in better aerated zones — common around respiring, oxygen releasing plant roots and in large pores where the watertable fluctuates up and down — then reddish brown ferric compounds are formed, giving the soil its distinctive mottled appearance. The Wola themselves associate the mottles in *pa tongom* soils, which they call *huwguwk*, with plant roots, pointing out that long rusty coloured mottled veins sometimes contain pieces of root fibre; the mottles result, they say, from the rotting of the parent roots and the water held in the channels left in the soil. The reduced conditions also promote the accumulation of organic matter, which is characteristically high in the dark topsoil, the fine dense root mat frequently associated with the vegetation that occurs under these conditions further encouraging it.

The mineralogy of these soils reflects their origin from one of the above described soils, having present a similar mineral suite, of gibbsite, montmorillonite, metahalloysite and kaolinite among others. The CEC is good and fertil-

Table 10.8
The average nutrient status of *tongom* gley soils
(n = 5 topsoils & subsoils; nos. in brackets = σ)

	pH	P (ppm)	K (me/100 g)	Ca (me/100 g)	Mg (me/100 g)	Na (me/100 g)	CEC (me/100 g)	C (%)	N (%)	C:N ratio
topsoil	5.12 (0.73)	32.42 (24.69)	0.83 (0.30)	25.0 (13.85)	2.58 (0.91)	0.95 (0.48)	30.30 (6.25)	13.22 (4.06)	1.09 (0.42)	16 (2.0)
subsoil	5.22 (0.40)	8.12 (5.72)	0.50 (0.33)	20.78 (12.44)	1.75 (0.99)	0.63 (0.26)	20.66 (3.45)	2.67 (1.96)	0.4 (0.46)	10.4 (6.8)

ity is moderate (Rutherford and Haantjens 1965:95, Bleeker 1983:39,68). Acidity varies widely, although topsoils are acid on the whole, with pH values around 5.1; they have low base status (Table 10.8). At Wola region altitudes these soils have high nitrogen values, the limiting nutrient under many agricultural regimes. Exchangeable potassium levels are good and phosphorus availability is fair given rates of organic matter fixation. Land use is restricted on these soils. When cultivated they largely support wet-loving crops such as taro and sedge. The Wola rarely dig ditches to drain them to grow other crops. They support a range of natural vegetation uncultivated, from forest to tall grassland and semi-swamp vegetation. They have little development potential unless drained, they may sometimes be fenced in and used to graze cattle.

Peaty Soils

***Iyb uw damiy* peat soils** (Hemist, Saprist; Histosols): When wet conditions become severe, gleyed mineral soils give way to peaty organic ones; under marshy conditions of year round saturation and waterlogging organic matter accumulates and peats develop. The *iyb uw damiy* (lit. water is rots) are characterised by layered organic horizons of varyingly decomposed plant material, the strata not always easily distinguished. Peaty soils occur in two contrasting topographical situations in the Kerewa-Giluwe region: on mountain summits (*hat maenda*) above 3000 metres as alpine peat soils, and below this altitude in swampy very poorly drained depressions (*hat suwl*) and on waterlogged seepage sites as bog soils. They cover only a small part of the region.

The alpine peats are shallow black to dark brown, well-decomposed peaty clay soils that overlie consolidated or unconsolidated almost fresh rock or grade into weakly developed thin clay mineral B horizons. Some show evidence of humus illuviation, clay content increase with depth and thin iron pan formation, which might be due to volcanic ash admixtures (Bleeker 1983:55). The *iyb uw damiy* bog soils of lower elevations are young and feature relatively little soil formation, comprising thick layers of black to brown peat of varying depth, which may range from well-decomposed, friable clayey peat, through to soft organic mud, all with very low bulk densities. These peat soils may comprise a series of layers in varying stages of decomposition down a profile, with the more decomposed vegetation in the deeper layers. They may also feature thin layers of in-washed alluvial sediment or tephra deposits, which might increase clay content, and result in them overlapping with *iyb dor tilai* alluvial soils.

The Wola are fully aware that it is the build-up and gradual decomposition of the vegetative material that accumulates in some poorly drained locations that results in these soils. One man explained that "where water has no path [to drain away], then leaves and wood build up and up over time and only rot slowly to give us this soil". Pedogenetic processes are markedly reduced in peaty soils. The more or less permanent saturation prevents air reaching peat deposits and the consequent lack of oxygen greatly reduces bacterial breakdown of organic matter; mineralisation by anaerobic bacteria is at a considerably lower rate than the supply of vegetative matter, which results in its steady accumulation. At higher altitudes precipitation is very high due to the very frequent low cloud cover and evapotranspiration rates are also low due to the cool climate, such that slopes may even remain water saturated; the low temperatures furthermore greatly reduce the activity of fungi which are responsible for mineralisation processes at high altitudes in the tropics (Mohr *et al.* 1972), hence there is a considerable build up of organic matter.

The fertility of these soils is moderate (Table 10.9). They are acid in reaction, at pH 4.9 to 5.9, and consequently of low base status. The accumulated organic matter results in a predictably high carbon content (a defining characteristic of the USDA Histosol order), although even at 15% to 40% this is not exceptional for this region where A1 horizon carbon contents are uniformly substantial. The high organic matter content again accounts for the high CEC. It is also reflected in high nitrogen values; the high C/N ratios that typify these soils, reportedly ranging from 8 to 25 or more (Bleeker 1983:58), suggest a low rate of nitrification and hence small denitrification losses. The available phosphorus levels are again low to moderate, as with other soils generally throughout the region, although exchangeable potassium levels are favourable.

Land use is restricted on these soils. The inaccessible alpine peats are beyond agricultural use, experiencing low cloud and frosts, and remain under montane grassland, featuring alpine species, and patches of mossy forest. The lower altitude *iyb uw damiy* bog soils support a range of hydrophytic natural vegetation, including sedges, reeds, shrubs and bog woodland. Some sites

Table 10.9
The average nutrient status of *iyb uw damiy* peat soil
(n = 2 topsoils & subsoils; nos. in brackets = σ)

	pH	P (ppm)	K (me/100 g)	Ca (me/100 g)	Mg (me/100 g)	Na (me/100 g)	CEC (me/100 g)	C (%)	N (%)	C:N ratio
topsoil	5.45	8.7	0.77	12.6	1.92	6.67	27.7	15.02	1.35	13
	(0.77)	(8.9)	(0.04)	(0.34)	(0.41)	(8.17)	(2.33)	(0.76)	(0.84)	(7.1)
subsoil	5.42	6.17	0.53	16.2	1.14	0.94	21.7	8.69	0.53	13
	(0.73)	(1.0)	(0.13)	(11.45)	(0.26)	(0.46)	(8.91)	(10.4)	(0.52)	(7.1)

under these Histosols may be suitable for taro cultivation, particularly around their margins, so long as not liable to total inundation, resulting in an *iyb mael* pond unfavourable to crops. They are beyond extensive cultivation without considerable reclamation work, notably the digging of drainage ditches to reduce water levels. The people living in the upper Mendi valley have drained and cultivated some of the peat soils in the lake Egari area, so inducing mineralisation in the topsoil layers (Radcliffe's 1986:67 Wimteh soil series), but by and large the Wola infrequently drain these soils, which anyway cover only a small part of their region (unlike elsewhere in New Guinea, in the Baliem and Wahgi valleys for instance, where people engage in extensive drainage measures, doing so since antiquity in some places — Heider 1970:42; Bleeker 1983:60–61; Steensberg 1980:85–87). The soils may have high chemical fertility when cleared and drained, as indicated by investigations of those under commercial tea and coffee cultivation on plantations in the Western Highlands (Drover 1973), and have development potential with appropriate capital investment.

Waip peat soils (Folist, Fibrist; Histosols): In some places vegetation accumulates to considerable depths under less saturated conditions, only rotting down slowly into a peaty organic soil. The Wola call such soils *waip*, a term they apply not only to peaty deposits but also to the litter horizon of decomposing vegetation that may cover the surface of any soil. It is less of a soil taxon to their minds than the other taxa described in this chapter, referring to the layer of rotting vegetation of variable thickness that builds up at the surface, which on mineral soils is necessary to the development of *suw pombray* topsoil, as it disintegrates and supplies the *iyba* 'grease' required for fertility. Indeed it is difficult to draw a line between a thick well decomposed litter layer and a peaty deposit, except that the latter customarily extends to a considerable depth.

The *waip* peat soils are young and feature little, if any soil formation, comprising thick layers of black to brown rotted vegetation to varying depths, largely comprising raw, unripened and open structured organic matter, of very low bulk density. These peats may comprise a series of layers, the extent of decomposition varying with depth, the most decomposed layers being furthest down. The Wola distinguish between the surface layers of relatively unrotted material where vegetative structures remain visible, which they call *iysh shor paenpaen* (lit. tree leaves bed-layer), and the deeper layers of more rotted peaty material, which they call *waip*. Any soil may feature these two layers. They accumulate to depth in only a few locations that favour build-up of vegetative material and where rotting down is gradual. They are common under *Dicranopteris* fern brakes and stands of *Cyathea* tree ferns.

These soils are acidic in reaction with a pH of 4.5 when rotted down (Table 10.10). Very high organic carbon levels characterise them, and high nitrogen levels too. The high C:N ratios suggest low rates of nitrification. Exchangeable potassium levels are moderate, but phosphorus low. If the Wola cultivate such sites they remove the recent peaty overburden, either rolling it up and throwing it off the site or, if it dries out sufficiently after cutting down

Table 10.10
The nutrient status of *waip* peat soil (n = 1 topsoil & subsoil)

	pH	P (ppm)	K (me/100 g)	Ca (me/100 g)	Mg (me/100 g)	Na (me/100 g)	CEC (me/100 g)	C (%)	N (%)	C:N ratio
topsoil	4.49	10.6	0.87	4.14	1.88	1.09	25.5	45.2	1.72	26
subsoil	5.81	0.3	0.18	4.02	0.63	0.67	29.7	11.1	0.71	16

the green top cover, by burning it. They then farm the organic soil, which as it dries out and decomposition proceeds shrinks, becoming more akin to a mineral *suw pombray* A topsoil horizon, and perhaps exposing with time and repeated digging a *payhonez* clayey AB horizon. These organic soils cover very small areas in their region.

CHAPTER 11

KEEPING UP WITH SOIL STATUS: THE IMPLICATIONS OF VARIABILITY

Soil, as everyone knows, is essential to plant growth, yet the Wola, who are highly skilful subsistence cultivators, maintain that assessment of it does not feature in their selection of garden sites. Their apparently offhand attitude to soil on potential cultivation sites is unexpected. According to them, an inspection of the soil before clearing it of vegetation for cultivation is not among the considerations that constrain and influence their choice of site, which include issues like cultivation rights as stipulated by their kin-founded land tenure system, site aspect and ease of enclosure, location relative to house and other gardens, and so on. Perhaps their apparently indifferent approach to soil use and management should not surprise us, for it seems to chime in with their contrary agricultural practices, of semi-continuous cropping within a swidden regime context.

It is possible to 'explain' away their apparently nonchalant attitude to soil inspection on the grounds that its validity is difficult to assess. The people know their local regions so intimately that they have no need deliberately to look closely at the soil at any place before deciding to cultivate it. They already know its status on those territories where they have rights of access to garden land by virtue of living there, constantly walking over them in the course of their daily lives. But my Wola friends insist that even if they found themselves in an entirely unknown part of their region (e.g. by virtue of affinal connections) they still would not closely inspect the soil before cultivating it.

Alternatively, we may try to account for their assertions by arguing that while they think that they do not look closely at the soil before establishing a garden, this is only their perception, and that they are unconscious of their assessment of soil resources (e.g. walking around barefoot that they are tactually aware of texture and structure, or seeing areas rooted over by pigs that they are aware of soil resources without need of digging to inspect them). Furthermore, we might suggest that the vegetation growing on a site may give an indication of the soil's worth, by its health and the prolificness of its growth, even the presence of certain species above others. But again the Wola largely deny that this is so. And field observations support their assertions; floristic composition is not apparently associated with their soil assessments nor different soils (see Chapter 9).[1]

I. SOIL ASSESSMENT

In an attempt to assess the status of Wola disclaimers about soil assessment, and evaluate generally the standing of local soil knowledge, this chapter reports on a survey of soil sites, and compares its findings with an investigation of indigenous soil appraisal. The sites surveyed occurred, by and large, on the local territories of two neighbouring *semonda* communities, situated adjacent to the Was and Ak rivers (Aenda and Ebay *suw* territories — see Map 5.2). It also included parts of the territories of other neighbouring *semonda* communities, particularly those with locations that had good examples of certain soil classes, such as sandy and gley soils, alluvial soils on extensive water-borne deposits (at lower altitudes in the vicinity of lake Kutubu) and clayey soils on volcanic parent materials. The soils inspected came from the major classes distinguished by the Wola, as documented in the previous chapter, to cover the range of soil resources available to them.

The survey sites, eighty-five in all, were not selected randomly, but with two aims in mind, related to the gardeners' unexpected disclaimers about soil appraisal. Firstly, to describe and classify the soils of the area according to a framework sympathetic to local perceptions and judgements, and secondly, to investigate and assess the affects of local land use and agriculture on the soils (Clarke and Street 1967; Wood 1979). In relation to the first aim, an effort was made to include in the survey sample representative examples of the different soil types identified in the area. And in relation to the second aim, sites were selected according to their vegetational cover and slope, in an effort to obtain representative coverage of different land uses, both past and present, and assess their impact on the soils. While trying to obtain a representative sample of the soils in the survey region, by structuring the selection of sites according to the above guidelines to ensure the necessary distribution, sites were otherwise selected in a casual manner depending to some extent on local people's ideas of where would be a good place to look at a particular soil that met certain of the above criteria. Another factor that influenced the choice of sites was that a wide geographical spread of sites from around the region was more desirable than bunching in one locality, where this was commensurate with

---

[1] In order to demonstrate any correlation that we might assume ought to exist between soil potential and vegetation (Allen 1982), we should need to make a comparison, by detailed logging, of plant development on different sites, it not relating to any easily seen macro-botanical feature such as species type. The Wola deny that they look systematically for the presence or absence of certain associations to assess soil fertility. They may clear any vegetational succession, from recent, weedy herbaceous regrowth and grasses, through to cane grass, secondary woodland and primary forest (see Chapter 7 on the composition of these communities). If the vegetation, whatever the succession, obviously looks spindly and poor however, they may take this as evidence of poor fertility (a location where *Miscanthus* cane grass has many dry, brown leaves, for example, may be taken as a sign of a poor soil).

the above objectives. A final factor was that landowners did not object to me digging profile pits on sites (a consideration that applied to gardens only).

The sites and soils were described using a standard survey form (after Hodgson 1976).[2] The data were subsequently analysed multivariately using the SPSS-X computing package, employing ordination by principal components, with factor and cluster analysis (Webster 1977). The analysis distinguished between different classes of soils according to all of the survey data, suggesting correlations that might help assess both the effects of soil status on local land use and of subsequent agricultural practices on the soils.[3] And it compared these findings with indigenous soil judgements and perceptions, to assess the validity of assertions that the soil does not significantly influence and guide cultivation decisions.

The "unearthing" of correlations between land use and soils, and local assessments of the cultivation potential of different soils, demands the classification of soils by profile and site. The procedure adopted is as follows. Firstly, different horizon types are defined, after assessing the variation in properties within horizons. Secondly, different profile types are distinguished, as comprising different sequences of these horizon types. Thirdly, these soil profiles are correlated with the site data to generate an all-embracing classification of soil resources. The strategy overall produces a classification of entire soils, as comprising sequences of defined horizons occurring in specified site locations. Finally, this soil resource classification is compared with land use and local assessments of soil-worth in an attempt to trace connections between cultivation practices and soil status.

Horizons, Profiles and Sites

Any pattern to the variation in properties within horizons was assessed initially in a series of cross-tabulations of those pairs of soil properties that might reasonably be expected to show some correlation in their variation. While

---

[2] The first page of the form described the site (relief, drainage, land use and so on). The second and third pages recorded horizon details for a soil profile dug centrally on the site (depth, colour, structure, and so on). The nominal depth of the profiles was set at 50 cm, as that most relevant to an assessment of soil-related factors affecting agricultural use of the land, but this was often exceeded. The fourth page included an assessment of soil-site relationships (soil variation across the site), and a local description of the soil and assessment of its agricultural potential. The latter description and assessment was supplied by those who assisted me, plus any other individuals who happened to be on site during the survey, of whom there were invariably one, two or more. They were asked to comment on the agricultural potential of the soil and assess its status on a six-point scale (as very good, good, middling, poor, very poor, or waterlogged), a ranking procedure with which the Wola are familiar, having a series of linguistic markers to indicate the relevant gradations (e.g. *ebay ora* very good, *ebay* good, *ebay sha* middling, *kor* poor, *kor ora* very poor and *suw pa* waterlogged) — see Sillitoe 1979:116 for comments on this manner of assessment, which the Wola use frequently when ranking anything, achieving relatively fine distinctions.
[3] For further analytical details, including cluster dendrograms and soils data, see the *Papua New Guinea Journal of Agriculture, Forestry & Fisheries* Vol. 36.

some evidenced a degree of correlation, the results were not conclusive, indicating that the horizons do not vary internally in any straightforward manner with certain properties varying regularly in relation to one another. The majority of properties showed a random distribution and no correlation. This lack of success was not surprising because if an obvious variation in properties was present within horizons I should have expected local people to have commented upon it. A more complex classification has to be anticipated than one where only a few key parameters control class ascription. The properties responsible for the larger part of the variation within horizons was assessed by principal components analysis (the first three factors accounted for the greater part of the variation that occurred within each horizon), and a classification generated using cluster analysis (an agglomerate method which creates a hierarchy from a similarity matrix by the nearest neighbour strategy; four clusters identified per horizon). The **horizon variation and classes** were as follows:

Horizon 1 Differentiated largely by stoniness, thickness of horizon and depth of rooting, and differences in colour and texture, as follows:

cluster 1: thick (6cm), dark reddish brown, virtually stoneless horizon with many fibrous roots throughout.

cluster 2: thin to thick (av. 4cm), very dark reddish brown, very slightly stony horizon (small rounded chert stones) with abundant woody, fibrous and fleshy roots throughout.

cluster 3: thin (2cm), very dark reddish brown, stoneless horizon with abundant woody fibrous and fleshy roots throughout.

cluster 4: thick (7cm), very dark reddish brown, very slightly stony horizon (very small and small rounded volcanic stones) with abundant woody and fibrous roots throughout.

There was not a great deal of variation within horizon 1 on this analysis (even between the two main groupings) and the lumping of all such horizons into a single class called *waip* by local people is understandable.

Horizon 2 Differentiated largely by mottling, together with structure, consistence and texture, plus thickness, colour, and porosity (although standard errors indicated considerable overlap between clusters in some of these properties — e.g. colour), as follows:

cluster 1: thick (20.5cm), brownish black, scarcely mottled silt loam; slightly sticky, moderately plastic and friable; blocky and granular structured, medium packed acidic horizon of middling porosity.

cluster 2: average thickness (14cm), dark brown scarcely mottled silt loam; slightly sticky and plastic and very friable; granular structured loosely packed acidic horizon of high porosity.

cluster 3: thick (17.3cm), brown, faintly mottled silty clay loam; slightly sticky, moderately plastic and friable; granular and angular blocky structured medium packed acidic horizon of low porosity.

cluster 4: variable (av. 8cm), greyish brown, unmottled silty clay loam; moderately sticky and plastic, and very friable; granular or angular blocky structured medium packed alkaline horizon of low porosity.

The differences between these horizon 2 clusters is again not large, as the standard errors indicated, and the lumping of all such horizons by the local people into a single category called *suw pombray* is reasonable (which they may qualify as necessary according to stoniness (*pombray araytol*), wetness (*pa pombray*), sandiness (*muw pombray*) etc. — see Chapter 10). The principal difference between the two main groupings is structural: a loosely packed soil of low density, poor coherence and highly porous crumb structure on the one hand, and a more packed soil of higher density and coherence, and less porous more aggregated blocky structure on the other.

> **Horizon 3** Differentiated largely by texture, consistence, organic matter status and colour, plus mottling and horizon thickness, together with stoniness, soil water, structure, porosity, and roots, as follows:
> **cluster 1:** average thickness (23.3cm), bright/yellowish brown, mottled, very slightly stony (small subrounded chert), mineral, sandy clay loam; slightly moist, moderately sticky, very plastic and firm; fine through coarse angular blocky or angular blocky and granular structured horizon of low porosity and few roots.
> **cluster 2:** thin (6cm), brown, unmottled, slightly stony (very small to medium angular chert), humose sandy clay loam; very moist, moderately sticky and plastic, and friable; fine and medium granular and/or angular blocky structured horizon of low porosity and roots common.
> **cluster 3:** average thickness (28cm), dark reddish grey, unmottled, stoneless, humose silty clay loam; wet, slightly sticky, moderately plastic and very friable; fine through very coarse angular blocky structured horizon of low porosity and many roots.
> **cluster 4:** thick (41cm), dull brown, unmottled, stoneless, humose sandy loam; slightly moist and sticky, moderately plastic and very friable; fine and medium granular and angular blocky structured horizon of middling porosity and few roots.

The differences between horizon 3 sub-types are more marked than for the previous two horizons, although they may be more apparent than real given the skewed distribution. The principal differences between the two major groupings relate to colour, stoniness, structure and rooting: on the one hand a more red-brown, nearly stoneless horizon with larger more aggregated peds and deep to fine roots, and on the other a more yellow-brown slightly stonier horizon with smaller less aggregated peds and shallower to very fine roots. This dual division parallels the identification of two principal types of transition horizon between topsoil and subsoil by the Wola (when they identify any at all). They distinguish between and name these horizons according to the dominance of properties from either the topsoil (horizon 2) above or the subsoil (horizon 4) below as *pombray sha* or *hundbiy sha* (qualified further for stoniness and so on by terms like *araytol* etc. where necessary). Their discrimination rests largely on colour differences.

> **Horizon 4** Differentiated largely by mottling and stoniness, texture, rooting, colour and porosity, plus soil-water and root depth, as follows:
> **cluster 1:** orange/bright yellowish brown, few to fine faint mottles, slightly stony (very small through medium subrounded chert), silty clay; slightly moist, low porosity, with few shallow (41cm) woody and fibrous roots.

cluster 2: brown, few to medium to distinct mottles, moderately stony (small and medium rounded volcanic), sandy clay; slightly moist, moderately porous with few deep (57cm) fibrous and fleshy roots.

cluster 3: greenish/bluish grey, few fine distinct mottles, almost stoneless, silty clay; wet, low porosity with deep (64cm) fibrous roots common.

cluster 4: light grey, very many to coarse prominent mottles, very slightly stony (very small through medium subrounded and subangular volcanic), sandy clay; very moist, low porosity with few shallow (39cm) woody and fibrous roots.

The cluster analysis separated some (though not all) of the anaerobic gleyed horizons from the aerobic ones, which parallels the local named discrimination, between *pa tongom* gleyed horizons and others which are aerobic. It is difficult to see any other similarities between cluster groups and local distinctions. According to this agglomerative analysis there is little variation in horizon 4 at most locations. When we consider the extensive range of properties by which we are differentiating between horizons, these horizons are more similar than dissimilar. Further differentiation between horizons depends almost exclusively on stoniness, paralleling another distinction made by the local farmers, who regularly suffix a horizon designation with the word *araytol* or stony when the soil has a considerable stone content.

When the horizon clusters were amalgamated together into profiles a total of twenty-eight combinations resulted, which varied considerably in occurrence (50% of the classes were represented by one site only). The classification was too cumbersome, having too many classes defined by too many fine distinctions. The following steps are taken to rationalise it to a manageable scheme:

1) Ignore horizon 1 because under cultivation it is a transitory horizon, it is lost entirely when a site is cultivated; furthermore, when present, this horizon evidences very little variation.
2) Ignore horizon 3 because it is a transition horizon that is often not present at all, and is infrequently identified by the local people.
3) Divide horizon 2 into two clusters, as justified by the cluster analysis.
4) Divide horizon 4 into two clusters, again as justified by the cluster analysis.

Five of the eight possible rationalised horizon combinations occurred together as observed profiles, three comprising two horizons and two only one. The two-horizon profiles were a pair of aerobic topsoil/clay profiles and one gleyed profile, and the single horizon ones were alluvial and skeletal profiles. These five **profile classes were** as follows:

profile a: coherent, more clayey topsoil with aggregated less porous structure; over a slightly stony and moist, faintly mottled, firm aerobic orange clay subsoil. The most abundant profile in the region (no. sites = 58).

profile b: looser, more silty topsoil with crumbly porous structure; over a slightly stony and moist, faintly mottled, firm, aerobic orange clay subsoil. The second most abundant profile in the region (no. sites = 20).

profile c: coherent, more clayey topsoil with aggregated less porous structure; over a wet and distinctly mottled, very sticky anaerobic greenish-grey subsoil. A waterlogged and gleyed profile. An uncommon profile in the region (no. sites = 3).

profile d: coherent, more clayey topsoil with aggregated less porous structure, extending to considerable depths (50cm). A recent alluvial profile. An uncommon profile in the region (no. sites = 2).

profile e: thin organic horizon 1 (ignored); over a slightly stony and moist, faintly mottled, firm aerobic orange clay subsoil. An uncommon profile in the region (no. sites = 2).

The generation of an overall classification of soil resources as perceived by the local people required the combination of the above profile classes with site parameters. This demanded the establishment of some patterned variation between the sites surveyed. Again, a series of preliminary cross-tabulations of those parameters that might reasonably be expected to show some degree of relatedness in their variation evidenced surprisingly little patterning, other than the obvious, so the factors featuring in site variation were assessed by principal components analysis, and site classes were created by cluster analysis. Two major site groupings emerged, variation between them centring on differences in altitude and parent material (variation between the two clusters that comprised each of these groupings centred largely on differences in aspect and slope, and to a lesser extent drainage, relief and vegetation). The number of sites in each cluster varied considerably. The skewedness of their distribution reflected the overriding importance of higher altitude locations to the Wola people. The greater part of their territory is in the highlands, the vast majority live and cultivate here. It was decided to combine all the low altitude sites (<1500 m) into a single group of varied sites and further divide the higher altitude ones to overcome this problem, giving five site classes. This **site variation and classes** were as follows:

Sites: Variation due principally to differences in site relief and drainage, altitude and vegetational factors, plus outcropping of rocks to a lesser extent.

site i: higher altitude (2015m), east-north-east facing, steeply sloping, normally drained interfluve site on limestone, possibly supporting secondary regrowth (no. sites = 23).

site ii: mid-altitude (1815m), east-north-east facing, steeply sloping, normally drained lower valleyside site on limestone, probably supporting grass or crops (no. sites = 30).

site iii: higher altitude (1983m), north-west facing, steeply sloping, shedding interfluve site on limestone, possibly supporting secondary regrowth (no. sites = 16).

site iv: mid-altitude (1770m), west-north-west facing variably moderately steeply sloping, receiving (with some run-off) lower valleyside site on limestone, possibly supporting secondary regrowth (no. sites = 11).

site v: low altitude (1230m), variably north through west to south facing, level to steeply sloping, receiving (with some run-off), plane valleyfloor or bench site on volcanics or alluvium supporting secondary regrowth or crops (no. sites = 5).

In summary, the classification divides sites by an interrelated series of parameters: altitude, local relief, drainage and slope. Higher altitude sites occupy up-slope locations on interfluves and upper valleysides, from which, steeper sloping, drainage is good due to rapid run-off. Mid- to low-altitude sites tend to occupy the lower down-slope locations on valleysides and floors which are more gently sloping and from which drainage is more likely to be poor.

Soil Resources and Local Assessments

The final step in this investigation of soil and site survey data requires the combination of soil profile and site classifications to generate an overall **soil resource classification**, to compare with local people's perceptions of, and agricultural use of, their land. When combined, fourteen of the possible twenty-five soil profile and site combinations were represented, some considerably more often than others (occurring from one to twenty-one times — Table 11.1).

We are now in a position to assess the status of Wola assertions regarding their non-inspection of soil before gardening it, by comparing the computer-generated soil resource classes with local assessments of the cultivation value of the soils inspected and the use to which they actually put the land. A crosstabulation of local assessments of soils against the soil resources classes revealed no apparent relationship. It appears that none of the soil resource classes painstakingly distinguished correlates with any local assessment class, the very best and worst soils are likely to occur in almost all the soil resource classes. We cannot say that any of the classes in which the majority of soils observed fall represent better or indifferent arable soils. The exceptions have too few sites to allow us to draw reliable conclusions. But there are some trends noticeable upon inspection which merit comment (Table 11.2 — soil profile and site classes separated to make comparison easier).

The soil profile class **a** 'coherent topsoil over orange clay' predominates over the entire spectrum of assessment classes, except for **6** 'waterlogged' soils. But the ratio to class **b** 'loose topsoil over orange clay' changes as assessment falls, from 4:1 for classes one and two to 2:1 for classes four and five. In other

Table 11.1
The soil resource classification

Soil profile classes	Site classes				
	(i)	(ii)	(iii)	(iv)	(v)
a	16	21	11	8	2
b	7	8	3	1	1
c		1		2	
d					2
e			2		

Table 11.2
The distribution of local assessment and land use classes between the soil profile and site components of the soil resource classes.

	SOIL RESOURCE CLASSES									
	Soil Profile					Sites				
	a	b	c	d	e	i	ii	iii	iv	v
Local assessment class:										
1. very good	10	2		2		2	5		2	5
2. good	15	3				7	5	3	3	
3. middling	6	2				1	6	1		
4. poor	12	3				4	8	5	1	
5. very poor	14	6			2	9	5	7	1	
6. waterlogged	1	1	3				1		4	
Land use class:										
1. never cultivated	12	2	2		1	5	2	5	4	1
2. long abandoned garden	20	4				8	9	3	3	1
3. recently abandoned garden	10	6		1		5	6	4	1	1
4. sweet potato garden (inc. fallow)	14	8				5	13	2	1	1
5. taro garden	1	1	1				2	1		
6. mixed garden	1			1		2				

words, any soil resource class featuring loose topsoil is more likely to be judged an agriculturally poor soil. The predominant sites in all local assessment classes, except for **6** 'waterlogged soils', are **i** 'high altitude ENE facing up-slope sites' and **ii** 'mid-altitude ENE facing down-slope sites'. The feature they have in common is aspect, and this complies with Wola assertions that the best arable land faces the sun to the north-east. The predominance of sites with this aspect reflects the non-random selection of the survey sample, which strived to accommodate all local land usages; hence it includes a large number of cultivated or once cultivated sites, which have this favoured aspect. But aspect alone is not sufficient to give a soil agricultural promise, and sites facing in a north-easterly direction cover the entire range of assessment from good to poor. The distribution of the next site class **iii** 'high altitude NW facing up-slope sites' between good and poor assessment classes further illustrates the importance of aspect, for these sites are twice as likely to occur in the poor classes four and five as they are in the good classes one and two.

The **v** 'low altitude' sites occur without exception in the first-class local assessment category. Although they are too small a sample to draw firm conclusions from, they reflect Wola assertions that soils occurring at lower altitudes are better, that crops grow more quickly there. They say that they avoid them because they fear sickness (notably malaria) and also because they are adjacent to the region of the Foi people whom they fear as sorcerers. Higher temperatures contribute to the perceived increase in agricultural productivity, but people attribute it in part to the occurrence of larger areas of favourable land too. This lower region has larger areas of recent alluvial soils than higher altitudes, and soils surveyed in this profile class **d** fall into the best local assessment class one. The soil profile class **e** 'thin organic layer over mineral horizon' predictably falls into the worst assessment class five. And profile class **c** 'gleyed soils' predominates in local assessment class six of waterlogged soils suitable for taro. All the soils occurring in this latter assessment class fall as expected in the down-slope locations of site classes **ii** and **iv** where drainage is more likely to be poor.

There is scant correlation between this chapter's computer-generated soil resource classes and the previous chapter's indigenous soil taxonomic classes (Table 11.3), which is perhaps expectable where the local classification scheme gives prominence to only a few of the observed properties used to construct the soil resource classes, such as colour (e.g. *pombray* and *hundbiy*, 'dark and bright brown'), texture and stoniness (e.g. *muw* and *araytol*, 'sandy and cherty'), and so on. Furthermore, the Wola are adamant that their soil classification scheme bears no relation to potential soil productivity, which is borne out in a comparison of local assessments of soil potential with indigenous soil classes (Table 11.4). The way in which they classify soil is not significant regarding their judgement of its agricultural potential (except for the *pa* gley class of soils thought suitable for taro). But the soil resource classification formulated here is no more successful at predicting the arable worth of soil as judged by local farmers. It is the uncommonly occurring alluvial, gleyed and skeletal soils that display a certain predictability regarding their local assessment; the majority of soils show scant pattern. Besides, it seems dubious to argue on the basis of local assessments of soil potential, when the people themselves both deny that they inspect the soil before they decide whether or not to cultivate it, and it is suggested that they are possibly unaware of any pedological judgements they make. We can overcome this objection to some extent by comparing the soil resource classes against land use.

The use people make of land is equivalent to their assessment of its worth in practice (although this approach has shortcomings too, we cannot simply equate cultivation of land with good soil, its condition possibly having deteriorated under cultivation from what it was before clearance; nor can we equate recently abandoned plots with poor soil, factors other than soil-related ones prompting abandonment). Again, the initial impression of a crosstabulation is that no land use predominates in any of the soil resource classes, which suggests that none of them are indicative of good or poor arable soils (Table 11.2). But there are once more some trends worthy of comment.

The soil profile class **a** 'coherent topsoil over orange clay' again predominates over the entire spectrum of land uses. There is no indication that a larger proportion of the soils in this class are more likely to be under cultivation than under secondary regrowth or primary forest, they are fairly evenly spread. The profile class **b** 'loose topsoil over orange clay' is not so evenly distributed, the figures suggesting that it is likely to be cultivated, with a ratio of 2:3 sites cultivated:uncultivated, compared to 1:3 sites cultivated: uncultivated for soil profile class **a**. This contradicts the trend of the local soil assessment comparison where any class featuring loose topsoil is more likely to be assessed as

Table 11.3
Wola profile classification compared with soil profile clusters

Wola Soil Classification	Horizon Cluster Sequence for Soil Profiles																																						Total
pombray	1																																						1
muw pombray		2																																					2
pombray+hundbiy	10	6	3	3	1	2	1	3	1	1	1																												32
pombray+hundbiy araytol	2	1																																					3
pombray araytol+hundbiy			1			1																																	2
pombray araytol+hundbiy araytol	4	3	1									1																											8
pombray+hundbiy+hundbiy araytol							1																																1
pombray+hundbiysha+hundbiy																	1																						1
pombray araytol+hundbiysha araytol+hundbiy araytol	1		1						1																														3
pombray+hundbiysha+tongom					1																																		1
waip+pombray+hundbiy+tongom								1																															1
waip+pombray+hundbiy				4						1																													5
waip+pombray+hundbiy araytol								1					1																										2
waip+pombray araytol+hundbiy araytol				1																																			1
hundbiysha araytol+hundbiy araytol	1																																						1
dowhuwniy+hundbiy araytol																																	1					1
waip+tiyptiyp																																			1			1
pombray+tiyptiyp	1																	1																					1
waip+pombray+tiyptiyp					1																																1		2
pombray araytol+tiyptiyp										1				1													1			1					1				5
pombray+tiyptiyp+kolbatindiy						1																																	1
pombray+iybmuw		1			1																																		2
waip+pombray+iybmuw	1													1																									2
pombray+pombraysha+haenhok																										1													1
pombray+pa tongom	1					1	1				2																												1
waip+pa pombray+pa tongom																																						1	4
																																							1

Table 11.4
Indigenous profile classification compared with local assessment of soil agricultural potential

Wola Soil Classification	Wola Assessment Classes					
	1	2	3	4	5	6
pombray			1			
muw pombray	2					
pombray+hundbiy	3	8	3	9	9	
pombray+hundbiy araytol	1	1			1	
pombray araytol+hundbiy			1		1	
pombray araytol+hundbiy araytol	1	4	2		1	
pombray+hundbiy+hundbiy araytol				1		
pombray+hundbiysha+hundbiy				1		
pombray araytol+hundbiysha araytol+hundbiy araytol	1					2
pombray+hundbiysha+tongom	1					
waip+pombray+hundbiy+tongom				1		
waip+pombray+hundbiy	1	1		1	2	
waip+pombray+hundbiy araytol		2				
waip+pombray araytol+hundbiy araytol						1
hundbiysha araytol+hundbiy araytol					1	
dowhuwniy+hundbiy araytol						1
waip+tiyptiyp					1	
pombray+tiyptiyp						2
waip+pombray+tiyptiyp	4	1				
pombray araytol+tiyptiyp				1		
pombray+tiyptiyp+kolbatindiy				1	1	
pombray+iybmuw		1			1	
waip+pombray+iybmuw				1		
pombray+pombraysha+haenhok			1			
pombray+pa tongom						4
waip+pa pombray+pa tongom						1
TOTALS	14	18	8	18	22	5

agriculturally poor. This may be explained, at least in part, by the larger proportion of **b** profile class soils coming from recently abandoned gardens (at a ratio of 1:2 recently abandoned to other classes, compared to 1:4 for **a** profile class soils). This suggests that a larger proportion of them might be judged as agriculturally tired soils on which cultivation has been abandoned to allow them to regenerate. It is also possible that some of the soils under cultivation were showing signs of declining productivity, and with a larger proportion of soil profile class **b** soils under cultivation there is an increased chance of them showing signs of becoming productively poor soils.

There is a wider spread of different land uses across the various site classes than across the soil profile classes. There is perhaps a weak trend for down-slope sites to be preferred over up-slope sites for cultivation. The aspect preference is again evident, with a greater proportion of north-east facing sites under cultivation and north-west facing ones under natural vegetation. The land under primary forest and never cultivated covers the whole range of soil-site classes, indicating that none of the soil resource classes identified is either the product of cultivation nor is so preferred above others and widely used as to occur less commonly under virgin forest. It also suggests adequate availability of those soils adjudged suitable for cultivation by the Wola. The low altitude soils also show an even spread over the land use range. The position with wet soils and sites under taro is not so clear-cut as might be anticipated, the multivariate analysis not assigning all wet soils to the gley class, some evidencing more properties akin to another class.

The conclusion of this comparison of local assessment and land use against the computer-generated soil resource classification is that there are surprisingly few correlations between them, where we might have expected some we have found remarkably little. We cannot tell the local Wola people, using the soil resource classification painstakingly built up by analysing data on their soils, what they might unwittingly look for in selecting a site for cultivation. The analysis has strived to identify a series of soil and site types, to uncover some deep hidden pattern underlying them that might relate to soil potential and cultivation use, and it has largely failed for the vast majority of soils to reveal anything other than a few trends, some of which are fairly weak and tenuous. In some regards this may be deemed a negative result, the investigation be thought to have failed. But only to those who assume the superiority of Western environmental scientific method, that it should come up with a more insightful and informed interpretation of soil resources than the people who for generations have successfully derived their livelihood from cultivating them. If one believes that the knowledge of local people is valid and informed, the study is a success. Regardless of efforts to generate an objective and scientifically grounded appraisal of soil resources, it has failed to go beyond what the local people know.

The survey analysis has vindicated Wola assertions that they have no need to inspect soil closely before cultivating it, not because they already know its status but because of the strikingly uniform nature on the whole of the soil resources available to them. The outward uniformity of the soils, which is central to an understanding of local pedological lore, can be attributed in part to their young age (Wood 1987). Inceptisols vary relatively little compared to soils of older orders having existed for an insufficient time to diverge markedly under the varying pressures exerted by different local environments. Volcanic rejuvination of soils (see Chapter 5) contributes further to their youth and uniformity. There is not consequently a great deal to choose between the majority of soils of the Wola region by readily observed properties (i.e. those not involving laboratory analysis). While the foregoing analysis distinguished between some of the grossly different soils (such as recent alluvial, gleyed and skeletal profiles), these only cover a small part of the region. The larger part of it comprises the 'dark topsoil/clayey subsoil' classes of soils which the analysis consistently grouped together into stable clusters (when it divided up the horizons that comprise these soils into cluster groups, it made some very fine distinctions, which neither local people nor soil scientists would consider significant). We can see that the Wola cultivate the majority of their crops on very similar soils (the notable exception is wet-soil-loving-taro), which cover by far-and-away the larger part of their country. In the light of this evidence their assertions no longer seem so remarkable, the soils of their region being, by and large, so similar that close inspection would be pointless.

Local Appraisal: Variable Soil Properties

Nonetheless not all soils are the same when it comes to cultivation. The Wola do assess soils, if not before gardening, then certainly when they are under cultivation. In this event, how do they judge the worth of any soil, and why is this a post-cultivation process? The Wola assess the agricultural potential of their soils according to a few properties which they take to be critical to their productivity. They relate to topsoil largely, focussing understandably on the horizon in which crops largely root and grow. Nevertheless, while it is the status of the dark *pombray* topsoil that is critical in the appraisal of agricultural potential, the Wola recognise that the subsoil can influence the character of the topsoil, especially if the latter is thin and the former near the surface. The properties central to the appraisal of a soil's productive status include its depth, strength, stoniness, colour, *iyba* 'grease' content, and water state. The properties featuring in **local soil appraisal** are as follows:

> The **depth** of topsoil, which may be assessed as *onduwp* (lit: much) or *g^{e}nk* (lit: little) or qualified versions of these words, is important as the medium in which crops are recognised to grow well. Although there is really no lower limit to the thickness of topsoil acceptable in a garden, the thicker it is the better, and if the subsoil shows through in places the site is likely to be abandoned.
> The **strength** of the soil, and assessment of it, relates in part to concerns over soil depth because the clayey subsoils are judged too strong for good crop growth. By strength the Wola are referring to the consistence and friability of the soil. They talk of soil as *buriy* (lit: strong) or *tomiy* (lit: weak) or as a qualified version of these terms, and assess it as a handling characteristic. A *tomiy* soil may have a good, friable, crumb structure, on the other hand it may be damp and soft, which are not necessarily good agronomic features. If a soil is *buriy* strong its agricultural usefulness is low because they say roots and tubers have trouble penetrating it, the mechanical resistance to their growth results in stunted development and poor yields.
> The **stoniness** of a soil only becomes critical when it exceeds a certain percentage, hindering cultivation and acting to increase soil strength, impeding adequate root development. Some stones are judged beneficial to a soil. They act to warm it up the Wola say, heating in the sun and retaining the absorbed heat longer than soil alone, so promoting the growth of crops which prefer a warm soil to a cool one. Stones also promote porosity, creating cavities and points of weakness in the soil which roots can exploit, and so off-set soil compaction. They break the soil up according to the Wola and promote friability, a process which they refer to as *kolkol kay* (lit. crumble say). And some stones, especially *araytol* chert or large limestone outcrops, promote the development of *iyba* 'grease'; people say that the rocks 'give grease' to the adjacent soils, which may as a consequence be

cultivated 'over and over again' (the idea that stones as well as rotting vegetation can augment a soil's fertility is an intriguing one in the context of shifting cultivation where to-date we have only considered as relevent the regeneration and decomposition of natural vegetation). The *iyba* may originate from seepages adjacent to outcrops.

The **colour** of topsoil is also significant in assessing the likely fertility status of a soil. The darker the soil the better. It does not necessarily have to be black to be judged good, dark brown is acceptable too, indeed it may be judged best, black often indicating a soil too wet ideally to suit sweet potato (although tolerable for taro cultivation). A light brown soil the Wola consider to be less productive, the pale shade frequently resulting from mixing with brighter coloured, structurally more coherent, and stronger clay subsoil. Topsoil colour reflects organic matter content, darker soils having higher organic matter status. It also relates to *iyba* 'grease' levels.

The *iyba* (lit: blood or sap) or *hobor* (lit: fat or grease) **content** of a soil derives primarily from rotting plant matter. It is assessed by the soapy silty feel that organic matter imparts to soil, the greasier the better. It 'dries' out under cultivation, less rotting plant material being returned to the soil, which becomes *suw kabiy* (lit. soil dry). This 'drying' may continue until the soil becomes 'grease' exhausted or *iyba na wiy* (lit: *iyba*-'grease' not resides) — it is interesting to note that a weak, sick person is also *iyba na wiy*, that is someone without blood. The growing crops take up the *iyba* until little remains. The only crop that can continue to yield on a considerably *iyba* depleted soil is the staple sweet potato. When the garden is abandoned, the rotting of vegetation deposited by the regrowth will replenish the *iyba* 'grease' levels of the topsoil.[4]

The **water state** of a soil, its *iyb* content, is critical for the healthy growth of crops. The majority require a moist aerobic soil; the staple sweet potato cannot tolerate conditions too wet. The notable exception is taro which thrives in a waterlogged soil. Women also dislike wet soils because they are heavier and stickier, making them difficult to work with digging sticks. While the distinction of waterlogged *pa* sites from others is straightforward, differentiating between moister and drier better drained soils is not easy. When under natural vegetation soils tend to be wetter, and the extent to which they will dry out and improve when cleared and exposed to the sun is difficult to assess.

An element of chance features in all of these appraised properties, which relates to Wola assertions that they do not inspect soils before cultivating them. They are all subject to change once a soil is under cultivation. The depth of topsoil is liable to diminish somewhat due to erosion, notably on slopes in newly planted gardens where the soil is exposed and scarcely protected by

---

[4] There is some parallel between this concept and the early notion of European science of 'juices of the earth'. It is also noteworthy that recently the Wola have asserted that their gardens are 'drying up, the soil becoming like ash' because of the oil well recently established at lake Kutubu, maintaining that it is draining away the oil from under their region. The association of *dez* mineral oil with *suw iyba* soil grease is a new one, dating from the pumping of oil in the early 1990s, which is when they say soil conditions started to deteriorate.

vegetation (see Chapter 6). And loss of fine soil particles, leaving the larger stones behind, can increase stoniness beyond the point where it imparts beneficial qualities to the soil, hindering cultivation and crop growth. The soil's strength is thus likely to increase and diminish yield potential, especially if the subsoil is exposed and subsequently mixed with the topsoil during cultivation. It is not only the incorporation of clayey subsoil that increases *buriy* strength, some topsoils can become excessively dry and hard when exposed to the sun for a considerable period of time under cultivation, and if they have a non-granular structure this can render them unsuitable for further cultivation.

The change in soil water content under cultivation is difficult to judge, but it usually falls. Until the sun has 'looked on' the soil, as the Wola put it, they cannot be sure of its water state under cultivation; it is possible that the soil might rapidly become too dry and 'strong'. Furthermore the water state can be adversely affected during cultivation of a site. When establishing a garden people are careful to keep off the site after prolonged heavy rain for fear of puddling it with their feet into a *suw mondow* state of liquid mud; for the same reason they take care clearing areas where there is *gaimb kolowmon*, a thick black layer of rotten water-filled cane grass stems, which trodden in will puddle and degrade the soil's structure and render it unsuitable for cultivation.

The organic matter related *iyba* content is certain to decline under cultivation (see Chapter 12) — a fall in carbon content being long associated with fertility decline and site abandonment under shifting cultivation (Nye and Greenland 1960; Sanchez 1976; Brams 1971; Zinke *et al.* 1978). Again rate of depletion is not easy to estimate, although some locations are customarily recognised as more likely to retain respectable *iyba* levels than others (such as folds and down slope locations), but these may be too shaded and not see enough sun for optimum crop growth, sufficient to reduce water content, warm the soil, and allow maximum photosynthesis.

Agricultural Potential: Developmental Soil States

While the Wola have no series of appraisal class terms that they can apply to a soil before cultivation, to label its agricultural potential, they have a clear idea of what comprises a good, bad or indifferent agricultural soil. But they talk only in generalities using these criteria, not specific predictions. A good topsoil for example, should extend to a fair depth, ideally with an abundant store of *iyba* organic matter, have a reasonable water content, and perhaps a modicum of stones. Structure features too in any soil assessment, the Wola having a keen sense of this, referring to small sized crumb aggregates as *suw ombo*, and larger blocky ones as *suw kompae* or *suw hul* (lit. soil bones). A good soil has a loosely packed, non-coherent, porous crumb structure, such a friable soil is *suw kolkol bay* (lit. soil crumbles does). A commonly heard phrase for such a friable granular soil is *dowhuwniy nonbiy* (lit: sweepings like), which likens it to the crumbs of rubbish, grit and dirt people periodically sweep from houses and houseyards. We can perhaps sense in part what they are looking for here in the cluster division of horizon 2 (the horizon that most often equates with *pombray* topsoil), into loosely packed soils of low density, poor coherence and highly porous crumb structure, and more packed soils of higher density and coherence, with a less porous more aggregated blocky structure.

These criteria, by which local people assess the agricultural potential of a soil, cut across their soil classification classes (see Chapter 10) although they may be used to qualify a class, by referring for example to *pombray buriy* 'strong topsoil' or *pombray iyba wiy* 'topsoil with *iyba* grease'. These criteria relate to a series of soil states distinguished by the Wola rather than to soil classification. They serve to demonstrate further how appraisal of soil potential is largely a relative issue for them, closely associated with time and use. It is an ongoing process, the course of which is not readily predicted, and it is based on observed soil performance under cultivation, occurring during and after land use rather than before it. During the survey work, when asked for an assessment of a soil's worth, people regularly looked at how any crops were growing, whether they were flourishing or not, before pronouncing on the issue. While the soil state classes distinguished by the Wola relate to soil assessment, they do so post-cultivation only. Indeed these soil state classes are less soil use assessment classes than steps in a broad developmental sequence that soils may follow under cultivation. They refer largely to *suw pombray* topsoil. These **soil state** classes are as follows:

> **suw ka** (lit: soil raw) is either soil under long standing natural vegetation or newly cleared soil that has not been cropped. It is agriculturally untried or 'raw' soil. It has a good *iyba* 'grease' content but its final water state and strength are difficult to judge.
>
> **suw hemem** is the best soil state, and few soils achieve it. It occurs where a considerable depth of vegetation waste accumulates, notably at the foot of slopes and on small flat areas. It is often human-made as a result of the build up of decaying vegetation and topsoil along a fence line at the bottom of a slope. It rots down to produce a thick layer of soft, dark, *iyba*-'grease'-rich topsoil in which crops flourish. It is common, as a consequence, to see a variety of crops growing at the foot of the slope in an established garden adjacent to the fence where *suw hemem* accumulates, the remainder of the site given over almost exclusively to sweet potato. It must not be too wet a locale, the good crumb structure characteristic of this soil state cannot develop on excessively wet soil.
>
> **suw huwniy** is a soil state achieved in some gardens following exposure to the sun. It is a good soil for sweet potato. It is not as soft as *suw hemem*, comprising coarser crumbs, and it is relatively deficient in *iyba* 'grease'. But it is porous and *tomiy* weak, so tubers can penetrate and grow well in it. It only occurs following the break up of the topsoil, when women have heaped it up into mounds for sweet potato. And the more times it is cultivated the better the granular *huwniy* structure may become for sweet potato cultivation. If a soil develops into the *huwniy* state, the time that it remains cultivable relates to the depth of the topsoil. If it is considerable, and the garden slope gentle such that erosion losses are negligible when the soil is exposed under a

newly planted crop, then it can remain in this state indefinitely and support a sweet potato garden for decades. A common strategy with these gardens is to work in a rotational manner around them, leaving an area to rest for some months under *bol* grass (*Ischaemum polystachyum*) to replenish its *iyba* 'grease' levels (see Chapter 13), a practice called *suw hombshor* (lit: soil share-out, i.e. share out its use, to conserve a modest organic matter content).

suw taebowgiy is the worst soil state. The soil is *buriy* strong, hard and cloddy. Sweet potato tubers find it hard to penetrate and grow in it, and weeds can compete effectively with the crop. Another sign that a soil is maybe in this poor state is many *pondiyp* ground orchids (*Spathoglottis* spp.), although these plants occur too irregularly to be a sure indicator of fertility status. The soil is entirely deficient in *iyba* 'grease'. Any tubers that grow are small and stringy, and may be so poor as to become what the Wola call *hokay haeriy*, that is bitter tasting with a blotchy flesh that turns an unpalatable grey colour when exposed (like a green apple turns brown when cut open). Indeed another name for this soil state is *suw haeriy* after these unappetising tubers; it may also be referred to as *maend*, intimating that like plants infected with *maend* rots (see Chapter 9) crops do not flourish in such soil. When a garden soil becomes *taebowgiy* or *kor* (lit. bad) it is time to abandon the plot. The time that soils take to reach this poor condition varies, from one or two years under cultivation onwards.

suw pa is waterlogged soil. It is unsuitable for any crops other than wet-loving taro and skirt sedge, although a range of other crops may be planted on any higher ground, notably around the base of trees where the transpiration-driven-suction of large roots has somewhat dried out the topsoil and bound it together. Waterlogged soil does not follow the above development sequence but remains *pa*. Nevertheless it quickly becomes tired under cultivation. Taro is a heavy user of *iyba* 'grease' supplies and these soils are cropped once only and then allowed to regenerate their natural vegetation cover and *iyba* 'grease' fertility.

This review of the rationale behind indigenous soil assessments gives further credence to Wola assertions about feeling no compunction to investigate soil before cultivating it. They can hardly assess the favourability or otherwise of any soil beforehand, if its character only becomes evident under cultivation. It is necessary to clear a site and allow the soil to dry out somewhat in the sun to appreciate better its potential. It also becomes more apparent following its break up and mounding, when the extent to which it may develop the favourable porous granular *huwniy* structure becomes clear. The extent to which its *iyba* 'grease' reserves might be conserved is also largely unknown, although certain localities are more well-inclined to this than others, such as down-slope and in folds where the best *hemem* soils are likely to form.

The locations where favourable soils are likely to develop are limited on any site. It would be pointless to assess an entire garden from one of these favoured locations alone, when they make up only a small part of its area. Likewise the assessment of any factors — for example topsoil depth or colour, the criteria that might be thought readily determinable — is not feasible because they can vary greatly over short distances within a garden. There is little point in checking them in one or two places when they will probably differ everywhere else. The same applies to the assessment of stoniness. The gardeners themselves exploit these site micro-variations as they become apparent when they plant their crops, siting taro on particularly wet spots such as seepages, and a variety of crops such as greens, pulses and cucurbits along the bottom of slopes and in folds, where the topsoil is likely to be deeper and more *iyba* organic rich.

The soils of the Wola region not only display a considerable homogeneity overall, are similar in a broad typological sense as the multivariate analysis demonstrated, but also, in the variations they do manifest, vary in a largely unpredictable and continuous manner, both in the way some of their properties respond to cultivation and spatially as distributed across garden landscapes. Where soils are generally so similar, and variations between them crucial to crop growth not readily perceived, even canny and experienced farmers like the Wola are hard put to make reliably informed comments about their possible potential. They cannot be at all sure about their behaviour. After all, some soils they maintain, progressively improve the longer they are under cultivation, and the only way to find out is to garden them. This is the exact reverse of the current image of soils under shifting cultivation regimes, as soon running down and forcing a change of site.

The attitude of these New Guinea highlanders to soil appraisal and cultivation appears yet again to contradict accepted wisdom with respect to tropical farming systems. Regarding soil inspection, their disinterest when selecting a garden site contrasts with the behaviour of people elsewhere who look at vegetation and soil, and even reportedly subject it to tests like tasting it or pushing in implements to check stickiness (Allan 1967:94; Conklin 1957:36–37; Allen 1982). The implications of their stance in relation to garden site selection are intriguing, particularly with respect to any assumed jockeying between homesteads for access to the most productive locations. Regardless of their outward uniformity, the yields obtained off soils at different locations are not similar, but may vary some twentyfold for sweet potato tubers, so there is a strong incentive for households to secure rights to the better productive places, particularly those, it might be surmised, with socially ambitious men struggling to handle more pigs than others.

II. SOCIAL STANDING AND SOIL RESOURCES

Among the Wola, the achievement of high social status, and the epithet *ol howma* man-of-renown, demands success in the management of wealth in the ceremonial exchange transactions which feature prominently in their society (see Chapter 1). The Wola esteem pigs highly as wealth transactable in this all-important sphere, like they do throughout the highland region. Consequently, men of renown must handle more pigs in exchanges than others. But pigs differ crucially from other, inanimate valuables: unlike them they have to be continually 'produced', that is herded and fed daily with cultivated produce, notably sweet potato tubers. The implication is that *ol howma* men of renown somehow use the influence that accrues to their social standing to gain access to, and exploit, more or better productive resources than others, which in this context implies cultivation of more productive garden land, gaining higher crop returns from their labour inputs. Those with more fertile land will clearly have an advantage in the production of pig wealth. This is an interesting proposition in the light of the discussion of Wola soil appraisal, particularly the way in which they say that the arable properties of soils develop under cultivation. It is also intriguing, given the egalitarian political constitution of Wola society, to ask how successful men might contrive to gain access to sites that have proved their worth, if indeed they can.

The proposal that the socially ambitious strive to gain access to better arable land to secure a competitive advantage over their peers in the production of wealth 'on the trotter' may be tested by looking for correlations between men's social standing and the potential productivity of the land their households cultivate, here using data collected on 294 gardens in the Was valley (Map 5.1), cultivated by men of differing social status.[5] Social standing is assessed initially according to exchange-earned *ol howma* social status, and subsequently by age and kin group affiliation, to see if there is any relationship between social position and natural resource use. The productivity of land cultivated is assessed according to a few critical site factors and measures of soil fertility (Hodgson 1976).[6] These site and soil parameters are analyzed for corre-

---

[5] The data on garden sites and soils used in this investigation of social status may be found on the internet at the following address (URL): http://www.dur.ac.uk/~danøps/soil.html.

[6] The soils cover a range of the Inceptisols that occur in Wolaland (see Chapter 10), variably affected by falls of volcanic ash. (See Chapter 12 for further technical details, also Radcliffe 1986 for analytical data on similar soils in the nearby Mendi valley.). Altitude was measured with an aneroid altimeter, aspect with a prismatic compass, and slope with a simple protractor-mounted level. Surface topography was described according to a series of nine broad classes (anticline, convex slope etc.), and the vegetation cleared at the time of cultivation was taken to be that currently surrounding the garden, as verified by local people. In each garden surveyed a soil profile pit was dug and the depth of topsoil measured and the number of horizons recorded; those assisting in the work also identified the horizons present and I noted the indigenous classification (see Chapter 10). The colour of the topsoil was assessed subjectively, not using Munsell charts, according to one of four gross classes: black, dark brown, brown or light brown (while not scientifically accurate the results are sufficient for the trends sought here). And a composite sample of topsoil was collected by gathering ten or more handfuls of soil at random across garden sites, for bulking together and field analysis of nutrients using a Sudbury soil test kit. This kit is intended to help farmers and horticulturalists assess soil fertiliser needs. It tests for 'available', not total plant nutrients, and the results are related to recommended fertiliser applications (Sudbury Technical Products Ltd. pers. comm.). While these field tests are considerably less accurate than laboratory analyses of soil fertility and can only be approximately correlated with them over a range of values, they indicate relative differences in fertility between sites sufficient for this chapter's purposes. Those present during the survey were also asked to assess the agricultural worth of the soil, according to one of five classes, from very good through to barren. The familiarity of the Wola with such ranking exercises, who use a series of adjectival markers to indicate gradations, has been noted previously. A similar ranking order was used in the assessment of men's social standing, taken up later. In assessing soil worth, the following classes were used: *suw ebayda* (lit. soil good-very), *suw ebay* (lit. soil good), *suw ora* (lit.soil only — i.e. average soil), *suw kor* (lit. soil bad), and *suw taebogiy* (lit. soil barren).

lations with the different indicators of social standing, taking them as dependent variables that should vary in some patterned manner between social groups if the hypothesis of a connection between social standing and garden productive status is valid (a statistically significant correlation is $P < 0.05$).

Social Status Compared with Site and Soil Status

There are few significant correlations between men's *ol howma* social status and differences in the site and soil characteristics of the gardens cultivated by their households (Table 11.5). The evidence suggests that there is little relationship between the conditions that prevail in their gardens and any social influence which men might be presumed to have by virtue of their social standing, whether high or low. (The men claiming title to the gardens surveyed, whose households cultivated them, are arranged according to their *ol howma* standing into nine groups, numbered one through to nine. The status of the men had been assessed previously in a survey of their communities, in which people were asked their opinion of the *ol howma* standing of the men known to them. In status group one are the men of highest status, whom the majority of people in their locale would acknowledge to be of *ol howma* status, and in status group nine are the youngest men (all under twenty-one years in age) whom no one would consider to be of any social consequence. The men in the status groups from seven downwards no one thought merited reference regarding *ol howma* standing, and they might be considered as a single large group of low social repute. They are grouped according to age, with the eldest first. There are eight men in each of the first seven groups, and ten in each of the last two groups (to achieve some balance in the number of gardens surveyed in each sample group[7])).

The mean aspect of gardens is similar over the entire range of social status groups, although with a considerable spread (as the standard deviations indicate). These data affirm again that the direction in which Wola farmers prefer their gardens to face is more-or-less east, because it maximises exposure to the sun, particularly in the morning, an important consideration in a mountainous region which is frequently overcast in the afternoons. The longer direct sunlight reaches crops daily the greater photosynthetic opportunity and the more rapid their growth. Although gardens with a favourable aspect will be more productive and hence preferred, men of higher social standing do not monopolise these sites.

The slope of gardens also affects their potential productivity. Erosion risks and possible crop damage increase with slope steepness, especially

---

[7] The size of the samples was as follows: status group 1 = 52 gardens, status group 2 = 43 gardens, status group 3 = 37 gardens, status group 4 = 31 gardens, status group 5 = 33 gardens, status group 6 = 25 gardens, status group 7 = 27 gardens, status group 8 = 24 gardens, and status group 9 = 26 gardens.

Table 11.5

Mean garden site and soil factors compared with gardeners' social status (values in brackets = standard deviations; some percentage totals exceed 100 because some gardens occur in two classes; degrees of freedom 8, 128/282)

		Social Status Groups									Statistical Significance		
		1	2	3	4	5	6	7	8	9	CV	F	Sig
SITE:													
Aspect (degrees)		102 (97)	101 (91)	86 (92)	114 (113)	92 (91)	93 (87)	87 (81)	81 (88)	86 (94)	98.3	0.39	NS
Slope (degrees)		19 (8.3)	20 (9.3)	22 (9.3)	20 (7.1)	20 (10.0)	19 (8.4)	18 (10.3)	20 (9.9)	23 (9.9)	45.4	0.85	NS
Slope Form (% gardens)	even	56	46	49	52	50	46	62	48	64			
	syncline	29	17	19	13	19	21	15	36	18			
	anticline	6	21	19	7	19	17	8	16	14			
	concave	4	10	5	7	6	4	15	4				
	convex	2	2	11	3	9	4			5			
	stepped	4	5		7		8	4	8				
	rocky	6	2	3	7		4	8					
	irregular		5		13	3							
	undulating	4			4	9							
Altitude (m)		1996 (94)	1776 (91)	1816 (89)	1762 (55)	1813 (98)	1835 (135)	1785 (69)	1840 (80)	1818 (93)	5.1	1.15	NS
Vegetation Cover (% gardens)	1y forest	6	3	8	6	13	4	23	4	14			
	2y forest	21	14	16	16	19	25	8	4	41			
	tree/cane	17	5	22	16	16	17	4	27	9			
	cane	50	64	49	53	42	46	57	61	32			
	grass	6	14	5	9	10	8	8	4	4			
SOIL:													
Depth (cm)		16.2 (11.8)	20.5 (15.8)	21.8 (15.3)	18.8 (12.8)	15.8 (9.3)	16.0 (11.7)	13.5 (9.5)	19.4 (16.6)	19.5 (14.1)	73.6	1.24	NS
Topsoil Colour (% gardens)	black	10	7	14	19	9	13	7	8	9			
	d. brown	37	44	32	34	38	25	31	46	45			
	brown	28	25	32	31	38	29	31	35	32			
	l. brown	25	24	22	16	15	33	31	11	14			

Table 11.5 *Continued*

		Social Status Groups									Statistical Significance		
		1	2	3	4	5	6	7	8	9	CV	F	Sig
Number Horizons		2.1 (0.5)	2.2 (0.5)	2.1 (0.5)	2.1 (0.6)	2.1 (0.5)	2.3 (0.5)	2.2 (0.6)	2.1 (0.5)	2.1 (0.7)	25.0	0.85	NS
pH		5.3 (0.3)	5.2 (0.2)	5.2 (0.2)	5.2 (0.2)	5.1 (0.2)	5.2 (0.4)	5.3 (0.3)	5.3 (0.3)	5.4 (0.6)	5.8	0.91	NS
N (ppm)		116 (27)	127 (17)	126 (26)	123 (24)	120 (34)	136 (21)	106 (30)	104 (46)	87 (32)	24.0	3.08	S
P (ppm)		6.8 (4.5)	7.2 (4.7)	7.2 (4.6)	7.9 (4.4)	7.2 (4.8)	7.3 (5.4)	9.3 (5.6)	7.6 (4.7)	6.4 (4.7)	65.1	0.31	NS
K (ppm)		155 (126)	167 (89)	160 (117)	153 (60)	144 (116)	151 (71)	145 (114)	234 (184)	246 (173)	71.3	1.16	NS
Local Soil Assessment (% gardens)	v. good	4	13	13		9	9	8	8	5			
	good	51	45	43	47	44	56	54	24	63			
	fair	29	22	30	40	28	22	23	40	16			
	no good	8	17	6	10	13	13	15	16	5			
	barren	8	3	8	3	6			12	11			
Sweet Potato Yields (kg/ha)		11955 (5117)	9200 (3337)	17529 (8966)	10733 (10941)		8330 (4188)		15933 (8890)		62.7	1.38	NS
Garden Type (% gardens)	mixed veg	6	2		3	6	8	4	8				
	abn m. veg	6	2	3	9	6		4		14			
	sweet pot	51	67	71	79	66	67	58	81	64			
	abn s.p.	30	19	16	9	22	17	27	11	14			
	taro	6	2	5			4	7		4			
	abn taro	7	8	5			4			4			
Times Cultivated (% gardens)	1	43	36	35	50	44	29	46	31	68			
	2	23	29	30	22	25	17	15	35	18			
	3	6	12	16	9	6	17	8	8	5			
	4	9	9	3	13	3	4		4				
	≥5	11	2	11	3	13	33	19	11				
	≥10	8	12	5	3	9		12	11	9			

when gardens are newly planted and lack established cover (see Chapter 6). And topsoil depth frequently decreases due to runoff losses. Too gentle a slope on the other hand, or a level site, results in increased soil wetness with reduced drainage rates, even waterlogging, detrimental to the yield of many crops, notably sweet potato, which cannot tolerate conditions too wet (see Chapter 4). The mean slope of the sites surveyed was about 20° for all social status groups, again there is no indication that men of a certain standing have gardens with slopes that differ significantly from any others, better optimising the balance between erosion and drainage constraints. The data on slope form likewise do not suggest that the slope topography of gardens differs with gardeners' social status. The most common slope form across social status groups is relatively even and featureless. Individuals of lesser social standing are not more likely to have sites with less favourable surface topography, where the ground is particularly broken up, uneven and irregular.

Altitude is pertinent regarding productivity, exerting an influence over temperature. The warmer it is, the faster plants grow and yield. The altitudinal range of gardens in the Wola region — with a lapse rate of between 0.5°C and 0.7°C every 100 metres (McAlpine *et al.* 1983) — is sufficient to have a noticeable influence on crop growth. Furthermore, the danger of frost damage and crop loss, although infrequently of severe occurrence, increases with altitude (see Chapter 3). Nevertheless, there is no indication that men of higher standing are more likely to have their gardens located at more favourable elevations. The mean altitude across social status groups is around 1800 m, with no significant variation.

The vegetation cleared from garden sites could conceivably have a bearing on their subsequent productivity (Clarke & Street 1967). But there is no pattern evident between vegetation cleared and men's social status. Any pattern revealed would have been difficult to interpret unambiguously. While virgin forest sites might be assumed to have a higher potential fertility, the effort required to clear them is considerable and they are frequently inconveniently located some distance from homesteads. The fertility assumption is not always valid either, some sites cleared from secondary regrowth giving comparable crop yields. The large proportion of gardens cleared from cane vegetation, across all social status groups, evidence the relative worth of these sites for gardens. The problem with cane or grass sites is the scarcity of suitable timber for fencing, which men sometimes have consequently to transport from considerable distances away, at great effort. Anyway, whatever the relative advantages and disadvantages, the Wola region has large areas of both caneland and forest, sufficient to meet any preferences men might have for establishing cultivations from particular vegetational successions.

The variations in soil properties between gardens parallel site characteristics, scarcely correlating with *ol howma* social status. The Wola say that depth of topsoil is an important determinant of garden productivity. When the topsoil is too thin, less than about 10 centimetres deep, they point out that the dense clayey subsoil near the surface becomes incorporated into the tilled layer during cultivation, and productivity declines. The depth of topsoil averages about 18 cm on the sites surveyed, across social status groups, with a fair spread about the mean.[8] The variation in soil depth between groups is not significant, showing no pattern related to social renown. Indeed men of highest renown have one of the lowest average depths of topsoil in their gardens, instead of one of the highest as predicted.

The number of horizons has less obvious potential implications for crop yield. A possible connection with garden productivity, is that sites with only one horizon usually have a considerable depth of topsoil extending down beyond 50 cm or so. They are the best sites, with large topsoil reserves. Those sites with three or more horizons might also be marginally better in some situations because they often have a transitional horizon between crumbly topsoil and firm clayey subsoil, moderating the change from one to the other and reducing the poor prospects of a shallow depth of topsoil. Nevertheless, the number of horizons observed in profiles shows no variation across social status groups; they consistently average around two. This is perhaps to be expected. The Wola do not dig holes on prospective garden sites to check horizon sequences, details of which, beyond topsoil depth, water content, and such like, are largely irrelevant to them agronomically. The sequence of horizons in a profile also relates to the local soil classification scheme (see Chapter 10), and predictably there is no pattern to the occurrence of different local soil classes between social status groups. The total number of different horizon combinations, and hence soil profile types, identified in the survey was thirty-one, which gives a cumbersome, random cross-tabulation, in which no relationships are apparent. There is no tendency evident for men of a certain social standing to cultivate their gardens on a particular range of soil horizon combinations.

The colour of topsoil is an indicator used by Wola cultivators to assess the likely fertility status of a soil. A darker soil is better; black to dark brown being acceptable. They consider a light brown soil to be less productive, frequently resulting from mixing with brighter coloured, more coherent, and less fertile clay subsoil. Again, there is little indication that topsoil colour correlates with gardeners' *ol howma* status. The occurrence of light brown soils actually declines between status groups one and five, and the occurrence of black and dark brown soils is variable across all status groups, with no suggestion that a larger proportion of the gardens cultivated by the households of men of higher social standing have darker topsoils. The colour of a topsoil is also a reflection of its organic matter content, darker soils having a higher organic matter status. Organic matter is an important source of some plant

---

[8] The measurement of topsoil depth from a single profile pit dug in each garden is open to question given the marked variation that may occur in topsoil depth across a single site. The statistics gain a certain credibility, sufficient for comparative purposes here, if the several pits dug in the gardens of the men comprising each status group are taken as a series of random depth samples.

nutrients, particularly in soils like those of the Wola region which have high volcanic ash contents, the amorphous minerals of which give rise to supply problems with some nutrients (see Chapter 12). The Wola acknowledge the importance of organic matter content to potential soil fertility not only in their colour appraisal of soil but also in their concept of *suw hobor* soil grease, assessed by the silky texture imparted to soil by organic matter. The silkier and more soapy-like the feel of a soil they say the higher its *suw hobor* organic-matter-derived content and the potentially more fertile it will prove under cultivation. A light coloured soil indicates a decline in organic matter content and falling *hobor* 'grease' levels.

The fertility status of the soils surveyed, as measured in a series of simple field tests, again shows scant relationship with gardeners' social status. Nitrogen is the only property, of all those investigated, to vary significantly between social status groups. But the pattern of the variation scarcely complies with the hypothesis that men of higher *ol howma* standing will tend to cultivate the sites of higher fertility status. The men in group one, of highest renown, cultivate soils with lower average nitrogen contents than any others except those in the three lowest ranked groups, and the peak for the highest nitrogen content soils occurs with group six, none of whom are of renowned *ol howma* status. Furthermore, it is the ratio of nitrogen to potassium that is noteworthy regarding the yield of sweet potato tubers, not absolute levels of potential nitrogen availability (see Chapter 12). The average supply of potassium in the soils tested shows hardly any variation between social status groups, except for the last two groups where it evidences a rise (caution needs to be exercised in interpreting this apparent increase, the standard deviations indicating a considerable spread of results). According to these data, the younger men, those of lowest social standing, tend to cultivate soils with lower average N:K ratios at about 0.4, which is thought preferable regarding sweet potato tuber yield than the higher ratios of around 0.8 of the soils cultivated by men of higher social status.

The pH of the soils tested reveals a more or less steady average value across social status groups too, around the overall mean of 5.2. The strongly acidic reaction of the soils cultivated by the Wola has serious implications regarding the availability of plant nutrients, making deficiencies probable (Landon 1991:115–117). It may particularly exacerbate the phosphorus deficiencies common to volcanic ash soils, promoting fierce fixation (see Chapter 12). The even spread of pH values suggests that men share this problem equally. The phosphorus contents of the soils cultivated are very low overall and reveal no variation according to social status. The standard deviations indicate a range of soil phosphorus availabilities, some sites considerably more deficient than others. They suggest that farmers, regardless of status, find it difficult to predict fertility status before clearing sites, just as they maintain, fertility depending in large part on the availability of phosphorus, a major limiting plant nutrient in these soils, particularly for crops other than sweet potato. This apparent inability to forsee fertility has implications, discussed later, for the absence of correlations between social standing and garden productive potential.

Age Compared with Site and Soil Status

While the garden site and soil parameters show no correlation with gardeners' *ol howma* social status, there is some suggestion in the data of a possible relationship with age. The last three social status groups, arranged according to the cultivators' ages, show some possible trends for certain characters (e.g. in soil fertility). There is no clear connection between *ol howma* status and age, except that men of high social esteem are mature, usually in their late thirties to fifties (Sillitoe 1979). Younger men have yet fully to develop their exchange networks and negotiating skills, while older men are losing their networks and astuteness, others considering them chancy exchange partners who may die without their kin subsequently fully acknowledging or meeting their obligations.

It is plausible that the potential productivity of garden sites might relate in some manner to the age of those who cultivate them. Mature men are more experienced cultivators and might be assumed to be more skilful in selecting garden sites. They are more likely to know the locations of those sites cultivated some years previously which yielded particularly well, and are sufficiently rested to sustain a further round of cultivation. They may have seen them cultivated when younger or heard from older relatives about their good yields when under cultivation. It is reasonable to suppose consequently that they are more likely to choose sites of higher fertility, if this is at all possible. Furthermore, we may suppose that those gardens cultivated long-term, apparently for generations, are particularly productive, as sites repeatedly put under crops over many years, and that older men are more likely to have taken possession of such enduringly productive sites through inheritance. But the data on sites and soils scarcely bear out this age hypothesis, there is little more patterning to them than when ordered according to *ol howma* social status (Table 11.6). (The men claiming rights to the gardens surveyed are ordered according to their ages into ten groups for comparative purposes, ranging in size from five to nine men, and containing from ten to forty-four gardens.[9])

The site characteristics of aspect, slope, surface topography, altitude and vegetation before cultivation show no significant variation according to gardener's age. There are no trends indicating potential productivity differences

---

[9] The size of the samples is as follows: <18 years = 5 men & 10 gardens, 18–21 years = 5 men & 13 gardens, 21–24 years = 8 men & 19 gardens, 24–27 years = 9 men & 30 gardens, 27–30 years = 9 men & 37 gardens, 30–35 years = 8 men & 38 gardens, 35–40 years = 9 men & 44 gardens, 40–45 years = 8 men & 43 gardens, 45–50 years = 8 men & 38 gardens, and >50 years = 7 men & 22 gardens.

Table 11.6

Mean garden site and soil factors compared with gardeners' ages (values in brackets = standard deviations; some percentage totals exceed 100 because some gardens occur in two classes; degrees of freedom 9, 127/281)

		Age Groups (yrs)										Statistical Significance		
		<18	18-21	21-24	24-27	27-30	30-35	35-40	40-45	45-50	>50	CV	F	Sig
SITE:														
Aspect (degrees)		112 (133)	85 (81)	82 (84)	86 (82)	102 (100)	87 (94)	94 (90)	105 (102)	94 (90)	106 (92)	98.7	0.26	NS
Slope (degrees)		22 (11.3)	24 (9.8)	21 (10.1)	20 (7.6)	20 (9.1)	20 (7.9)	21 (8.8)	18 (9.1)	21 (11.1)	20 (7.9)	45.5	0.69	NS
Slope Form (% gardens)	even	80	54	56	47	51	49	57	50	50	46			
	syncline	20	23	22	20	22	11	18	24	24	32			
	anticline	23	17	17	14	14	18	12	11	9				
	concave	6		8	3	5	5	5	9					
	convex		8	6	3	3	11	7		8	5			
	stepped			6	7	5	5		7	8	5			
	rocky				7	5	3	2	2	5				
	irregular						5		2	3	9			
	undulating				3		11		2					
Altitude (m)		1850 (89)	1817 (105)	1813 (89)	1811 (106)	1807 (106)	1757 (96)	1775 (82)	1798 (77)	1811 (103)	1831 (56)	5.2	0.83	NS
Vegetation Cover (% gardens)	1y forest	20	8	5	10	5	8	11	9	5	5			
	2y forest	20	46	5	17	24	13	16	12	32	9			
	tree/cane	20		37	17	14	21	9	16	8	9			
	cane	40	38	53	46	54	45	48	51	55	81			
	grass		8		10	3	13	16	12		5			
SOIL:														
Depth (cm)		27.3 (14.7)	13.5 (10.8)	19.0 (17.0)	20.0 (12.3)	14.9 (10.5)	21.1 (16.6)	18.9 (13.3)	15.4 (11.0)	16.1 (11.3)	19.8 (15.3)	73.4	1.40	NS
Topsoil Colour (% gardens)	black	10	15	0	10	11	19	9	9	8	14			
	d. brown	60	39	47	33	38	38	39	37	42	9			
	brown	20	31	42	30	32	30	25	33	31	32			
	l. brown	10	15	11	27	19	13	27	21	19	46			

Table 11.6 Continued

		Age Groups (yrs)										Statistical Significance		
		<18	18–21	21–24	24–27	27–30	30–35	35–40	40–45	45–50	>50	CV	F	Sig
Number Horizons		1.9 (0.8)	2.1 (0.5)	2.2 (0.6)	2.0 (0.5)	2.2 (0.4)	2.1 (0.6)	2.0 (0.5)	2.3 (0.5)	2.1 (0.5)	2.1 (0.6)	24.7	1.27	NS
pH		5.4 (0.5)	5.3 (0.6)	5.3 (0.3)	5.2 (0.2)	5.2 (0.4)	5.2 (0.2)	5.2 (0.2)	5.3 (0.3)	5.2 (0.4)	5.2 (0.2)	5.9	0.35	NS
N (ppm)		77 (38)	108 (26)	99 (50)	123 (27)	125 (22)	122 (28)	109 (38)	122 (30)	126 (16)	125 (19)	24.6	2.07	S
P (ppm)		5.0 (6.0)	7.3 (4.4)	6.9 (5.0)	8.5 (4.8)	8.1 (5.3)	6.5 (4.6)	7.1 (5.0)	8.4 (3.9)	6.7 (4.6)	6.8 (4.7)	64.9	0.05	NS
K (ppm)		332 (222)	157 (103)	269 (185)	161 (85)	150 (92)	147 (114)	150 (140)	176 (121)	146 (78)	148 (78)	69.4	1.95	S
Local Soil Assessment (% gardens)	v. good	13		6	10	6	11	14	7		9			
	good	61	50	33	62	42	45	25	69	42	55			
	fair	13	25	39	17	33	33	36	14	33	27			
	no good	13		11	7	11	11	18	7	17	9			
	barren		25	11	4	8		7	3	8				
Sweet Potato Yields (kg/ha)				12625 (8970)			16561 (9541)	8549 (4523)	10483 (11016)	15041 (7434)	11323 (5737)	62.7	2.42	NS
Garden Type (% gardens)	mixed veg	30		11	3	8					13			
	abn m.veg				7	5		5	5	5	5			
	sweet pot	60	61	84	67	62	84	66	62	55	50			
	abn s.p.		31	5	13	22	16	27	21	21	22			
	taro	10			3	3		2	5	8	5			
	abn taro		8		7				7	11	5			
Times Cultivated (% gardens)	1	70	69	37	53	40	21	45	30	47	45			
	2	30	15	42	10	22	31	27	16	32	18			
	3		8	5	20	8	11	7	14	8	5			
	4			5	3	3	11	5	12	5	5			
	≥5				7	24	15	2	21	3	18			
	≥10		8	11	7	3	11	14	7	5	9			

ascribable to variations in age. There is little divergence attributable to age at all. The older men for example, do not appear to cultivate markedly less steep nor topographically less rugged sites, physically somewhat less demanding to work, as a concession to their advancing years. The weak trend evident for older men to clear a smaller proportion of their gardens from forest and a larger proportion from cane grass is difficult to interpret because of the heavy physical work demanded by fencing in caneland; furthermore it is debatable whether cane is physically less hard work to clear than forest.

There are perhaps some trends evident between the soil data and age, although they challenge the hypothesis that older men will cultivate the potentially more productive sites. Regarding topsoil colour, younger men cultivate a smaller proportion of sites with light brown soil of lower organic matter content, men over 50 years in age have a considerable percentage of their gardens on such light coloured soils. According to the field tests, the nutritional status of soils shows some variation with age. There is a weak downward trend in pH, the soils in older men's gardens tending to be marginally more acid (although the large standard deviations for young men compromise this trend). But there is no evident impact on phosphorus availabilities. There is a statistically significant variation in the nitrogen and potassium availabilities of the soils cultivated by different age groups. The younger men tend to cultivate soils of lower nitrogen but higher potassium status which, as noted, tends to give them more favourable N:K ratios for sweet potato production than older men. This again controverts the idea that more mature men will have selected the potentially more productive garden sites.

Kin Group Affiliation Compared with Site and Soil Status

The investigation turns now to the social groups that structure access to land in Wola society in a bid to uncover possible correlations between men's social position and their exploitation of natural resources in cultivation, having uncovered no apparent trends in the data when they are compared with either the social standing or the age of gardeners. It explores the possibility that the members of different kin groups may have assess to land resources of differing quality. The social groups that order land tenure in Wola society are the small kin-constituted *semgenk* corporations, localised groups which, although they have a patrifilial recruitment bias, comprise collections of bilaterally related kin. They control access to tracts of territory, over which they claim corporate rights (see Chapter 1). Those who have an acknowledged extant social connection with a *semgenk* group, validated through active wealth exchange relations with kin members, whether resident locally or not, can clear and establish gardens on its territory (except for sites to which other relatives have prior claims by virtue of previous cultivation, either by themselves or close kinspersons).

It is plausible that some *semg^{e}nk* groups have access to better tracts of cultivable land than others. The Wola say that some groups originate from the first settlers to arrive in any area, while others descend from later arrivals, usually non-agnatic relatives of the original colonists, whose descendants have subsequently founded independent *semg^enk* groups. It is possible that the first-comers laid claim to the best tracts of cultivable land, and the later established *semg^enk* have had to make do with the less promising areas.[10] While some redistribution of land may have occurred over the generations with intermarriage between *semg^enk*, this may not have been extensive, because the larger *semonda* social corporations — made up of *semg^enk* that have contiguous land holdings interdigitated across the same region — are ideologically exogamous, which reduces rates of intermarriage such that any early disparities in the quality of land resources held might be anticipated to remain evident.

Another possibility is that the quality of land available to individuals varies depending on the number of persons eligible to share in the resources of the *semg^enk* to which they are affiliated, who are acting on their rights and cultivating on its land. The *semg^enk* groups vary considerably in size, due to the vagaries of demography, some expanding rapidly in numbers while others languish. It is conceivable that those associated with smaller groups cultivate more productive land resources per head than those affiliated to larger ones who have to share out their available land among a larger number of households. The areas of tracts claimed by *semg^enk* do not vary proportionally with the size of their active memberships, a small group will on balance have a larger land resource area from which its households may choose sites for cultivation.

A review of the garden site and soils data by *semg^enk* kin group affiliation, to assess the third social hypothesis that there is some relationship between kin group association and the quality of the land houshholds cultivate, reveals even fewer possible correlations than those weakly evident in the comparisons with social standing and gardeners' ages (Table 11.7). (The first three *semg^enk* groups listed comprise one *semonda* and the other five another neighbouring *semonda*. The first five listed are said to descend directly from original male ancestor settlers in the region of the Was river valley concerned, and the last three from latecomers related to the first arrived groups through distant female genealogical connections and marriage. The number of men affiliated to the *semg^enk* groups and activating cultivation rights to land on their territories range from two to fifteen.[11])

---

[10] This assumes that the peopling of the region followed local, sometimes mythological accounts; if its populating parallelled that evident today in some less densely settled regions of the Southern Highlands, restriction of access to cultivable land may have only occurred as numbers of people increased (Sillitoe 1994).

[11] The size of the *semgenk* samples is as follows: Ind = 15 men & 51 gardens, Piywa = 3 men & 8 gardens, Kolomb = 9 men & 32 gardens, Mayka = 10 men & 46 gardens, Maenget = 9 men & 32 gardens, Puwgael = 2 men & 13 gardens, Wenja = 14 men & 51 gardens, and Huwlael = 14 men & 51 gardens.

Table 11.7

Mean garden site and soil factors compared with gardeners' kin group affiliation (values in brackets = standard deviations; some percentage totals exceed 100 because some gardens occur in two classes; degrees of freedom 7, 117/283

		Kin Groups								Statistical Significance		
		Ind	Piywa	Kolomb	Mayka	Maenget	Puwgael	Wenja	Huwlael	CV	F	Sig
SITE:												
Aspect (degrees)		83 (71)	110 (108)	174 (131)	89 (91)	77 (71)	73 (65)	74 (80)	98 (91)	93.6	4.57	S
Slope (degrees)		19 (8.0)	18 (9.4)	20 (6.7)	20 (9.4)	20 (9.5)	23 (9.8)	21 (9.7)	20 (10.3)	45.6	0.46	NS
Slope Form (% gardens)	even	56	88	38	48	53	69	49	53			
	syncline	15		22	22	27	8	26	25			
	anticline	19	13	19	9	15	15	12	13			
	concave	4		6	9	6	8	6	6			
	convex	2		9	4			6	6			
	stepped	7		6	7			2	2			
	rocky	2			4	3		4	8			
	irregular			3	4			2				
	undulating	4			4	9			6			
Altitude (m)		1793 (117)	1806 (125)	1811 (103)	1797 (90)	1825 (63)	1791 (94)	1795 (93)	1808 (85)			
Vegetation Cover (% gardens)	1y forest	15		9	15	3		4	6			
	2y forest	11	13	13	11	9	23	27	30			
	tree/cane	5		13	25	20	8	10	21			
	cane	54	74	59	43	65	69	46	41			
	grass	15	13	6	6	3		13	2			
SOIL:												
Depth (cm)		14.5 (13.1)	16.5 (12.1)	17.3 (9.8)	18.0 (13.1)	25.0 (16.5)	15.4 (13.6)	18.9 (12.9)	17.5 (13.0)	73.0	1.94	NS
Topsoil Colour (% gardens)	black	11	13		11	12	15	13	12			
	d. brown	34	25	37	46	47	39	29	35			
	brown	32	25	33	32	26	31	33	27			

Table 11.7 Continued

					Kin Groups					Statistical Significance		
		Ind	Piywa	Kolomb	Mayka	Maenget	Puwgael	Wenja	Huwlael	CV	F	Sig
Number Horizons		2.2 (0.5)	2.4 (0.5)	2.3 (0.6)	2.1 (0.5)	2.0 (0.7)	2.2 (0.4)	2.1 (0.6)	2.0 (0.3)	24.7	1.11	NS
pH		5.2 (0.4)	5.0 (0.3)	5.1 (0.2)	5.3 (0.3)	5.3 (0.2)	5.3 (0.3)	5.2 (0.2)	5.3 (0.4)	5.8	0.90	NS
N (ppm)		120 (22)	125 (21)	123 (21)	112 (35)	110 (41)	130 (11)	118 (37)	122 (27)	25.8	0.56	NS
P (ppm)		7.0 (4.7)	6.7 (4.5)	7.8 (4.5)	7.7 (5.0)	6.7 (4.8)	7.4 (4.9)	7.7 (4.6)	7.3 (4.9)	65.3	0.13	NS
K (ppm)		112 (71)	106 (27)	148 (83)	202 (154)	223 (171)	220 (64)	164 (106)	162 (106)	69.9	1.97	NS
Local Soil Assessment (% gardens)	v. good	10	25	3	11	9		6	6			
	good	44	51	60	48	53	38	53	33			
	fair	30	12	19	28	19	31	25	43			
	no good	10	12	9	9	13	31	14	8			
	barren	6		9	4	6		2	10			
Sweet Potato Yields (kg/ha)		9871 (4109)		13815 (7026)	16732 (11921)	14400 (5794)	8700 (5515)	6200 (794)	9220 (5621)	62.7	1.23	NS
Garden Type (% gardens)	mixed veg	2		6	4	6		2	2			
	abn m.veg	66	13	6	9	3	8	6	6			
	sweet pot	24	62	67	59	70	53	77	58			
	abn s.p.	4	25	18	17	9	8	21	24			
	taro				9	6	8		2			
	abn taro			3	2	6	23		8			
Times Cultivated (% gardens)	1	44	12	43	51	41	39	29	49			
	2	19	12	27	19	38	31	25	22			
	3	9	25	3	4	6	15	11	15			
	4	6		3	9		15	6	8			
	≥5	13	39	9	11			23	4			
	≥10	9	12	15	6	15		6	2			

The only statistically significant correlation concerns site aspect, which is the result of a single marked variation among the first-settled *semgᵉnk* and nothing to do with arrival status, nor numbers of *semgᵉnk* members actively cultivating a group's lands. The arrival status of the *semgᵉnk* groups bears no apparent relation to any of the documented site nor soil parameters. No noticeable differences are evident between the five original colonising kin groups and the three more recently arrived ones, on the right of the table. Regarding the size of *semgᵉnk*, there are perhaps some weak trends evident. The small *semgᵉnk* groups of Piywa and Puwgael have a larger proportion of their gardens on sites with even topography, avoiding difficult terrain. They also clear a larger proportion of their gardens from cane grass and avoid primary forest, although the implication of this is difficult to evaluate, as pointed out. The soil samples collected from the gardens of the small *semgᵉnk* have high available nitrogen contents too, although only Puwgael *semgᵉnk* has a correspondingly high potassium availability, giving an N:K ratio favourable to sweet potato production.

Soil Assessment, Crop Yields and Social Standing

The up-shot of this close investigation of some site characteristics and soil parameters assumed to influence garden fertility and crop productivity is that surprisingly few correlations have emerged with sociological factors relating to either gardeners' social standing, age or kin group affiliation. We have found only a few weak trends where we might have expected to uncover some relationships plausibly related to household productive potential, especially pig herding capacity. These apparently negative findings would come as no surprise to the Wola themselves. They emphatically deny that some households are able systematically to exploit better land resources than others. They maintain that men do not use any influence that accrues to them because of their *ol howma* renown, nor any experience and knowledge they have by virtue of their age, nor any supposed advantage that accrues to them as a consequence of their *semgᵉnk* connections, to secure potentially more productive garden sites for their households to cultivate. The burden of this investigation is that the assertions of the Wola are correct when examined against measured features relating to land productivity. An exercise in indigenous soil assessment further confirmed the consistency of their assertions, as did a comparison of sweet potato tuber yields against the three measures of social standing.

When asked to rank the soils surveyed according to five classes ranging from 'very good' through to 'barren', Wola appraisals show no correlation with social status, age or kin group affiliation (Tables 11.5 to 11.7). Regarding social standing, the men in status group one, who are undeniably of *ol howma* renown, cultivate a considerable proportion of 'barren' assessed soils and a smaller than average percentage of 'very good' soils. The soils cultivated by the men comprising status group two score better overall on the ratio of 'barren' to 'very good' sites. The soils gardened by the men in group nine, the lowest in social esteem, also compare favourably with those cultivated by status group one men. The proportion of 'barren' and 'no good' soils cultivated by both is nearly equal, while status group nine has the edge regarding soils adjudged to be 'very good' and 'good'.

Age and local assessment evidence a similar random pattern. While a fair proportion of the soils cultivated by young men in their twenties are judged 'barren', for instance, cultivation of these soils is by no means confined to them. They cultivate a goodly percentage of 'very good' soils too, comparable with their older relatives. The trend with those soils assessed as 'good' illustrates clearly the absence of any correlation between soil status and age, the percentages going up-and-down irregularly between age groups with no hint of an age bias.

Kin group affiliation and local soil assessment has the same random quality. The members of the three later established $semg^enk$ (Puwgael, Wenja and Huwlael) appear not to garden a larger proportion of 'barren' soils overall, although the members of one of them cultivate a high percentage of soils thought to be 'no good' (a doubtful increase, the garden sample being relatively small compared to others). They also cultivate together fewer 'very good' and 'good' assessed soils, although the differences are not of a magnitude to suggest that they are disadvantaged in terms of access to soil resources. When considered against the size of $semg^enk$ groups, the trends are scarcely more persuasive. The members of the two smaller kin groups (Piywa and Puwgael) cultivate no sites assessed 'barren', and those affiliated to Piywa garden an above average proportion of 'very good' soils, but members of Puwgael cultivate no soils adjudged 'very good' and an above average percentage of 'no good' ones, which taken together with the relatively small sample of gardens from both groups compromises any apparent trends.

The ultimate test of these findings on garden site location and potential soil productivity is the extent to which crop yields support them, for in the final analysis it is these which really matter. A review of data on sweet potato tuber yields shows that jockeying to secure access to the potentially more productive garden locations would not be misplaced. The outward homogeneity of many of the soils cultivated by the Wola does not translate into relatively uniform yields across gardens. The extent of the variations in yields between the best and worst sites is considerable, ranging from 1700 kg ha^{-1} to 32,500 kg ha^{-1}, with a mean yield of 12,248 kg ha^{-1}. Differences in yields between sites of this magnitude indicate that access to the more productive sites would considerably boost someone's productive capacity with no change in labour inputs.

A comparison of sweet potato yields from different gardens with the social status, age and kin group affiliation of the men whose households harvest from them corroborate the foregoing findings with respect to variation in natural resource exploitation by different men, with no significant correlations (Tables 11.5 to 11.7).[12] The highest mean yields were recorded for gardens harvested from by men of middling *ol howma* social status, and the second highest from those of lowest exchange-ranked standing, with men of highest renown in third place. Regarding age, younger men in the 30-to-35 year age group evidence the highest tuber yields, with the under-30s in third place behind the 45-to-50 year age group. These yield trends again compromise any suggestion that more mature men are likely to achieve higher crop yields because they have greater experience and exploit better land resources. There is a certain trend evident between kin group affiliation and tuber yields, although not statistically significant, with the newer arrivals having lower mean yields than the original colonists. No comparison with *semg^{e}nk* size is possible given the small sample size. Whatever the trends evident, the large standard deviations caution against reading anything into the variations between the means; they suggest that a larger sample would tend to reveal relatively little variation in the mean yields from the gardens of men of different social standing, however assessed.

Production, Exchange and Natural Resource Exploitation

The findings reported here indicate that there are no pervasive correlations between men's social backgrounds and the likely productivity of the land resources they exploit (*conta* Waddell 1972:216; Allen & Crittenden 1987). This seems alien, even improbable to us, conditioned by our culture's capitalistic assumptions to expect individuals to use and manipulate whatever social advantage they have to secure the most for themselves, materially or otherwise. Furthermore, it is not only antithetical to what we may assume is axiomatic to human nature, but it is also apparently at odds with overt manifestations of self-oriented behaviour in Wola social life. These people, as I understand their social values, invest considerable importance in allowing individuals relative freedom of political action, and self-interest, be it individual or for a small circle of kin, is an overt aspect of their social behaviour (Sillitoe 1979). The conclusions reached here about natural resource use seem

---

[12] The yield data come from gardens largely under sweet potato, some from gardens throughout the entire harvesting period of a crop (i.e.both *waeniy* and *puw* stages — see Chapter 8 on these different stages) and other data from a single clear harvesting of mounds (i.e. the *waeniy* stage only) — they are roughly comparable, the *puw* stage not giving substantial yields like *waeniy*. The sample sizes were as follows (some *ol howma* status groups and men's age groups are amalgamated because of the uneven spread of the data collected, to give better balanced samples): status group 1 = 7 gardens, status group 2 = 5 gardens, status group 3 = 6 gardens, status groups 4 & 5 = 6 gardens, status groups 6 & 7 = 7 gardens, and status groups 8 & 9 = 6 gardens; <30 years = 4 gardens, 30–35 years = 4 gardens, 35–40 years = 6 gardens, 40–45 years = 6 gardens, 45–50 years = 8 gardens, and >50 years = 9 gardens; Ind = 7 gardens, Kolomb = 8 gardens, Mayka = 8 gardens, Maenget = 4 gardens, Puwgael = 2 gardens, Wenja = 3 gardens, and Huwlael = 5 gardens.

to contradict this interpretation of their society. What is it that apparently prevents men from self-interestedly using whatever influence or knowledge they have to secure for their households the most productive land resources available?

The Wola say that they find it difficult to assess the fertility of the soil on garden sites before cultivating them, as described. The spread of some of the data presented here complies with their assertions. They maintain that they do not inspect the soil on a site before clearing and cultivating it because they are unable accurately to judge its potential, which anyway will probably vary unpredictably across any site, and under cultivation. The absence of any correlation between *ol howma* status, age or *semgenk* affiliation and the potential productivity of garden sites, as measured by the characteristics reported on here, corroborates this initially improbable seeming assertion. In some regards, finding a good site depends somewhat on luck. Men's inability to judge finely the worth of soil before cultivating it considerably reduces any scope we might think they ought to have to secure the potentially better locations for their households' use by virtue of their social position.

When someone finds and cultivates a productive location, the land tenure system allows him, and his descendants, to garden the area indefinitely, with intervening fallow periods of appropriate duration to maintain soil fertility. A man and his direct descendants have a prior right to cultivate any site which he, or his direct ancestors, have gardened previously, and no one can usurp them whatever his social standing and *ol howma* derived influence. The only way in which a person can secure cultivation rights over an area that has proved to have a particularly fertile soil is to inherit it, or to persuade those with rights to it to allow him to cultivate it. So in some senses, men are dependent on what their forebearers cultivated for access to some areas of potentially good land, particularly those under long-term cultivation. Even if social standing did play some part in men's initial selection of these sites, their direct ancestors may not be of the same standing themselves; the sons of *ol howma* may not aspire to that status but nonetheless inherit rights to their fathers' previously cultivated sites.

Furthermore, soil fertility is not a constant property but subject to change depending on cultivation history and soil management. A onetime productive soil may be degraded if poorly managed. Inheritance of cultivation rights to a previously fertile site does not guarantee good crop yields, although it is probable if appropriate management strategies have been followed. Also, men are not entirely dependent on rights to previously cultivated sites for garden land. They may prevail upon relatives or friends to allow them to cultivate areas that they have prior rights to but no plans to use in the forseeable future (although this can give rise to complex disputes over land tenure in following generations). There is no indication that individuals use whatever influence accrues to them through their success in wealth exchanges to pressure others to allow them access to more promising locations, by placing them in their debt or whatever. Men can also claim and clear virgin land on the territories of the *semgenk* to which they are affiliated. In the Was valley this pool of unused land is large. The drawbacks to cultivating such areas include the heavy physical labour demanded to clear them and their frequently inconvenient locations, situated considerable distances from settlements, perhaps in demon lurking locales. Whatever, the problem of identifying locations with promising soil resources remains.

Regardless of social position, whether judged by *ol howma* status or age or kin group connection, men are hard put to use whatever influence, knowledge or opportunity they have to select sites with the most promising soils for cultivation. But there is more here than men's inability reliably to assess the potential of a location's soil resources before cultivation. Site characteristics are known before clearance (aspect, slope and so on), yet there is scant correlation between them and the sociological factors. The prime sites, whatever their soil status, are not more often cultivated by men of a certain social category, who seem unable or disinterested in securing some control over them by virtue of social esteem, age or whatever. The system of land tenure reinforces their attitudes further, rights to land setting limits on any scope they have for manoeuvre.

It is as if men, who largely select sites for clearance and cultivation, are not too bothered about finely judging issues pertaining to land productivity. Their inability reliably to assess edaphic conditions doubtless plays a part. If they cannot be sure of a site's soil fertility, this perhaps discourages them from paying close attention to site characteristics, which will come to nought if the soil proves relatively poor. Their consideration of site factors is also probably influenced equally by social circumstances (such as proximity to homesteads and rights to previously cultivated areas), as it is by environmental ones (such as aspect and slope — so long as these are within tolerable limits). But beyond these issues, site selection reflects something about men's general attitude to production.

It is of paramount importance to the acephalous political constitution of Wola society that men do not pay too much regard to production, but concentrate their energies on the sphere of wealth exchange (Sillitoe 1988). Production is necessary to supply them with things to exchange, which regarding garden productivity relates primarily to the rearing locally of pig wealth. While marxism has fallen wall-like as a fashionable theory in the social sciences, its central tenet remains unchallenged, that those who control productive capital required by others, whether it be land or whatever, have a potent source of political control over their lives. From this perspective, it is critical to the continuance of the egalitarian Wola polity that no one aspires to secure control over the productive resources involved in pig or other material production. It would compromise the stateless order by placing a facet of political power, currently diffused throughout the social fabric, within reach of some small interest group, which could be antithetical to the continuance of the egalitarian polity.

The findings of this investigation are one manifestation of the consequences of this prescription. The soil contributes indirectly to the Wola political system, for men's inability to judge its fertility and response under cultivation (some soils improving, others worsening with cultivation) helps thwart any ideas some 'tyro-Machiavellians' may have of consolidating the more productive areas under the stewardship of their households. Soil status obviates consolidation of social status, thwarting efflorescence into something politically more hierarchical. Those men of renown, the *ol howma* of Wola society, who handle more pigs than others in exchange transactions are not able to do so because they have access to or control better natural productive resources. There is nothing to choose between the potential crop productivity of their gardens and those of other men. It is access to female labour, not better land resources, that allows these men to keep and handle larger numbers of pigs than average. Either they have female relatives, usually wives, capable of managing larger than average pig herds, who earn themselves the appellative *ten howma* for their admired abilities, or they can call on the services of several female relatives beyond average men's needs, or both.

The dependence of ambitious men on women's productive co-operation has critical implications. It is important to stress that access to things produced does not imply control over the resources used in their production. It is crucial to the continuance of headless Wola 'government' that individual men cannot aspire to control them. The attitude of the Wola to marriage is instructive here. When they marry, men say that they 'share women'; they refer to marriage as *ten tol bay* (lit. woman share do). They do not assume exclusive rights to the labour of females they marry, and certainly do not control them (whatever some males may say in public); they share relations with them, with their affinal kin. They can only exert some influence over their activities, up to a point. At any abuse of what everyone agrees are acceptable demands on their labour, women can turn to their kin for protection of their rights, even return home to them (Strathern 1972).

Esteemed men who manage more pigs than average in exchange contexts are able indirectly to gain access to greater productive capacity than others. It is their female relatives — both close, like wives and mothers, and more distant, such as the spouses of other male relatives — who interface between them and the productive process, as the herders of pigs. And the redemption of pig wealth from the everyday female productive domain to the public male-dominated exchange domain requires the transaction of shell or other wealth through the women involved to claim exchange title. It is a licit and symbolic return to their affinal relatives, acknowledging that they share in the productive efforts of their female kin, notably as guarantors of demands on their labour. The partnership between men and women, and the relationship of men to men through women, is an important aspect of the insulation of the productive from the exchange domain central to the acephalous Wola polity. It may seem improbable to go from soil to marriage, but such is the nature of anthropological enquiry. Both play a part in structuring pig production, ensuring an egalitarian order.

Garden Type, Time Cultivated and Status

Whatever the sociological implications of these findings, the validity of the conclusions reached above about men's social standing and the agricultural potential of their garden sites might be questioned, the random sample notwithstanding, on the grounds that the data were collected from different types of gardens, under cultivation for varying periods of time. If there is any systematic variation in the type of garden men cultivate, or the number of times they cultivate sites, according to social status, age or kin group affiliation, and there are any significant differences between such sites, this could compromise the comparisons made. We are not comparing like with like, although arguably all the sites surveyed were food gardens, whatever the differences between them, and we are interested in the overall relationship, if any, between social position and garden potential productivity. An inspection of the site and soil data according to garden type and times cultivated compared with the social categories used in the foregoing analysis suggests however that there are no such regular variations that might seriously jeopardise the conclusions reached (Tables 11.5 to 11.7). Any discrepancies are relatively minor, the random selection of the sample gardens gives a data base that falls within acceptable limits for the social comparisons attempted.

According to garden type, sweet potato gardens still under cultivation a year or so after the survey, predominate proportionately for all classes under the three social analyses — whether differentiated according to status, age or kin group — followed by sweet potato gardens abandoned within the year. Neither the distribution of sweet potato gardens nor that of mixed vegetable gardens shows any apparent pattern across social categories. It is difficult to see anything other than a random distribution when gardeners' social position is compared with the number of times they have cultivated garden sites. There is a tendency for the proportion of gardens cultivated by any of the social groups to decline with time under cultivation, sites cultivated for the first time tending to comprise the highest percentage. This probably reflects the fact that while some sites prove to be repeatedly cultivable, others do not,[13] resulting in a fair turnover of sites cleared, a goodly proportion of them under cultivation once only.

There is something of a trend evident for taro gardens. Men of higher social status are more likely to cultivate these, which is probably a reflection of the use of this crop in food exchanges, sometimes involving social gatherings

---

[13] Not necessarily because of edaphic shortcomings, other reasons like difficulty enclosing site, distance to it, and so on may also be prominent.

in gardens with a distribution of corms. It is men of a certain standing, who take their exchange commitments seriously and aspire to some renown, who are more likely to cultivate this crop for distribution among kin. Likewise there is a tendency for older men to cultivate more taro gardens. A possible explanation is that the exchange of taro tubers at small social events is going out of fashion and is unpopular with younger men. It is their older relatives who largely keep the custom alive (the distribution of taro at gatherings is not customarily perceived as a pursuit of older men). The number of taro gardens involved is small compared to the sample size overall and these trends are unlikely seriously to bias the conclusions reached, particularly given the lack of any marked variation in properties between different types of gardens.

There is no indication that men of higher *ol howma* social renown tend to monopolise those sites that prove suitable for longer term cultivation, as might be anticipated. Nor is there any suggestion that older men acquire a larger proportion of those sites cultivated many times, through inheritance or by virtue of having farmed for longer, although there is a tendency for men under twenty-one to have fewer gardens cultivated several times, reflecting the fact that they have had less time to consolidate their long-term holdings. Similarly, there is no suggestion in the data distribution that the members of any *semgenk* kin group are disadvantaged regarding access to land. The kin groups are inclined to have similar spreads of sites under cultivation for different times (for example, while the smaller and later founded Puwgael group has no gardens cultivated over four times, and might be thought to be either advantaged, having access to good virgin tracts and no need to keep sites under long term croppping, or disadvantaged, having few sites suitable for long term cultivation, neither supposition merits debate because both Piywa, the other small group, and Wenja and Huwlael, the other two later founded ones, have cultivated a considerable percentage of their garden sites more than five or ten times).

The inclusion of different types of garden, under cultivation for varying periods of time, in the comparison of social standing with site productive potential is supported further by the small variation between them in their site and soil properties. This evidence also raises some intriguing questions about the Wola farming system, particularly the behaviour of soils under cultivation. The variations are not of the order that might be anticipated, which is intriguing, for it raises again issues about the environmental implications of subsistence cultivation in this region.

CHAPTER 12

OUT OF THE SOIL: FERTILITY UNDER CULTIVATION

The common image of subsistence farmers in the tropics has people shifting their cultivations frequently from one site to another because of problems with weed control, disease build-up or fertility decline. The Wola practice such shifting cultivation, but they sometimes decide not to move on and may cultivate the same site for decades. They farm some areas of land semi-permanently within, as described, the broad context of a shifting cultivation regime. The balance of environmental evidence, observation of large numbers of gardens over many years, and the declarations of local people, indicate that soil fertility is the major constraint on production, the incidence of weed, disease and pest infestation rarely rising to levels sufficient seriously to limit yields (Waller 1985; Bridge & Page 1984; Thistleton & Masandur 1985). A look at the previous chapter's status data reveals some intriguing trends relating to these issues. The variations in sites and soils, between different types of garden and periods under cultivation, are not of the kind that might be expected, which questions again suppositions about the assumed behaviour of soils under subsistence cultivation in this montane region of the tropics.

I. SOILS UNDER CULTIVATION

The sample of sites investigated in the previous chapter included examples of the three major types of garden distinguished by the Wola (Sillitoe 1983), as follows: 1) gardens in which sweet potato predominates, sometimes as a monocrop, called *hokay em* (overwhelmingly comprising the largest area under cultivation, as reflected in the sample); 2) gardens in which taro predominates, called *ma em*; and 3) those gardens, often of small area and adjacent to houses, in which a range of crops are grown, called *em g^emb* (sometimes these gardens are subsequently incorporated into long-term sweet potato gardens, when adjacent to them). These three different types of garden may be cultivated for varying periods of time, some are abandoned within a year or so and not recultivated for some time (the subsequent time under fallow varying widely), others remain under cultivation for longer, some for considerable periods of time.

There is a very significant correlation between garden type and the length of time that a site may be under cultivation (Figure 12.1).[1] Both taro gardens

---

[1] The suffix 'abn' labels gardens abandoned within a year of the survey. The number of times the gardens had been cultivated approximates roughly to their time under cultivation in years (at least for the first one or two cultivations, after which short fallow breaks displace the correlation). The gardens are divided up into those cultivated from one through to four times, and then more than five and more than ten times.

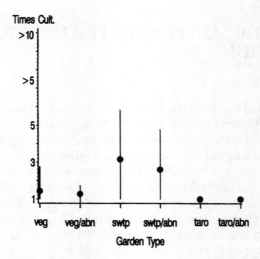

Figure 12.1 Comparison of garden type with time under cultivation.

and small mixed vegetable gardens are customarily cultivated only once before abandonment, whereas gardens under sweet potato may be cultivated several times, sometimes many times. According to the Wola, gardens put down a second time to taro or a varied intercrop will not give worthwhile yields, wherever the site and whatever the soil (except for the occasional exceptionally fertile pocket on some sites, such as may occur in protected folds, which may support a range of vegetables more than once in a garden's lifetime). Sweet potato gardens, on the other hand, may continue to give respectable, even improved, yields many times with appropriate management. It is the crop, not the site nor the soil that in part determines the probable fate of a garden. If a location put down to taro or mixed vegetables was planted with sweet potato instead it would probably have a different cultivation history.

It is not, the Wola assert, that they put the better locations under sweet potato and hence they have longer productive lives. A comparison of local assessments of soils with the types of garden cultivated on them supports their assertions. There is no apparent association of soils judged locally to be good or poor with particular types of garden, even when sites abandoned within one year of the survey are distinguished (Table 12.1). There is a tendency for taro gardens to be located on soils judged to be 'good' (although none were thought to be 'very good'), and a sizeable proportion of those abandoned within the year were considered 'barren' — a possible acknowledgement of their impending abandonment following harvest of the crop. The soils under mixed vegetable intercrops show a relatively even spread across assessment classes, the 'barren' and 'no good' soils unexpectedly making up a higher proportion of those gardens not nearing abandonment, and 'good' soils predominating proportionately under those approaching it, together with some 'very good' ones. The position with soils in sweet potato gardens, the largest sample, is more even between sites remaining under cultivation and those nearing abandonment. There is surprisingly little difference between the ongoing cultivated sites and those approaching the end of their productive lives within the year; while those nearing abandonment have proportionally more 'barren' and 'no good' assessed soils, the difference is not large, and the proportion of 'fair' to 'good' soils is very similar between the different gardens.

The time for which sites are under cultivation compared with local assessments of the agricultural worth of soils further corroborates the assertions of the Wola that there is no correlation between the cultivation status of gardens and their productive potential. On all sites, whether cultivated once or more than ten times, soils assessed as 'good' and then 'fair' make up the largest proportion under cultivation. While the percentage of soils judged 'barren' is comparatively high on sites cultivated over ten times, it is also high for those under crops for the first time, with fewer, if any 'barren' assessments in intervening time intervals. And soils judged 'very good' occur proportionally more often on sites cultivated for over five to ten times. These trends comply with Wola assertions about the absence of any correlation between site cultivation status and soil fertility, which contradict orthodox views about soils under cultivation, that without amendments they ought to show a steady decline in agricultural worth with time cropped.

The site characteristics and soil properties evidence relatively small variations between different garden types (Table 12.1), and what differences there are need to be judged against the skewed sample with its small number of taro and mixed vegetable gardens. While differences in aspect are significant, they are difficult to interpret given the large standard deviations; sites nearing abandonment are not more likely to face consistently in a different direction to others. Slope does not vary significantly between gardens; nor does it feature in abandonment apparently, steeper sloped sites not being more likely to be abandoned. The differences in altitude, although significant, are again difficult to interpret given the wide spread of data; there is some suggestion that higher altitude sites are more likely to be abandoned over others. Sweet potato gardens occur on a wider range of slope forms (although comprising the larger part of the sample, this is perhaps expectable); there is little evidence that taro gardens occur on sites of a particular surface configuration, indicative of the wet conditions favoured by this crop; and the small area of vegetable gardens makes it more likely that they will occur on even surfaces. Again, sweet potato gardens may be cleared from any vegetation, whereas taro and vegetable gardens are rarely established on grassy sites, and vegetable ones infrequently from forested locations.

While depth of topsoil is similar between differently cropped gardens, there is a consistent trend for sites approaching abandonment to have shallower depths of topsoil, reflecting perhaps some erosional losses when exposed under cultivation, the ensuing reduction in depth contributing to the decision to abandon. Differences in the colour of topsoil only relate tentatively to garden type, with gardens approaching abandonment somewhat more likely to have light brown soils, lower in organic matter content; dark brown and brown soils otherwise tend to dominate whatever the type of cultivation. The pH of sites differs significantly, sweet potato and small vegetable

Table 12.1

Mean garden site and soil factors compared with garden type and number of times cultivated (values in brackets = standard deviations; some percentage totals exceed 100 because some gardens occur in two classes; degrees of freedom 5, 131/285)

		GARDEN TYPE						STATISTICAL SIGNIFICANCE			TIMES CULTIVATED						STATISTICAL SIGNIFICANCE		
		mix veg	abn m.v.	swt pot	abn s.p.	taro	abn taro	CV	F	Sig	1	2	3	4	≥5	≥10	CV	F	Sig
SITE:																			
Aspect (degrees)		126 (118)	148 (129)	78 (79)	121 (99)	176 (145)	104 (106)	94.1	5.23	S	113 (79)	80 (67)	87 (109)	95 (75)	82 (84)	73	96.9	1.77	NS
Slope (degrees)		17 (7.8)	18 (10.8)	21 (8.8)	20 (8.9)	16 (12.1)	18 (11.2)	45.2	1.37	NS (9.5)	20 (9.4)	23 (7.0)	19 (7.7)	20 (8.8)	18 (7.9)	17	44.9	2.17	S
Slope Form (% gardens)	even	78	64	50	51	20	82			56	52	44	50	50	44				
	syncline		7	21	24	60				15	23	19	38	25	35				
	anticline	11		16	15	10				12	19	19	6	9	13				
	concave		7	6	7					8	3			9	13				
	convex	11		5	4	10				5	4	7	3	3					
	stepped		21	3			18			6	1	7							
	rocky			4	2		9			5	4	4							
	irregular			3	2					3		4	13						
	undulating			3	2					2	3	4		3	4				
Altitude (m)		1869 (66)	1875 (68)	1795 (84)	1821 (108)	1725 (90)	1839 (118)	5.0	2.6	S	1802 (96)	1800 (92)	1847 (106)	1775 (72)	1842 (29)	1769 (80)	5.2	0.75	NS
Vegetation Cover (% gardens)	1y forest		7	6	10	50	55				16	7							
	2y forest			16	23	20	18				25	20	14	12	6				
	tree/cane	11		17	14	30	27				12	21	14	6	12	17			
	cane	89	93	50	49						43	47	65	65	66	61			
	grass			11	4						4	5	7	17	16	22			
SOIL:																			
Depth (cm)		26.7 (23.9)	17.0 (11.3)	18.7 (13.3)	14.7 (11.6)	22.6 (14.2)	14.8 (9.1)	73.3	1.86	NS	17.9 (13.0)	18.3 (12.7)	16.7 (12.5)	15.1 (12.1)	20.1 (17.5)	18.2 (12.6)	74.3	0.34	NS

Table 12.1
Continued

		GARDEN TYPE					STATISTICAL SIGNIFICANCE			TIMES CULTIVATED						STATISTICAL SIGNIFICANCE		
		mix veg	abn m.v.	swt pot	abn s.p.	taro	abn taro	CV	F	Sig	1	2	3	4	≥5	≥10	CV	F	Sig
Topsoil	black	11		12	7	11	18				7	16	11	12	16	4			
Colour	d. brown	33	50	37	35	67	9				39	37	52	59	12	31			
(% gardens)	brown	45	43	31	29	11	18				38	21	4	12	50	39			
	l. brown	11	7	20	29	11	55				16	26	33	17	22	26			
Number		1.8	2.2	2.1	2.2	2.0	2.3	24.7	1.34	NS	2.1	2.2	2.1	2.2	2.2	1.9	24.7	1.28	NS
Horizons		(0.4)	(0.8)	(0.5)	(0.5)	(0.7)	(0.5)				(0.6)	(0.5)	(0.5)	(0.4)	(0.4)	(0.3)			
pH		5.3	5.0	5.2	5.2	5.6	5.7	5.4	4.54	S	5.3	5.2	5.2	5.2	5.3	5.3	5.8	0.50	NS
		(0.3)	(–)	(0.3)	(0.2)	(0.5)	(0.8)				(0.4)	(0.2)	(0.3)	(0.2)	(0.1)	(0.2)			
N		142	125	117	127	100	117	25.2	1.50	NS	116	117	126	122	128	113	25.7	0.49	NS
(ppm)		(14)	(–)	(31)	(19)	(39)	(38)				(32)	(31)	(33)	(20)	(22)	(29)			
P		9.7	2.0	7.7	6.6	3.7	9.7	62.6	2.00	NS	6.6	7.6	9.6	6.9	9.2	6.1	163.4	1.33	NS
(ppm)		(2.3)	(–)	(4.7)	(4.8)	(2.5)	(2.3)				(4.5)	(4.6)	(4.5)	(5.0)	(5.1)	(4.6)			
K		174	136	172	114	241	213	70.5	1.84	NS	162	163	159	233	167	157	69.9	0.60	NS
(ppm)		(125)	(–)	(123)	(62)	(166)	(125)				(120)	(123)	(122)	(118)	(94)	(134)			
Local	v. good	22	7	9	6						7	3	4	6	19	17			
Soil	good	22	57	46	44	90	46				45	55	48	50	40	39			
Assessment	fair	22	29	30	26	10	27				27	32	24	38	25	26			
(% gardens)	no good	12	7	11	15		9				12	6	24	6	16	9			
	barren	22		4	9		18				9	4				9			

gardens covering considerably narrower ranges, with increasingly acid conditions as abandonment nears, whereas the reverse occurs in taro gardens, probably due in part to changes in redox conditions on wetter sites with cultivation. The variation in other nutrients between different garden types is not significant, although sites nearing abandonment under all cultivation regimes evidence a fall in available potassium; the variable trends for available nitrogen and phosphorus are difficult to assess.

When garden site characteristics and soil properties are compared with the number of times gardens are cultivated, they too show surprisingly few correlations. Differences in aspect are not significant, extending over an almost constant range with time under cultivation; those sites that remain longer under cultivation do not face more often in any particular direction. The differences in slope, while not great, are significant, and indicate that sites cropped for several years tend to have somewhat less steep slopes. They also occur on slopes more restricted in their range of surface topography, suggesting that gardens cleared on less favourable terrain, such as irregular ground or places with many prominent rock outcrops, are less likely to be recultivated several times. These differences in slope form are unlikely to have a notable effect on soil erosion loss rates, although they may contribute to the cultivable life of a site.

Regarding vegetation, the more times a site is cultivated, the less likely it is to be cleared from forest, either primary or secondary, with tall cane grass and short coarse grasses increasingly cleared as time under cultivation increases. The depth of topsoil in gardens remains fairly constant in range, however many times they are cultivated, partially reflecting perhaps differences in slope. Topsoil colour shows no apparent variation explicable according to time under cultivation; the proportions of darker to lighter coloured soils do not change regularly with time. While pH may evidence a small fall after one cultivation (the large standard deviation range compromises this trend), it remains much the same from then on. And while the levels of available nitrogen, phosphorus and potassium fluctuate across time intervals, they do so randomly.

While this review of variations in site and soil parameters according to garden type and time under cultivation, vindicates the use of a mixed random sample in the foregoing socio-pedological investigation, its findings sit awkwardly with widely accepted ideas and evidence from elsewhere about what happens to soils under similar subsistence regimes. They are expected to undergo a steady, sometimes a dramatic, decline in fertility (see Chapter 2; also Clarke 1971; Wood 1979, 1984 in New Guinea contexts). They confirm again the idea that the Wola live in a place against time, soils apparently not manifesting expected changes in their properties under cultivation. There are some trends discernible in the data as gardens near the end of their cultivable lives that comply with the conventional view — like a decline in topsoil depth, likelihood of lighter coloured soils with lower organic matter, a tendency for soil acidity to increase and available potassium levels to fall, as abandonment approaches. But the same trends are not evident with increasing time gardens are under cultivation, with no striking changes in soil properties (the only noteworthy differences are those in slope and vegetation cleared).

The unexpected behaviour of soils under cultivation, contrary to accepted assumptions, prompts the question: what is happening to soil fertility under cultivation in the Wola region? Perhaps the soil tests conducted in the field were too approximate and lead to erroneous conclusions. Or perhaps they do signify some important differences between the behaviour of soils under cultivation in the highlands of Papua New Guinea and those under shifting regimes elsewhere, as intimated by local soil appraisal. If so, they merit further investigation, for they point to the need of a radical reinterpretation of local cultivation practices and soil management, away from current emphases on environmental degradation. This chapter addresses changes in the nutrient status of soils under cultivation for varying periods of time, including those intriguing sites that are under semi-permanent cultivation without the benefit of outside soil amendments, in an attempt to understand how the Wola apparently avoid the soil constraints reported under shifting cultivation, which normally oblige traditional farmers in the tropics regularly to shift their cultivations.

The soils investigated are *hundbiy* clays of the Tropept sub-order, the most widely cultivated in the western Wola region. Soil samples were collected from one hundred and ten locations on Aenda *semonda* territory in the Was valley (Map 5.1). They were selected, so far as possible, from sites of similar slope, altitude, aspect and so on, to minimise spatial environmental variations and 'noise' factors when time cultivated and cropping history were of interest. They ranged from sites never cultivated and under primary montane rainforest, to gardens under cultivation for varying time periods, to sites abandoned for different periods. Eleven different classes of site are distinguished according to cultivation status or land use history, each comprising a sample of ten topsoils (Table 12.2).[2]

---

[2] All of the soil samples were composites. Subsamples of soil were collected at random across the sites (from 0 to 10 cm depth for topsoils). Between 10 and 20 spots were sampled at each location (depending on swidden size), while walking about it. The soil collected was mixed up, and a coned and quartered sample bagged after drying in the sun. The samples were subsequently ground to pass a 2 mm sieve before analysis. All of the soils were analyzed for the principal fertility indicators — pH, phosphorus, exchangeable cations, organic carbon and nitrogen — and some of them were tested for other measures — such as boron, base saturation, aluminium and sodicity. The experimental procedures used in analysing the soil samples were: pH: measured electometrically in 1:2.5 soil:distilled water suspension; pH (NaF): measured in 1 g soil: 50 ml NaF suspension (2 mins after mixing); extractable Al: measured in unbuffered 1N KCl; P: measured by Olsen bicarbonate extraction; P retention: soil shaken with standard phosphate solution and filtered, and P remaining in solution measured colorimetrically, giving soil retained P by subtraction; exchangeable cations: extracted in 1N ammonium acetate at pH7 followed by atomic absorption spectrophotometry; Kc: reserve potassium extracted in boiling nitric acid; CEC: summation of extractable cations plus exchange acidity (measured by depression of pH in CH_3COONH_4 extract); C: measured by the Walkley-Black dichromate oxidation method; N: measured by the Kjeldahl digestion method; ECe: electrical conductivity measured in 1:5 soil:water extract; B: soluble B determined in hot water extract. Analytical variability between the laboratories that performed the soil analyses contributed in some measure to the variation recorded (soil laboratories are notorious for varying in their analyses for the same nutrient), but a statistical check indicated that overall this source of variation was not significant for these data. The soil analysis results may be found on the internet at the following address (URL): http://www.dur.ac.uk/~dan øps/soil.html.

Table 12.2

Measures of topsoil chemical fertility compared with site land use status
(values in brackets = standard deviations; n = 110 sites; degrees of freedom = 10, 99)

	Virgin Sites	SITE STATUS: VIRGIN THROUGH CROPPED TO FALLOW VEGETATION						Fallow Sites: Years Under Fallow				STATISTICAL SIGNIFICANCE		
		Cropped Sites: Times Cultivated												
		1	2	3	4	≥ 5	≥ 10	<1	≥ 1	≥ 5	≥ 10	F	P	
pH	5.02 (0.66)	5.39 (0.92)	5.08 (0.47)	4.94 (0.40)	5.01 (0.67)	5.32 (0.53)	5.25 (0.28)	5.28 (0.61)	5.22 (0.41)	5.1 (0.31)	4.86 (0.27)	1.14	0.34	N.S.
Phosphorus (ppm)	17.1 (12.1)	17.3 (16.2)	12.2 (13.4)	7.1 (4.5)	7.8 (6.3)	7.2 (5.5)	5.7 (5.2)	13.9 (17.4)	7.2 (7.9)	6.6 (6.0)	7.2 (6.8)	1.71	0.09	N.S.
Potassium (me/100 g)	1.04 (0.57)	0.56 (0.20)	0.68 (0.30)	0.63 (0.36)	0.44 (0.49)	0.27 (0.13)	0.28 (0.17)	0.87 (0.96)	0.58 (0.28)	0.79 (0.97)	0.46 (0.25)	2.09	0.03	S.
Calcium (me/100 g)	13.9 (11.2)	12.2 (6.4)	14.1 (9.4)	8.7 (7.4)	15.3 (11.7)	16.7 (15.8)	8.1 (3.7)	19.7 (23.8)	12.4 (9.4)	8.0 (5.3)	7.3 (7.7)	1.23	0.29	N.S.
Magnesium (me/100 g)	3.04 (2.53)	3.36 (2.25)	3.25 (2.23)	1.90 (1.31)	2.11 (1.54)	2.14 (0.70)	2.35 (0.98)	2.65 (2.48)	2.04 (0.92)	1.66 (0.96)	2.45 (2.39)	0.95	0.49	N.S.
CEC (me/100 g)	31.3 (5.2)	27.4 (9.4)	31.0 (8.3)	27.1 (5.6)	27.4 (12.1)	29.0 (4.9)	22.6 (7.4)	28.0 (7.4)	31.6 (6.4)	23.3 (7.1)	26.9 (6.9)	1.58	0.13	N.S.
Carbon (%)	25.9 (6.6)	17.6 (8.6)	17.3 (6.4)	14.9 (5.3)	16.8 (10.9)	12.8 (5.9)	11.6 (6.4)	14.5 (6.8)	16.7 (5.2)	11.9 (4.1)	17.5 (6.0)	3.55	0.001	S.
Nitrogen (%)	1.41 (0.62)	0.96 (0.29)	1.06 (0.41)	0.86 (0.39)	0.99 (0.32)	0.93 (0.28)	0.74 (0.26)	0.93 (0.42)	1.12 (0.43)	0.79 (0.29)	1.08 (0.46)	2.23	0.02	S.

The objective is to assess whether soil fertility, as measured by the properties analyzed, varies significantly according to site cultivation status whether it was a virgin location, or land cultivated a specified number of times, or somewhere left under fallow regrowth for some interval (statistically significant = $P < 0.05$). The results were assessed by the ANOVA procedure. The cultivation status of the soil (time cultivated/abandoned) was equated with different treatments, and the soil properties with dependent variables, assumed to vary with time under cultivation. A regression curve ($y = a + be^{-ct}$) was fitted to the data means where sites differed significantly by their cultivation histories.

Soil Nutrient Status and Land Use History

The analytical investigation indicates that while organic matter and the elements nitrogen and potassium all decline significantly with time under cultivation, they settle at new equilibria which do not signify deficiencies. The other nutrients show no significant variation over time, but phosphorus levels are markedly low throughout. Soil conditions suggest possible availability problems with some nutrients. These soil trends accord with the wide variety of crops observed under cultivation early in the life of gardens, followed by sweet potato, which plays a significant part in Wola manipulations of the swidden agricultural regime, long established gardens being virtually under it exclusively. This crop is able to continue yielding adequately long-term with the subsequent decline in nutrient availabilities, notably because nutrients remain in favourable ratios for tuberisation and because of its tolerance of low phosphorus conditions. It is the ability of this crop to cope with changes in the fertility status of soils under semi-continuous cultivation which accounts in considerable part for how these farmers are able to avoid the problems that normally prompt site abandonment under tropical subsistence agricultural systems.

Acidity: The topsoils are, in general, acidic in reaction, averaging 5.13 overall (although measuring pH on re-wetted dry soil samples is likely to depress it, by up to 0.5, suggesting that the soils may not be as acidic as some of the extremely low values suggest). The topsoils cover a considerable range from pH 3.8 to 6.9. The pH of subsoils is marginally higher, with an overall mean value of 5.23. Soil pH serves as a general indicator of fertility status, acidity influencing the availability of most soil nutrients. While the results suggest a sub-optimal pH range for maximum availability of most nutrients, moderately acid conditions are reportedly favourable to many tropical crops, including sweet potato (Landon 1991:281).

There is an initial rise in pH on sites cultivated for the first time, reflecting the burn off of natural vegetation cleared from sites, increasing cation levels. Thereafter an increase in acidity is evident in the trend of mean pH values, until sites are recultivated a fourth or consecutive time, when it declines again. The increase in acidity is understandable given the increased effects of leaching the longer a site is kept under cultivation (an increase in soil acidity has been reported frequently under shifting cultivation — Siebert 1987; Brinkman & Nascimento 1973; Sanchez 1976; Nye & Greenland 1960). But the fall after several cultivations is more problematic, it could reflect the tendency for people to leave these sites longer under grass and weeds between recultivations, and sometimes to burn off the resulting growth of coarse grasses, the increase in readily available cations being sufficient to boost pH again. When sites are abandoned, the data indicate a steady decline in pH the longer they are left under natural regrowth, which probably reflects the ever larger proportion of available cations taken up by the standing natural vegetation as its biomass increases and locks up evermore nutrients. The subsoil pH values also cover a considerable range, although predictably no time effect is evident under cultivation.

In view of the importance of pH as an all-round measure of soil fertility, the trends evidenced in these data give few reliable clues as to what is happening in these highland New Guinea soils under cultivation regarding nutrient availability, if anything they suggest that what changes there are will be of a relatively small magnitude. Any evident time effects regarding topsoil acidity are relatively small, the spread around the means is considerable (sites under cultivation for the first time, for example, almost span the entire pH range). The variation in pH between sites shows no significant difference with time under cultivation. The only significant change in pH occurs on abandoned sites, where the decline evidences a straight line relationship with time under natural vegetation (Figure 12.2).

A common problem with acid soils like these in the tropics is that aluminium becomes soluble and is present on soil colloid exchange sites, sometimes reaching concentrations toxic to the majority of plants, damaging roots and interfering with nutrient uptake. Aluminium saturation in excess of 60% of total exchangeable cations indicates toxic levels. None of the aluminium values recorded for the few topsoils tested were sufficiently high by this crite-

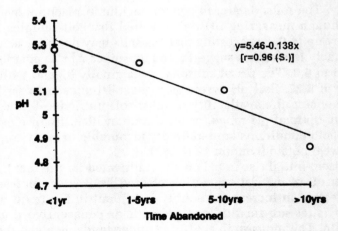

Figure 12.2 The relationship between soil pH and time sites abandoned.

rion to suggest toxicity, with a mean of 37 ppm. Calcareous parent materials not rich in aluminium may partly explain the low levels, although volcanic ash, deposited to varying extents across the region, gives an amorphous alumino-silicate gel which could elevate levels. The relatively high organic matter content of the soils (see below) might be a contributory factor, organic matter binding aluminium to itself in an unexchangeable form, so keeping levels down regardless of the relatively acid conditions. There is an apparent increase in aluminium levels over time, although the small number of samples tested for aluminium, and their skewedness relative to time under cultivation, make the correlation dubious. Nonetheless, an increase in aluminium concentrations, like hydrogen ions, with time under cultivation and greater exposure to leaching by rainfall, is quite feasible, although never approaching levels toxic for most crops.

The measurement of pH in sodium fluoride gives an indication of the contribution of volcanic ash to a soil, or more accurately the amount of poorly ordered material in the clay fraction. This is significant because non-crystalline minerals, like allophane, derived from volcanic ash, can have marked effects on soil properties, notably by giving rise to variable charge and increased phosphate fixation. Values above 10.0 suggest that poorly ordered minerals make a significant contribution to soil properties; the overall mean pH in NaF for the few topsoils tested is 9.9. (The small and skewed sample makes any comparison with time under cultivation dubious, and cultivation would not anyway change the distribution of the mineral fraction.)

Phosphorus: The extractable or 'available' phosphorus levels in the soil solutions were perhaps more variable than expected, although on the whole they were low. Anything less than 10 ppm is generally regarded as deficient; the overall mean value for the topsoils is 9.9 ppm and for subsoils 4.1 ppm. This low 'available' phosphorus corroborates the contention that ash-derived, poorly ordered minerals are having a significant influence on soil properties, leading to strong fixation (Gebhardt & Coleman 1974; Parfitt & Mavo 1975). The measurement of 'phosphate retention' gives an estimate of the relative unavailability of phosphorus to plants due to fixation by the soil. Phosphorus retention values above 90% are regarded as evidence of very high rates of fixation, and the mean value for these topsoils is 93%.[3]

When we compare 'available' topsoil solution phosphorus with the cultivation history of soils, there is some patterning evident. Although the variation is not statistically significant when we take all soils together (both those under cultivation and those abandoned under secondary regrowth), it is when we take only those under virgin forest and crops ($F = 2.23$ and $P = 0.05$). The trend is for topsoil phosphorus availabilities to decline with time under cultivation (Figure 12.3 traces the likely rate of fall). This parallels the

---

[3] Phosphorus sorption isotherms, which give a more accurate assessment of phosphorus fixation, constructed for similar soils elsewhere in the Southern Highlands region confirm these observations of very high phosphate fixation capacities (Radcliffe 1986:114–18; Floyd et al. 1987, 1988).

situation reported for similar soils elsewhere, with an increase in phosphorus levels upon burning followed by a decline (Sanchez 1976; Watters & Bascones 1971; Siebert 1987). The slight rise evident in topsoil phosphorus levels with the first cultivation may reflect the boost received following the destruction of the natural vegetation upon clearance. The marked fall in phosphorus levels from one through to three cultivations is noteworthy, correlating closely with Wola cropping practices, where the range of crops grown falls off considerably after one or two cultivations as sites pass under a virtual sweet potato monocrop. There is no significant variation in the subsoil supply of this major plant nutrient over time; it is consistently low. It seems likely, given these results, that phosphate is a limiting nutrient in these soils, particularly after one or two cultivations, although given sweet potato's relatively low demand for phosphorus this may be a fairly minor constraint on its yield with other nutrient limitations as outlined below. Also, somewhat off-setting these limitations, organic derived phosphate is probably a notable source of phosphorus in these apparently deficient soils (Mueller-Harvey et al. 1985, 1989 stress the influence of soil organic matter on crop phosphorus availabilities). It is likely that plants depend partly on organic matter mineralisation to supply them with some of their phosphorus needs, particularly given the high soil organic matter contents (see below).

Exchangeable Cations & CEC: Regarding the exchangeable cations, potassium is required in the largest amounts by crops, and is particularly important to the yield of sweet potato. If soil potassium levels of less than 0.2 me/100 g are taken to indicate deficiency (and 0.1 me/100 g as the absolute minimum for agricultural production), the availability of exchangeable potassium is low

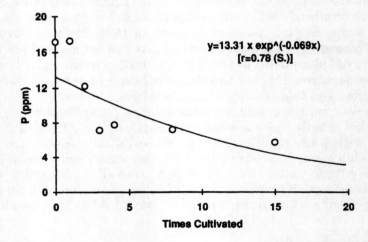

Figure 12.3 The relationship between topsoil phosphorus content and site cultivation status.

to deficient on some sites, suggesting that potassium supply may on occasion be a major constraint to crop yields, particularly for sweet potato, over the yield of which it has a marked influence. The mean topsoil potassium level is 0.60 me/100 g, indicative of adequate to good levels overall; and the mean subsoil level is a deficient 0.11 me/100 g. If we take 0.4 me/100 g as an adequate level, then some 56% of the topsoils have adequate potassium supplies, the remainder having sub-optimal levels, although only 8% are deficient (<0.2 me/100 g).

The potassium constraint is further reflected, as might be anticipated, in measurements of the ion as a percentage of base saturation. If present in adequate amounts, exchangeable potassium should account for a minimum of 2% to 3% of total exchangeable base saturation, a level reached in none of the small sample of soils in which measured, with a mean of 1.2%. The measurement of 'reserve' potassium (Kc), which indicates the amount of unavailable potassium locked up in minerals and waiting to be weathered out, serves as a measure of soil potassium reserves present over and above those available, and results were also low for the few soil samples on which measured, with a mean value of 0.025 me/100 g, indicating that the sub-optimal supply of potassium may be an intrinsic feature on some of these soils, an on-going constraint on crop production, not solely due to cultivation and removal of potassium in harvested crops, nor leaching losses. (Neither of the small, skewed samples of results for percent saturation nor reserve potassium show any time related variation).

There is evidence of a statistically significant difference between sites in topsoil potassium availability and time under cultivation. This univalent cation is more readily lost by leaching, and cumulatively in crop removals. The trend is for available topsoil potassium reserves to decline steadily under cultivation, until they reach an equilibrium point where presumably the potassium weathered out of parent material balances that lost in leaching and taken up by crops. Under natural vegetation, on the other hand, potassium is cycled tightly around the ecosystem with little leaking out (Edwards 1982). It is possible, as available potassium reserves near the new equilibrium, after several years under cultivation, that they are approaching crop deficient levels, an important trigger regarding the abandonment of some sites, prefiguring a marked decline in yields (mean potassium values after ≥5 years cultivation, at 0.27 me/100 g, and ≥10 years at 0.28 me/100 g, are approaching the deficiency thresholds). A regression indicates the likely rate of potassium decline under cultivation, and the possible time scale leading up to abandonment (Figure 12.4).

A sharp increase in available potassium at the start of cultivation, followed by a marked decrease, is commonly observed under shifting regimes, suggesting that potassium is leached at a faster rate than other cations (see Chapter 2). Any increase immediately after clearance could be the result of the burning of the vegetation cut down on newly cleared sites, this practice releasing potassium from the plants and returning it to the soil in ash. Such a

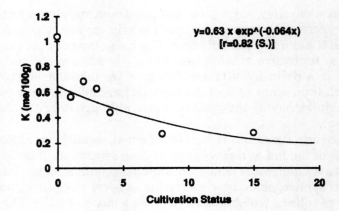

Figure 12.4 The relationship between topsoil exchangeable potassium and site cultivation status.

rise is not evident in these topsoil data, although the variation in potassium data between gardens cultivated the same number of times, some sites recording considerably higher levels, might be related to this phenomenon, for if a soil sample included material from below recent fire sites this could markedly elevate potassium values. The rise evident in topsoil potassium levels on abandoned sites might reflect their gradual restoration to pre-cultivation levels, as potassium released is incorporated into the tighter natural mineral cycle.

The subsoil data, on the other hand, evidenced a possible burn effect, with an initial rise in available potassium levels over those under natural vegetation, followed by a decline thereafter under cultivation, until on long-term sites recultivated many times they rise again. These trends could reflect something of the leachability of this cation, potassium released by the initial burn of natural vegetation moving rapidly, with relatively few plant roots to intercept it, from the topsoil into the subsoil. Later, longer grassy fallow intervals with occasional burn-offs could result in increased potassium release which, again quickly leached into the subsoil, elevate levels once more. The evidence suggests that as topsoil potassium levels fall in long established gardens, some of the ramifying and extensive root network of sweet potato may explore and deplete the denser subsoil of what little potassium it has available. Any decrease in the thickness of the top horizon, due to run-off erosion losses on steep slopes (see Chapter 6), will promote this root exploration of the more compact less-favourable-for-rooting subsoil, by bringing it nearer to the surface, and within range.

The dominant exchangeable base ion in all soils is calcium (as under Eastern Highlands forest — Edwards & Grubb 1982). The levels of this cation are adequate on the whole for crop production, although there are considerable variations between soils. The topsoil mean is 12.4 me/100 g, and that of the subsoil 3.5 me/100 g. Again, the percentage base saturation distribution reflects the predominance of calcium, it comprising a considerable fraction, sometimes over 50%, of the cations present in soils. The levels of magnesium, the other important nutrient cation, are also adequate, with a mean topsoil value of 2.5 me/100 g, and subsoil value of 0.49 me/100 g. This ion comprises something like 8% of cation saturation. These results are perhaps only to be expected given that the tested soils overlie limestone. The ratio of calcium to magnesium is generally within an acceptable range, with a topsoil mean value of 5:1. Antagonism of magnesium on potassium and calcium nutrition is only likely to be significant where the Ca:Mg ratio is <1 (Tergas & Popenoe 1971), which is only probable on basic igneous parent material.

The distribution of these two cations shows no statistical correlation with soil cultivation status. Neither of them evidence any patterned change as gardens age, which matches, as anticipated, the trend evident with pH, base availability influencing acidity. They show no firing effect on their levels either, which is unexpected. The calcium levels in the topsoil move up and down irregularly within a relatively narrow range, whereas the magnesium levels evidence a noticeable decline after two cultivations, following which they remain broadly uniform over time. The general stability in the availability of these cations possibly reflects parent material buffering capacity, replenishing any ion leaching losses, at least in the shallower profiles. The lower cation exchange capacity of the subsoils partly explains the decrease in the concentrations of these exchangeable nutrient ions down profiles (see below). The low subsoil levels of both cations evidence a gradual rise with time under cultivation, possibly reflecting the slower leaching rates and hence more gradual accumulation of these divalent ions from topsoils when more exposed to leaching under crops.

The ability of soils to hold nutrient cations depends on charges associated with the surfaces of clay particles and organic matter. They may be permanent, where isomorphous substitution occurs in clay mineral lattices, or pH dependent, where soil solution conditions affect adsorption/desorption of hydrogen ions from minerals and organic matter. Variable surface charge, with its pH consequent effects on a soil's ability to retain ions, is prominent in soils like these which contain some volcanic ash, with short range order minerals like allophane and halloysite.[4] While base saturation indicates the extent of cation leaching losses, and hence is a guide to a soil's capacity to retain these elements above protons and aluminium ions (occupying the remaining exchange sites), it is parameter of doubtful usefulness for soils which contain volcanic ash because of their variable charge characteristics (being calculated as the sum of basic cations divided by CEC at pH 7). Nevertheless base saturation continues to enjoy some prominence, featuring in both FAO (1974) and USDA (1975) classification schemes, for distinguishing between more fertile

---

[4] Radcliffe & Gillman 1985 indicate the significance of variable components in some similar soils from the Southern Highlands and discuss the influence of soil pH variations on exchange capacities.

eutric soils and less fertile dystric ones. By this rough and ready criterion the topsoils vary in fertility from poor to very good, although there is no time related correlation between base saturation and garden age.

The cation exchange capacity (CEC) of a soil is a better measure of its ability to retain ions against leaching losses — although it too yields questionable results, particularly when measured at pH 7, somewhat overestimating CEC for acidic soils with their dependence on variable charge clays for part of their capacity (Radcliffe 1986:98–99 indicates a perhaps 50% overestimation). Any variation in soil pH affects effective cation exchange capacity. Regardless, the indications are that the CEC values of the soils are fair, even high for tropical soils generally, largely due to their allophane and high organic matter contents. The mean topsoil CEC value is 27.8 me/100 g and the subsoil mean value is 15.9 me/100 g (whereas effective CEC might be half or less these values). While the absolute CEC values are unrealistic, they accurately show relative differences between soils when compared according to their land use histories. They indicate little correlation with time under cultivation. We should not anticipate that cultivation would affect CEC, except in so far as it leads to increased oxidisation rates and a fall in carbon levels, and some CEC is derived from organic matter. The absence of any CEC trend again complies with the poor correlation found between pH and time under cultivation, as pH is a significant determinant of CEC in these soils with their prominent variable surface charge component.

Organic Matter, Carbon & Nitrogen: The organic matter content of these soils is a significant contributor to their respectable exchange capacities. It is high for the tropics. The mean topsoil organic carbon level is 16.1%, with a considerable range of some 13% in either direction. At low elevations in the tropics, where humidity and temperature are high year round, rapid organic matter decomposition, humification and mineralisation occurs, whereas at higher altitudes, like Wolaland, relatively cool temperatures reduce biological and chemical activity rates significantly, and favour organic matter accumulation. Allophane and its derivatives also further inhibit mineralisation. The formation of stable humus-allophane complexes and inhibition of microbial decomposition of organic matter by allophane and related compounds contribute to the high organic matter contents of soils containing some volcanic ash (Wada 1977). The upshot is that these soils have a considerable store of potential plant nutrients locked up in their topsoil humus fraction, which is slowly made available, being released into the soil solution as mineralisation gradually proceeds, although some may remain indefinitely immobilised. The subsoil carbon contents are predictably considerably lower with a mean of 4.1%, and evidence no time related variation.

There are many reports of a decline in topsoil organic matter content under shifting cultivation regimes in the tropics, sometimes of quite dramatic proportions, with the enhanced aeration and oxidation resulting from the disturbance of the soil under cultivation (see Chapter 2). The magnitude of the decrease relates to topsoil organic carbon content before clearance, the higher it is, the greater the depletion (Sanchez 1976:369). These topsoils evidence a

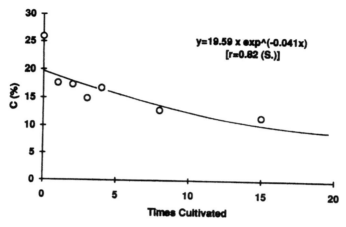

Figure 12.5 The relationship between topsoil carbon content and site cultivation status.

similar, though qualified decline, with a statistically significant difference between topsoil organic carbon and number of times sites are cultivated. (Loss on ignition was also consistently high on the samples measured and evidenced a decline with time). Topsoil organic carbon content gradually falls and levels off (Figure 12.5). This steady decline to new equilibrium levels is attributable to increased microbial respiration with soil aeration, and the return of smaller amounts of organic matter to the soil when under crops, and the loss of carbon as CO_2 gas with the burning of vegetation during cultivation, which may be repeated after initial clearance with the occasional burning of fallow grasses when pulled up.

But the fall is not sufficient to reduce soil carbon contents to markedly low levels, even after land has been repeatedly cultivated ten or more times (mean C value at ⩾10 cultivations is 11.6%). High humus contents have been reported elsewhere for Andepts under shifting cultivation, where farmed sites may be abandoned while the surface soil still contains up to 10% carbon (Popenoe 1959). Furthermore, the results within each time-under-cultivation class evidence a considerable spread, gardens under cultivation for the first time having a range of organic carbon contents which compare with those of plots cultivated ten or more times. While there is some decline in carbon levels under cultivation, the capacity of these soils to conserve organic and humified material is of such an order that depletion is not marked. They remain soils with, on the whole, relatively high organic carbon levels, whatever their cultivation history.

The nitrogen levels are also high, especially for the tropics. The mean topsoil nitrogen value is 0.99%, ranging from a hefty 2.5% to a moderate 0.2%; the mean subsoil value is considerably less at 0.33%. Again cool temperatures and mineralogical sequestration of organic matter are largely responsible for this store of nitrogen. Nitrogen, mineralised to ammonium and nitrified to nitrate, is normally subject, as an anion, to rapid leaching losses under high rainfall regimes, but here it is protected, kept locked up as organic nitrogen, and only gradually released into the soil solution as slow microbial activity brings about decomposition (Mueller-Harvey et al. 1985, 1989 stress the importance of soil organic matter to crop nitrogen availability in the tropics). Again, there is a correlation between soil nitrogen and time under cultivation; the mean topsoil value for sites cultivated once is 1.3 times greater than that of sites cultivated over ten times. The difference is significant; topsoil nitrogen gradually declining and levelling off (Figure 12.6).

The change in topsoil organic nitrogen with clearing is relatively small compared to carbon, as has been noted for soils affected by volcanic ash elsewhere (Sanchez 1976:370). Its steady decline to new equilibrium levels under cultivation is attributable to release of N_2 gas with the firing of vegetation, the take-up of some nitrogen by crops and hence removal from the system at harvest, some losses through leaching of nitrate anions, and expanding microbial populations immobilising a large proportion. The rise in subsoil nitrogen levels upon initial clearance of sites could evidence leaching losses from topsoils, the burning of natural cover increasing nitrogen contents, while reduction in root networks results in less efficient nitrate interception. It is possible that nitrogen, the element most commonly found to be limiting agricultural production — due to large plant demand and its susceptibility to heavy leaching losses — is limiting in this cultivation system, at least regarding the yield of some crops, particularly after a site has been cropped one or two times, with post-clearance nitrogen levels diminishing from the time of the burn onwards. The reason is not intrinsically low soil nitrogen contents but that organic matter decomposition rates are of an insufficient order to ensure adequate available nitrogen supplies.

The carbon:nitrogen (C:N) ratios, which indicate organic matter mineralisation rates, support this assumption. The gradual fall, attendant upon differ-

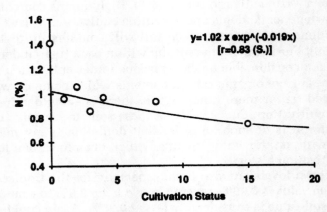

Figure 12.6 The relationship between topsoil nitogen content and site cultivation status.

ential rates of change in carbon and nitrogen levels under cultivation, suggests that the increase in biological activity following cultivation of land, when greater soil aeration results in increased oxidative activity, remains high no matter how long a site is cropped because of the very high organic matter contents, and that a large part of the potentially available nitrogen is immobilised as microbial tissue, while a fraction exists in a form resistant to decomposition (Sanchez 1976; Tergas & Popenoe 1971). While a range of factors, including moisture content, temperature, mineral environment, aeration, acidity and organic matter chemical composition all determine the activity of the soil micro-organisms responsible for the decomposition of soil humus, an adequate supply of nitrogen in comparison to carbon has long been recognised as necessary for optimal activity. If nitrogen is relatively scarce compared to carbon (a high ratio score), there will be competition between micro-organisms and plants for it, reducing the available supply of nitrogen to any crop with high rates of immobilisation.

A ratio of 10 or thereabouts signifies close to the end-product of breakdown, indicating well decomposed organic matter and little restriction on nitrogen availability. A C:N ratio above 18 indicates slow organic matter breakdown due to nitrogen shortages. The C:N ratios in these topsoils are on the high side. The mean ratio is 17.5,[5] extending over a considerable range from well broken down to scarcely decomposed plant residues, and suggests some inhibition of organic matter breakdown; a finding that corroborates with the above discussion on soil humus conservation. Nonetheless the high absolute nitrogen levels suggest that nitrogen supplies are probably sufficient to ensure that severe deficiencies do not occur and that the yield of crops which have relatively modest nitrogen demands may not be markedly limited. The mean subsoil C:N ratio is 13.4 which, while close to the decomposition end-product ratio, is of relatively minor agronomic significance given the lower carbon and nitrogen levels of this horizon and its small role in short-term crop nutrition. Neither the topsoil nor the subsoil C:N ratios show any significant time related correlation, although they tend to decline over time under cultivation, with the proportion of nitrogen to carbon in the soil increasing with time under tillage.

We should expect the C:N ratios to fall, given the differential rates of decline in soil carbon and nitrogen contents, as described. While carbon is lost from the soil in respired gas, microbial sequestration conserves a larger proportion of the nitrogen, which is in considerably smaller overall supply, and the ratio consequently falls, relatively less nitrogen being lost from the system. When we consider the wide distribution of the ratios in each time class, as indicated by the standard deviations, the pattern becomes less con-

---

[5] This compares with the ratio values of 14/16:1 reported by Mohr et al. 1972 for montane forests in S.E. Asia and by Jenny et al. 1948 of 11/17:1 in the Andes; Edwards & Grubb 1982 however report a lower C:N ratio of 9/11:1 under montane forest in the Eastern Highlands of Papua New Guinea.

vincing, but it remains plausible. High levels of microbial activity are feasible with the repeated breaking up of the soil, giving improved aeration and less saturated conditions. The pattern revealed, of a declining C:N ratio, resulting in improved availability of nitrogen over time 'under the digging stick', may also partially explain the unexpected claims of those who cultivate these soils that many of them improve agronomically the longer they are under cultivation, which is the reverse of the accepted notion that soil performance declines under swidden cultivation the longer continuous cropping proceeds, sometimes falling disastrously.

Salinity & Micronutrients: The soils present no salinity problems, which is expectable in a cool region experiencing high rainfall. The electrical conductivity (EC) measurements are very low (with mean values of 0.093 mS and 0.029 mS respectively for the few topsoils and subsoils measured), typical of soils that are effectively salt free (Landon 1991:158). The exchangeable sodium levels reflect this too, being very low (the mean topsoil exhangeable Na is 0.78 me/100 g, and for subsoils 0.68 me/100 g). None of these measures of salinity evidence any variation according to time under cultivation.

Finally, regarding micronutrients, these soils could experience some deficiencies sufficient to limit crop yields. The boron levels are low, with a topsoil mean around 0.5 μg/g soil for those few samples analysed, whatever the cultivation history of the site. It is thought that concentrations <1 μg/g are likely to lead to deficiencies in susceptible crops. Boron, an anion, is fairly readily leached like nitrogen from the soil, resulting in deficiencies in regions of high rainfall. It is possible that deficiencies might also partly occur because of a reverse process; strong adsorption being observed in soils with high allophane contents and associated minerals. Radcliffe (1986:122) is also of the opinion that local deficiencies of manganese could occur, and perhaps even of zinc and copper, though to lesser extents.

Soil Fertility Under Cultivation

According to this investigation of soil fertility under cultivation, the restricted availabilities of phosphorus and potassium are probably the major limiting factors on crop yield. Together possibly with boron, and nitrogen too. In summary, we may characterise the soils of the Wola region as acid with high organic matter contents, reflecting the cool, wet conditions of the Highlands and that the soils here contain some volcanic ash. The volcanically derived allophane and related minerals result in high fixation of soil phosphorus, the consequent low availabilty of which is a major limiting factor on crop production. These extruded minerals also result in soils with ion exchange properties which are depressed under acidic conditions. This depression of the variable component of the CEC — which is otherwise fair for the tropics with its considerable organic matter component — has no serious consequences for the metallic nutrients calcium and magnesium. But any resulting reduction in potassium retention is problematic because the general availability of potassium, like phosphorus, is low and limiting, and declines under cultivation, this univalent cation being more readily lost through leaching from the ecosystem than divalent ones, depending on clay types present. Organic carbon levels also fall under cultivation, and nitrogen levels too somewhat, although relatively less rapidly due to conservation. Associated sequestration leads to possible nitrogen shortages, which the high C:N ratios further indicate, even though total nitrogen supplies are overall adequate. A possible positive effect of the consequent long-term decline in C:N ratios is supplementation of available nitrogen supplies later. Of the micronutrients, boron levels are low, even deficient and could be limiting crop production too.

An appropriate low cost management recommendation for many tropical soils is to lime them to increase pH, racking up their CEC and increasing all cations held (Juo & Kang 1989). But this would improve these 'atypical' tropical soils little, having relatively little potassium present initially, although it might reduce the ferocity of phosphorus fixation. On the down side, it might also promote microbiological activity somewhat, which is reduced under acid conditions, so exacerbating problems of nitrogen availability. The variation in pH results between gardens of the same cultivation status, and the absence of any clear trend in pH over time under cultivation, suggests that the Wola have hit upon their own equivalent of liming in burning vegetation pulled up from brief grassy fallows to boost pH at intervals. The addition of rock phosphate might help too (although with no local source it would probably prove uneconomic with high bulk transport costs). Commercial agriculture, which is not likely in this rugged region, would doubtless go for additions of potassium and phosphate fertiliser, perhaps with nitrogen too (Floyd *et al.* 1987b), and it would also probably experiment with borax amendments. Industrial inorganic fertilisers are however inappropriate to contemporary farming conditions, being largely beyond the financial reach of the region's subsistence farmers.[6]

Regarding soil management, it is pertinent to ask what correlations, if any, exist between our scientific assesssments of soil chemical fertility and local assesssments of soil and site cultivation potential, even though the Wola traditionally have no concepts, so far as I am aware, that correspond to such fertility measures. The evidence suggests scant relationship between indigenous appraisal of resources as good, fair or poor, and chemical measures of fertility; farmers are apparently unlikely, using local assessment criteria, to select for cultivation soils with notably higher levels of nutrient availability over others (Table 12.3). The evidence suggests that local farmers may classify locations of similar fertility into polar opposite categories, some as 'very good' and others

---

[6] In a research report for the World Bank sponsored development project AFTSEMU, an economic review of the use of inorganic phosphate (triple superphosphate) and potassium (muriate of potash) fertilisers concluded that while the former gave a large yield response it was not cost effective, whereas the latter, while not giving such a yield response, was economic up to 170 kg K/ha (depending on soil base saturation) given its lower unit cost (Floyd *et al.* 1987b:21–23). The widespread use of commercial fertiliser is unlikely, as the report notes, and probably inadvisable, for the forseeable future in a region dominated by low income, subsistence agriculture.

Table 12.3

A comparison of local assessment of cultivation potential with mean site soil chemical fertility (value in brackets = standard deviations, n = 100 sites)

	LOCAL ASSESSMENT CLASSES				
	Very Good	Good	Middling	Poor	Very Poor
pH	5.22 (0.43)	4.96 (0.49)	5.40 (0.62)	4.90 (0.50)	5.18 (0.52)
Phosphorus	11.31 (9.20)	13.34 (16.42)	11.76 (10.47)	7.10 (10.34)	9.77 (13.77)
Potassium	0.60 (0.32)	0.72 (0.74)	0.54 (0.31)	0.55 (0.40)	0.74 (0.78)
Calcium	15.11 (11.16)	12.17 (12.75)	11.23 (7.87)	9.24 (10.25)	15.49 (21.49)
Magnesium	3.13 (2.33)	2.10 (1.32)	2.49 (1.42)	2.18 (2.07)	2.89 (3.36)
CEC	29.19 (8.32)	28.43 (8.01)	28.86 (5.53)	27.93 (8.34)	27.39 (8.39)
Carbon	14.10 (7.77)	17.56 (7.94)	17.50 (5.96)	15.86 (7.80)	17.38 (8.05)
Nitrogen	0.90 (0.48)	1.14 (0.40)	1.11 (0.35)	0.95 (0.36)	0.92 (0.37)

as 'very poor'. Places they adjudge 'poor' or 'very poor' are not markedly less deficient regarding one or more of the fertility measures, nor are those they assess 'good' or 'very good' notably better.

One of the reasons for the absence of any correlation is that the Wola do not look at the soil in isolation and comment on its possible fertility status, their assessments include a review of site factors too. (The distortion introduced becomes apparent if virgin sites are included in the comparison because many of these are adjudged potentially 'very poor', due to unfavourable aspect and so on, but are chemically very fertile — including them results in the very poor assessment class having soils of considerably higher mean chemical fertility than any others.) Again this raises the point that the assessment of soil fertility in the abstract is foreign to the Wola. It is a somewhat contrived exercise to ask them to comment on potential fertility from inspection of the soil alone, as their comments make abundantly clear (responses are commonly in the vein "I don't know about this soil, but if we clear the site and till and plant it, then we'll see"). This takes us back again to the practical nature of their knowledge, they start to cultivate sites and come to know their soil resources and assess their potential according to their response under use. While this chapter attempts to explain, according to our chemical understanding of soil fertility and related processes, the possible sequence of events that occurs following site clearance and subsequent approach of abandonment, this is knowledge which the Wola acquire and pass on through practice, very effectively given the nature of their soil resources, without need of any similar analytical discourse.

Table 12.4
A comparison of soil survey profile classes with mean site soil chemical fertility (values in brackets = ranges, n = 85 sites).

	SOCIAL SURVEY PROFILE CLASSES				
	a	b	c	d	e
pH	5.1	4.91	5.23	5.32	5.3
	(3.4)	(1.9)	(1.2)	(0.1)	(0.4)
Phosphorus	13.43	13.46	22.93	2.85	25.9
	(74.3)	(48.2)	(20.3)	(3.5)	(24.2)
Potassium	0.82	0.64	0.77	0.28	2.25
	(6.03)	(1.61)	(0.41)	(0.18)	(1.25)
Calcium	15.37	15.86	19.5	12.8	68.4
	(104)	(39.1)	(11.6)	(15.8)	(11.2)
Magnesium	2.89	2.68	2.20	1.69	12.92
	(14.68)	(6.06)	(1.33)	(1.96)	(2.92)
CEC	29.64	30.07	27.13	16.95	62.00
	(30.1)	(42.7)	(6.7)	(4.5)	(24.2)
Carbon	17.22	19.98	14.14	3.51	30.2
	(30.59)	(34.54)	(7.48)	(0.54)	(22.26)
Nitrogen	1.12	1.03	1.22	0.34	1.54
	(2.27)	(1.59)	(1.37)	(0.08)	(0.69)

The absence of correlations between indigenous assessments of arable worth and chemical fertility of soil also bears on the adaption of local knowledge to the region's soil resources. We established previously that it is difficult to distinguish between the soils that cover the larger part of Wolaland by observed features alone, and to use these to devise some scheme to predict possible soil productivity (see Chapter 11). A comparison of the profile classes compiled in the survey analysis with measures of site chemical fertility confirms these conclusions (Table 12.4). The differences between the two profiles (classes **a** and **b**) that predominate across the region, representing the soils on which the vast majority of cultivation occurs, are small for nearly all fertility measures. There are no readily seen features by which farmers can distinguish between any of these soils, that are reliable indicators of probable chemical fertility status, except perhaps for silty *suw hobor* 'grease' texture, indicative of high organic matter content (although whatever organic matter levels are, they will decline under cultivation). The striking fertility differences occur between the region's relatively infrequently occurring soil types, and are well known to local farmers who exploit them where available, as appropriate. Gley soils (class **c**) are particularly fertile, but the anaerobic conditions that prevail suit them to only a few crops without drainage, which the Wola resort to very rarely. Thin, high organic topsoils (class **e**), of a type that characterise abandoned house sites, are very fertile, as they are well aware, but of very limited occurrence and transient fertility. Recent alluvial soils (class **d**) are likely to prove markedly less chemically fertile, and are generally avoided anyway because of their precarious locations liable to periodic innundation and crop loss.

The sequence of soil-related events that follows the clearing of natural vegetation from a site and establishment of a garden is perhaps as follows. The burning of the vegetation cut down in clearing the site gives a critical, though short-lived boost to the availability of several elements, notably increasing pH (Table 12.5[7]). The decrease in acidity is marked at the time of clearance and planting, whatever the vegetation cleared from the site, whether montane rainforest through to coarse grasses. And the increase in nutrient availabilities is particularly noticeable with phosphorus, and potassium and nitrogen also, to lesser extents. The variation in the availability of the nutrients in the ash derived from the burning of vegetation from different communities has to be set against the different amounts of ash that the destruction of each yields (see Chapter 9). At a rough estimate the mass of ash resulting from the burning of *iyshabuw* forest is perhaps some 7 t ha^{-1}, from *obael* woodland and *gaimb* cane some 3 t ha^{-1}, and from well established *bol* coarse grasses some 1 t ha^{-1}. There is a noteworthy decrease in phosphorus availabilities as the biomass, and ash input, of the communities cleared declines. The trend with potassium is less clear-cut. The relatively high percentage nitrogen contents reflect the partial destruction of the plant material in the low intensity burn achieved under the region's wet climatic regime; the organic carbon content of the ash also indicates the substantial levels of partially burnt and charred material.

It is not that crops massively mine the nutrients made available by the initial burn and held in the virgin soil, for although these decline to new equilibrium points, they remain, even after years under cultivation, relatively constant at these decreased levels. Rather, conditions promoting the availability of critically limiting nutrients are encouraged, some of which, accumulated in the standing vegetation, are further released very briefly for uptake by the first crops, but their availabilities quickly fall to pre-burn levels, a proportion removed in harvested crops and the remainder becoming unavailable due to soil processes. The availability of limiting nutrients, some released via the ash of the burned vegetation, is sufficient to permit a wide variety of crops, several of them annuals, to flourish in newly cleared gardens, including a range of beans, green leafy vegetables, aroids and cucurbits. This crop diversity is short-lived however, parallelling the ephemeral nature of boosted nutrient availability levels following clearance.

---

[7] The data in this table on both ash and soils indicate the levels of readily crop available nutrients at the time of planting, and not total nutrient contents. The analytical methods followed correspond with those described earlier. Several sub-samples were collected at random across each garden site, bulked together and a sample taken for analysis. The ash samples come from fire sites, where a sufficient pile of burnt material remained for sampling purposes. The soil samples sometimes, though not invariably, included some ash where it was spread across the surface, either by the wind or the gardeners' activities.

Table 12.5

The average nutrient status of ash and topsoil samples collected from newly cultivated garden sites cleared from different vegetational communities (n total = 14 sites; nos. in brakets = σ).

Vegetation Cleared	Ash/Soil	No. sites sampled	pH	P (ppm)	K (me/100 g)	Ca (me/100 g)	Mg (me/100 g)	C (%)	N (%)
Iyshabuw montane forest	ash	4	9.11 (0.7)	754.5 (415.6)	9.13 (2.66)	94.82 (20.36)	29.63 (8.19)	7.42 (3.46)	0.35 (0.32)
	topsoil	4	6.47 (1.28)	49.8 (23.73)	0.96 (0.17)	17.5 (8.0)	3.52 (1.74)	22.41 (7.01)	1.41 (0.5)
Obael wood regrowth	ash	1	8.85	1164.0	8.97	82.34	79.8	6.02	0.45
	topsoil	1	5.83	34.0	1.22	26.85	6.13	14.9	0.99
Gaimb cane grass	ash	4	8.8 (0.83)	697.4 (767.4)	21.48 (14.42)	56.06 (24.75)	41.89 (26.0)	7.22 (1.13)	0.77 (0.1)
	topsoil	4	6.3 (0.8)	34.16 (34.26)	3.37 (1.68)	17.36 (12.53)	7.59 (5.21)	19.25 (7.67)	1.07 (0.16)
Bol coarse grass & herbs	ash	5	9.36 (0.78)	955.2 (291.5)	33.26 (22.12)	52.27 (22.97)	63.77 (31.07)	6.64 (3.01)	0.68 (0.38)
	topsoil	5	6.19 (0.43)	14.94 (7.14)	1.06 (0.51)	10.91 (9.64)	3.48 (1.68)	12.83 (0.92)	0.85 (0.18)

The wide distribution of the analytical results suggests that a closing caveat may be in order, for it intimates that soil nutrient availability may not be the key feature to the continued use or abandonment of all cultivated areas. The comments of the Wola suggest this too, for they often cite non-pedological factors when discussing reasons for garden abandonment — a common one is that it is becoming too difficult to keep pigs out, gardeners frequently giving up sites when they prove too arduous in labour terms to maintain enclosed and costly in pig depredations. The conservation, broadly speaking, of nutrients, sufficient at least for the cultivation of sweet potato for extended periods, further points to the significant part that non-nutritional factors may possibly play in swidden management. But soil nutrition takes on a prominent role when we consider other short-term crops, the demands of which have a knock-on effect on sweet potato, such that fertility indirectly influences the abandonment of some proportion of sites under this crop.

The consistently short-term nature of the intercropping that characterises swiddens when first cleared, giving way rapidly to sweet potato after one or two cultivations (except in some gardens with a few more favourable pockets of longer term fertility — folds, fence lines, and so on), suggests that it is nutritionally driven after the classic shifting cultivation paradigm. In order to ensure some variety in their predominantly sweet potato diet, and cultivate the required range of crops in sufficient amounts, the Wola are obliged repeatedly to clear new garden sites on which to cultivate them. They regularly clear new areas consequently, and bring them under cultivation. If the overall area under cultivation at any time is to remain more-or-less constant relative to the the number of mouths to be fed and the demand for food, some sites must inevitably be approaching the other end of the farming phase and abandonment. Some of these sites might still be able to support an adequate sweet potato crop, but with new areas successively being brought under cultivation there is no need to keep them under crops. In this event, the impetus behind the abandonment of some gardens might relate not to the availability of adequate nutrients for sweet potato growth but the decline in their availability sufficient to support a range of other crops, which although cultivated in considerably lesser amounts than sweet potato are nonetheless important to the Wola subsistence diet (see Sillitoe 1983 on the nutritional significance of these other crops). The sites which people keep long-term under sweet potato cultivation need not necessarily be markedly superior ones as measured by nutrient availability, others may be adequate too but superfluous to needs. They may be selected between on such grounds as ease of enclosure and its effectiveness, distance from homestead and other gardens, and so. Again, a range of non-soil factors may be playing a prominent part in land use patterns.

II. SWEET POTATO NUTRITION

The sweet potato emerges as occupying a key place in the Wola agricultural regime, whatever the motives behind site abandonment. It plays a crucial part in the region's singular agricultural system. Sites cultivated over again, maybe repeatedly many times, come to support it as a virtual mono-crop. Its fertility requirements are such that it can manage on the intrinsically low availabilities of nutrients. A review of sweet potato nutrition, in the context of the foregoing interpretation of soil response under long-term cultivation in the Southern Highlands, reveals what it is about the crop's demands that particularly suit it to these soils, and why it has become the region's staple, allowing the extended cultivation of land without external inputs, contrary to the general characterisation of tropical subsistence agricultural regimes, as necessitating shifting of cultivations (see Chapter 2).

It is necessary that inorganic nutrients are available in certain proportions one to another for optimum growth — some in relatively large amounts (macronutrients) and others in trace amounts (micronutrients) — any that are not will be limiting overall growth. But their levels are interlinked in complex feedback relationships. It has long been known for example, that optimum nutrient availability for maximum storage root production is not necessarily optimal for vigorous vine growth. The results of field experiments on the influence of different nutrient levels on sweet potato growth and yield are contradictory and difficult to interpret, as with many crops. They are further complicated by varying treatments, different cultivars researched and varying edaphic and other environmental conditions; if it is rainy for example, during the tuberisation phase, vine growth may continue and lead to a reduction in the proportion of dry matter diverted into tubers (Enyi 1977).[8]

The results of trials with inorganic nitrogen, phosphorus and potassium (NPK) fertilisers on growth and yield in the New Guinea highlands are inconsistent. Some trials report significantly increased yields, others little or no response, and yet others depressed yields. In the Bismarck Mountains, increased yields were observed with a combined NPK fertiliser, especially in currently cultivated and abandoned gardens, interpreted as indicating a decline in nutrient availability during the cropping period (Clarke & Street 1967).[9] At Goroka in the Eastern Highlands increased yields were obtained in trials with nitrogen and potassium but not phosphate (Kimber 1970). And in a series of trials at Aiyura responses to various treatment levels of NPK have been variable and scarcely significant, with fertilised plots often not giving

---

[8] Although high rainfall is reported to favour vine over tuber growth, increasing water availability experimentally has no effect on dry matter partioning to the vine, both vine and storage root dry weights increase as a result (Acock & Garner 1984). This suggests that the effect of high rainfall, in promoting vine over tuber growth, is unlikely to be the result of differences in water supply. High rainfall also affects soil aeration, and reduced storage root growth under high rainfall may possibly result from limited soil oxygen levels rather than induced vine competition.

[9] Besides a decrease in nutrient availability, other suggested possible causes of yield decline and eventual site abandonment include: increased pest infestation, increased weediness, and changes in soil structure (the density of soils under secondary forest [1.1] and old gardens [1.4] being higher than under primary forest [1.0]).

markedly better yields than unfertilised, and sometimes reduced ones (Bourke 1985).

Nitrogen & Potassium: While total dry matter content increases linearly with increased nitrogen supplies, experimental results disagree over its partioning between roots and vine (Bouwkamp 1985). According to some studies, increased nitrogen applications result in significant root yield reductions, too much nitrogen inhibiting root thickening, causing decreases in root yield against increases in foliage growth. (Nitrogen availability is associated with an increase in soil organic matter content, attributed to excessive vegetative vine growth). In other studies, storage root yields increased up to a point as nitrogen was increased, and then levelled off or declined, while vine growth responded positively to every rise in nitrogen level. These trends support the contention that rate of assimilate movement is more important for storage root development than increase in leaf area (see Chapter 2). But in other research, root yields have increased repeatedly for all increases in nitrogen, the highest nitrogen applications resulting in both the largest weights of root and vine.[10] The application of nitrogen generally increases root size but not number of roots. The possibility of microbial nitrogen fixation complicates the situation further, N fixing bacteria being suggested to explain how some sweet potato cultivars are capable of producing high yields on low nitrogen containing soils without fertilisation, where fertilisers do not significantly influence total biomass, storage root yields and foliage weights (Hill et al. 1990).

The effect of potassium on yield, the uptake of which is associated with root enlargement and increasing numbers of tubers, is likewise inconsistent (Bourke 1977). According to some studies, different potassium concentrations have little influence on yield, but others report that as potassium levels increase there is a linear yield response. And yet other studies report an initial increase in storage root yields with potassium additions, followed by a decline after a certain point (Constantin et al. 1977). There is doubtless an environmental component to these apparently conflicting findings. Soil potassium levels for example, have an effect on fertiliser response; when soil levels are above a certain threshold (<0.08 meq K 100 g^{-1} soil in one study — Nicholaides et al. 1985) sweet potato shows no yield increase. Trials in the Southern Highlands, suggest that economically optimum rates of potash

---

[10] According to trials at Keravat on New Britain, on young glassy volcanic soils under continuous cropping, nitrogen applications usually had a greater influence on sweet potato growth and yield than potassium, especially on grassland sites (although sometimes they depressed yields — the use of different cultivars may explain these inconsistent responses). It is surmised that nitrogen influences yield by increasing leaf area duration, in turn leading to a rise in mean tuber weight and yield, whereas the effect of potassium, which was of lesser importance here (except on a depleted forest soil where it significantly increased yields), is to increase the proportion of dry matter diverted to tubers, pushing up the number of tubers to a plant and, to a lesser extent, their mean weight (Bourke 1977, 1985). Nevertheless, yields under fertilisation were considerably lower than those obtained when the site was first cultivated (possibly because of nematode build-up).

application decrease with increasing soil potassium saturation, and above 15% are negligible (Floyd et al. 1987). The concentrations of other nutrients may also have an interactive effect; a decrease in plant tissue calcium and magnesium concentrations with increasing potassium additions suggests some antagonistic effect (Nicholaides et al. 1985). Nonetheless it is well established that of the mineral nutrients potassium has the largest effect on tuber yield. It increases total dry matter production by increasing net photosynthetic activity per unit leaf area. When potassium levels in leaves are high, so is the rate of photosynthesis, but as they fall so does photo-activity. It increases NAR by increasing the capacity of the tuberous root sink to accept carbohydrates transported from the leaves, by accelerating tuber respiration.[11] And it stimulates photosynthetic activity in the leaves by promoting the translocation of photosynthates into the root, reducing leaf carbohydrate levels, acting as a trigger for photosynthesis. The part played by potassium in increasing tuber number may also stimulate photosynthetic activity further by providing more carbohydrate sinks.

The 'law of the minimum' may also explain in part some of the apparently contradictory crop yield results. An increase in either nitrogen or potassium, the other held constant, reduces the number of storage roots (attributed to a reduction in the rate and/or duration of root bulking, and poor leaf area development). The evidence suggests that the ratio of nitrogen to potassium has an influence over tuber yield. It is necessary to keep the ratio of potassium to nitrogen high to maximise tuberous root yield. If one or other element is below its optimum ratio quota, it will act to hold down yield by its short supply, even though the other nutrient is available in amounts sufficient to support further growth. Nitrogen supply should be sufficient to give optimum, not excessive, growth of above-ground parts (to maximise efficiency of light interception per unit leaf area), and potassium supply should be sufficient to maximise sink capacity (promoting photosynthetic activity by depleting leaf carbohydrate levels). When the nutrients are supplied in adequate amounts, there is less evidence that vine growth competes with storage root growth. Under these conditions, those plants that evidence the greatest foliage growth also have the highest mean weight of fresh storage roots, vine growth associating positively with tuber growth. These findings question the idea that excessive vine growth need necessarily reduce storage root yields through competition for assimilates. While heavily fertilised plants may partition more of their total dry weight into vine and produce fewer storage roots, they give higher tuber yield overall because the tubers they produce are considerably larger. Less fertile conditions may enhance storage root initiation and favour tuber over vine growth, but the limitation of vine and foliage growth depresses yield.

---

[11] Also essential for sink activity is non-exposure of tubers to light; it is thought that light probably affects sweet potato tuberisation by interfering with hormone levels.

Whereas potassium suppresses excessive top growth, giving higher tuber yields, nitrogen promotes it, increasing dry matter production largely as foliage, to the possible detriment of root yield and quality (the increase in leaf expansion affecting the distribution ratio of a plant's above- to below-ground parts, increasing foliage above roots). The possible low nitrogen availability of Wola soils might consequently benefit sweet potato, the crop favouring prolific vegetative growth over tuber development when its levels are high (Bourke 1985). It is on potassium, of the three major plant nutrients, that adequate sweet potato yields particularly depend, and it appears that so long as it remains above some critical minimum level, New Guinea highlanders can go on cropping the same site indefinitely, whatever our preconceptions from elsewhere of non-amendment or shifting cultivation.

Phosphorus & Mycorrhizae: The phosphorus requirements and responses of sweet potato are ambiguous too, although there is general agreement that demand for this nutrient element is small. It is a crop that will tolerate soils particularly low in available phosphorus, which suits it to Wola soils, where phosphorus fixation is a major fertility problem (Wood 1984, 1991 reports a marked correlation in the neighbouring Tari Basin between gardens growing only sweet potato and low levels of soil available phosphorus). The crop's tolerance may be related to three possible mechanisms (after Djazuli & Tadano 1990):

1) its capacity to take up phosphorus from soils low in phosphorus (the ability of roots to elongate, proliferate hairs, or form mycorrhizal associations);
2) its tolerance of low tissue phosphorus concentrations (perhaps reflecting low internal requirements);
3) its ability to use phosphorus economically (when expressed as production of harvested organ to unit amount of phosphorus absorbed).

Trials frequently indicate that phosphorus applications have no effect on dry matter accumulation in sweet potato, on tuber yields or quality (Nicholaides *et al.* 1985; Hammett *et al.* 1982; Constantin *et al.* 1977). Other work however, reports increases in tuber yield with phosphorus fertilisation, and foliage growth too (Djazuli & Tadano 1990). While yet other research findings suggest that increasing available phosphorus depresses yields of both tubers and vines (Bourke 1977). There are several possible reasons for these conflicting results (Bouwkamp 1985). In many cases, little or no response to added phosphorus might be anticipated given that sweet potato does not require large amounts to achieve good yields (Scott & Bouwkamp 1974 estimate a total uptake of only 26 kg P/ha). Some crop responses may consequently pass undetected, sweet potato giving near optimum yields at very low soil phosphorus levels (Fox *et al.* 1974 found that soil solution phosphorus concentrations as low as 0.003 ppm gave 70% of optimum yield; and in a glasshouse experiment I obtained 60% of optimum plant growth at low phosphorus availabilities of 0.01 mMol P).[12] Lack of response to increasing phosphorus levels might also be attributable to soils already having sufficient available, adding more making no difference to growth and yield (Nicholaides et al. 1985; Hammett et al. 1982). Even at low concentrations, soils may have adequate supplies of available phosphorus for healthy sweet potato growth (Hammett et al. (1982) report no yield response to applied phosphorus when soil concentrations exceed 60 ppm).

Furthermore, vesicular-arbuscular (VA) mycorrhizae are often present on the fibrous root system of sweet potato. When these symbiotic fungi colonise roots they are thought to enhance uptake of phosphorus and other minerals, especially in low phosphorus soils, primarily by effectively increasing root volume.[13] The part played by VA mycorrhizae in sweet potato nutrition is undetermined. In some trials they stimulate growth (Ngeve & Roncadori 1985 report that they are as effective in promoting growth as an additional 800 ppm available P in the soil). In other trials, fungal infection has no effect, or even depresses yields (Mulongoy et al. 1988). Soil phosphorus content is critical to the magnitude of any mycorrhizally induced growth response. The frequency of root infection has repeatedly been found to increase as available phosphorus content of soils declines (Ngeve & Roncadori 1985 report that the growth response of mycorrhizae decreases when soil fertility increases beyond 100 ppm

---

[12] The experiment, conducted at Wye, involved fifty plants (grown from cuttings in perlite). Two cultivars were given five treatments (the phosphorus concentrations ranging between 0.5, 0.1, 0.05, 0.01 and 0.0 mMol). The results, analysed according to leaf area, vine length and dry matter production (foliage, roots and tubers), demonstrate the ability of sweet potato to tolerate low phosphorus availabilities, akin to the soils described here. While it is not possible to make precise comparisons between the experiment and sweet potato growth in the New Guinea highlands, the perlite and nutrient solution system not replicating the complicated phosphorus supply of the soil system, some broad generalisations are possible. The range of available P in the topsoils investigated was 0.2 ppm to 59.0 ppm, with 62% of soils analyzed having <10 ppm P. Regarding the ability of sweet potato to manage at low P levels like these, the experimental evidence suggests that yields 60% of the optimum may be possible on some sites where P availabilities are restricted, assuming selection of low P tolerant genotypes, and that yields considerably higher are feasible on some. Yields of this order are quite adequate for subsistence purposes and acceptable in low-input traditional horticulture. The two cultivars responded differently in some respects to P treatments (one a subsistence crop and the other a commercial one), indicating that genotype is important to P response and tolerance, the differences relating to expected level of inputs (Kanua & Floyd 1985). It is necessary to exercise care when generalising about nutrient responses of sweet potato, given the wide variablity exhibited by cultivars in adaptions to different environments. The breeding of new cultivars and formulation of extension advice needs to be approached with caution. It is probable that any local population, like that in the New Guinea highlands, will have selected cultivars over the generations well suited to their particular environmental conditions and agricultural arrangements, and while the cultivars may not prove high yielding with amendments (e.g. fertiliser applications), they may be relied upon to give adequate returns, sufficient for subsistence purposes.
[13] These fungi produce alkaline phosphatases that are possibly involved in phosphorus uptake and transfer, high host tissue phosphorus concentrations inhibit them colonising plants and producing enzymes (Djazuli & Tadano 1990).

available P in the soil, probably because higher tissue phosphorus concentrations affect membrane permeability and root carbohydrate allocation, reducing fungal-infection-attracting exudations). Differences in soil and other environmental conditions may explain some of the reported variations in sweet potato responses to mycorrhizae. Differences have also been observed between cultivars in extent of mycorrhizal infection (Mulongoy et al. 1988). And different species and strains of fungi, of which several have been reported colonising sweet potato roots (including *Glomus* spp., *Gigaspora* spp., and *Scutellospora* spp.), vary too in their colonising effectiveness

The sweet potato plant also has an extensive, rapidly growing, fibrous root system of dense architecture, penetrating downwards to soil depths over 2 m and laterally outwards for a similar distance (Kays 1985). This is doubtless significant in its tolerance of low phosphorus soils, greatly extending on the effective soil zone explored by the plant for nutrients. Besides depending on its ramifying root network, the sweet potato plant also relies under low phosphorus conditions on significantly increasing its phosphorus efficiency. It extends the growth duration of its leaves and maintains relatively constant leaf phosphorus levels (possibly by reduced root sink demand for phosphorus, for which the sweet potato tuber has a low internal requirement — Djazuli & Tadano 1990). Whatever, there is a limit to the low phosphorus levels that even sweet potato can tolerate, phosphorus being essential to any plant's cellular energy transactions, deficiency symptoms including stunted growth and small purple to brown hued leaves evidencing increased abscission. There must be some feature of New Guinean continuous soil cultivation practices that ensures phosphorus levels remain above this albeit low threshold.

Regardless of sweet potato's tolerance of low phosphorus conditions, yield responses in a series of trials in the Southern Highlands Province suggest that its low availability is the major fertility constraint on crop production, with potassium availability second (Floyd et al. 1987b, 1988). The nutritional status of soils, as assessed by phosphate sorption isotherms, best explains yield differences; the negative linear correlation between control yields and soil phosphate requirements confirm it. (Inadequate phosphate nutrition is a soil related problem: the rate of diffusion of phosphate to roots, as estimated by phosphate buffer capacity, determines phosphate availability, rather than the rate of phosphate uptake by roots.) On all soils except one (n = 9), phosphate fertiliser applications (up to 1000 kg P/ha) had a highly significant effect on tuber yields, whereas the effects of potash applications (up to 360 kg K/ha) were more variable.[14] And responses to potassium applications occurred only on soils which gave moderate (not low) control plot yields, which in turn correlated with level of phosphate fixation, and hence soil phosphate supply. In short, response to added potassium was reduced in soils

---

[14] In the Simbu region of the highlands, on similar Tropept and Hydrandept soils, highly significant sweet potato tuber yield responses were also obtained from phosphate applications of 50 kg P_2O_5/ha (Goodbody & Humphreys 1986).

of low phosphate status. A deficiency in soil phosphate may also curtail sweet potato response to other nutrients. It is the principal nutritional limitation on the production of crops other that sweet potato, like beans, cabbage and maize. Furthermore it appeared that indigenous mycorrhizal infections (involving *Glomus* and *Gigaspora* spp.) had significant effects on sweet potato phosphate nutrition. They were associated both with increased yield and reduced response to phosphate fertiliser (an increase in VA mycorrhizal infection diminished response to phosphate applications, and sites where fungal infection was higher had larger control plot yields than expected given their phosphate status). These interactions suggest that mycorrhizae could be potentially important to sweet potato yield in the region.

pH, Aluminium & other Nutrients: While sweet potato is not particularly sensitive to soil acidity, there is evidence that cultivars may vary in their responses to pH, and related exchangeable aluminium levels. Root growth may be enhanced, or only slightly reduced, at moderate (<250 uM) aluminium concentrations (for a range of possible reasons, including bound cation replacement, enhanced potassium mobilisation, inhibition of harmful microorganisms, or favourable interaction with other toxic elements), but above this level growth decreases significantly (Bouwkamp 1985). Aluminium tolerance varies between cultivars. It is thought that tolerance may relate to the extent that a plant does or does not attract aluminium to its developing storage roots. The more sensitive cultivars have greater cation exchange capacities than tolerant ones, and consequently attract and take up more aluminium. However, under conditions of no aluminium stress these sensitive cultivars grow better because they take up more ions, having greater cation exchange capacities.

Regarding other plant nutrients, there have been relatively few studies involving sweet potato. Calcium appears to be required in relatively small amounts (Scott & Bouwkamp 1974 suggest 60 kg Ca/ha). There is little evidence that natural soil supply significantly limits yields anywhere. It is likely that many soils, including highland New Guinea ones, are able to supply adequate calcium for good yields. The few studies conducted on calcium have focused on crop response to lime applications, sweet potato rarely responding to additions. The demand for magnesium is lower (Scott & Bouwkamp 1974 suggest 20 kg/ha), but deficiency problems are more common due to potassium antagonism (potassium applications on a soil low in available magnesium result in leaf deficiency symptoms, with interveinal chlorosis and the like, reduced top growth and, to a lesser extent, depressed root yields). The results of trials are again conflicting, some reporting moderate yield increases, others no responses.

The micronutient demands of sweet potato can be met by most soils. Boron concentrations, with their narrow range between deficiency and toxicity, may be injurious to both vine and root growth if in excess of 1 ppm. Applications rarely increase yield except on notably boron deficient soils. The application of boron with NPK fertilisers in a trial on volcanic soils on the nearby Nembi Plateau in the Southern Highlands depressed sweet potato yields, suggesting that boron, although low, is not deficient in these soils (Bourke 1985 — potash gave a large yield increase and phosphate a small but significant one, the reverse of other trials on similar soils in the region). Regarding manganese, concentrations of 0.031 ppm are reported optimal for early foliage and root growth, and applications on problem soils may increase yields. Zinc applications apparently give no yield response, whereas copper may sometimes increase yields.

This review of the nutritional requirements of sweet potato suggests that the crop's tolerance of low phosphorus conditions is particularly significant regarding its widespread cultivation in the New Guinea highlands. Phosphorus is one of the region's principal limiting nutrients, as is common for soils with volcanic ash contents, due to the marked fixation properties of the short range minerals that characterise them. One of the strengths of sweet potato is that it can tolerate low levels of phosphorus availability and still give acceptable tuber yields. While the soils of this region may have other fertility shortcomings, for example with nitrogen and potassium of the other macronutrients, and possibly boron too of the micronutrients, phosphorus unavailability is a major limitation. The supply of any of these nutrients, and occasionally perhaps some others, may be sub-optimal for sweet potato growth, but it is generally adequate for the modest yields expected by the subsistence farmers of the region. The soils have favourable potassium:nitrogen ratios and sufficient available phosphorus on the whole to give crop returns enough for Wola needs. The low phosphate tolerance and modest nitrogen demands of sweet potato support the hypothesis that potassium deficiencies most likely prompt site abandonment, where this occurs for fertility reasons, for while the levels of all three major plant nutrients — nitrogen, phosphorus and potassium — decline significantly in soils under cultivation, those of nitrogen and phosphate appear to settle at new equilibria sufficient to meet indefinitely sweet potato's minimal requirements. But this only goes part way to explaining how the Wola are able to sustain their long-term cropping of some sites without the addition of any external fertilisers. How do they manage to keep nutrient levels up to meet even their modest yield requirements?

SECTION VI

CONCLUSION

Tiriy tomb ez aesokemi. Ngo tomb mond aend sawokemi. Mond aend taengbiyp haek pundiy ukemi, hokay shor diy, imbil haeruw diy. Ngo tomb mond kolokemi diy hokay way bokemi. Hokay aend sen kor, bol diy taengbiyp diy, hokay shor aendaon onduwp uwkor. Hokay obun nay nokor. Den diy suw momonuw bolbol tomb hokay obun nay wiy, shor ngo pung bukor tomb.

First, they [gardeners] will pull up the grass. Then they will prepare the soil mounds. They will put pulled up grass, sweet potato leaves, and burnt vegetation in the mounds. Then they will heap up the mounds and plant sweet potato vines. Tubers grow well with plenty of grass and sweet potato foliage there. The sweet potato will consume them like food. When grass is mixed with the soil it feeds sweet potato, as its leaves rot.

Mayka Sal

CHAPTER 13

MOUNDS OF SOIL: GROUNDS FOR SOUND LAND RESOURCE MANAGEMENT

The ability of the Wola highlanders to maintain some of their gardens under long-term cultivation, within the broad context of a shifting strategy, raises intriguing questions about the nature of their agricultural practices, notably their sustainability. While they may move to clear new sites, when crop yields fall below certain tolerable levels, and sometimes for other reasons too, and so overcome any productivity shortfalls, they do not frequently have recourse to this option in some locales, keeping gardens under cultivation for decades, even generations. One of this book's aims is to account for this, within the broad context of a comprehensive ethnographically focussed review of the natural environment. This chapter draws together the many themes addressed to comprehend how, given the constraints of their montane environment, they can practice semi-permanent agriculture, adding no outside amendments to sites.

Regarding staple sweet potato yields, which are central to the agricultural system, a wide range of factors may limit yields in the Southern Highlands of New Guinea, among them deficiencies in soil fertility (Floyd *et al.* 1987, 1988), virus diseases (Waller 1985), root-knot nematode infestation (Bridge & Page 1984), leaf scab (Waller 1985; Kanua & Floyd 1985), insect damage (Thistleton & Masandur 1985), and the genetic potential of cultivars (Kanua & Floyd 1985). The balance of environmental evidence, general observations of many gardens over several years, and the comments of local people, suggest that soil fertility is the major constraint on sweet potato production, the incidence of disease and pest infestation rarely rising to levels sufficient seriously to limit yields.

The nature of the region's soils, their response to clearance and cropping, together with their management under cultivation, are taken as central to understanding the processes that allow the continuance of this broadly sound agricultural regime, one which is sustainable and not degrading of land resources. The chemical properties of the Wola region's soils which limit crop nutrition and production are, in summary: 1) low levels of phosphorus availability and high rates of phosphate fixation; 2) depressed cation exchange capacities and lowered availability of exchangeable cations, particularly potassium; 3) acid conditions, with concommitant high aluminium levels possibly threatened and total base saturation reduced; and 4) low available nitrogen levels likely with the high organic matter contents and resulting high C:N ratios. The physical properties of the soils are, by contrast, generally favourable to crop production, with their high organic matter contents, low bulk densities, and fair topsoil aeration and good drainage.

Management of Soil Resources

Regarding crop production, the supply of all three major plant nutrients potentially limit yield. After one or two cultivations of a site nitrogen, potassium and phosphate levels all fall below the nutritional requirements of many crops for healthy growth. Sweet potato is one of the few which has the capacity to continue producing tolerable yields under these particular nutritionally depleted conditions, and has consequently become the region's staple. The nutritional needs of sweet potato identified as suiting it to these soils are a relatively low phosphorus requirement and a preference for a highish potassium:nitrogen ratio.[1] The data on phosphorus nutrition demonstrate that sweet potato can continue to grow adequately under particularly low phosphate conditions, such as occur in soils where properties restrict its availability through fixation. And the fall in nitrogen under cultivation, if it outpaces that in potassium, may, up to a point, favour sweet potato tuber production. If the potassium decline is too precipitate, or extractable phosphate too inaccessible, yields will inevitably decline unacceptably, for even sweet potato demands a certain minimum supply of nutrients, below which it is unable to give tolerable tuber yields.

Plate 13.1 A newly planted sweet potato mound (*mond*).

---

[1] On similar soils (Tropepts & Hydrandepts) in the north Simbu region of the highlands, Goodbody & Humphreys (1986) report that pH, available phosphorus, phosphate retention and available potassium all show positive effects on the first harvest yields of sweet potato.

The evidence suggests that phosphate availability and potassium supply are probably the principal nutritional limitations in the majority of gardens (Floyd *et al.* 1987, 1988; Goodbody & Humphreys 1986; Wood 1984, 1991). While sweet potato may yield adequately on soils relatively low in extractable phosphorus, even potassium so long as its ratio relationship with nitrogen remains favourable (which is likely with the rapid depletion of available nitrogen by the other nitrogen demanding crops cultivated on a newly cleared plot, together with leaching losses and increased microbial immobilisation following clearance), minimal levels of these nutrients must nonetheless be maintained for tolerable tuber production to continue. In this event, a question posed at the opening of this study remains partly unanswered: how can subsistence farmers in the Wola region, and by extension throughout the highlands of New Guinea, sustain their 'bone gardens' under semi-permanent cultivation without adding external amendments to the soil? The answer is to be found in the traditional method of soil management practiced across the region, which features the building of soil mounds and the use of vegetation pulled up from gardens as compost when they are recultivated. This incorporation of herbaceous and grassy regrowth as compost is an interesting example of the important role weeds may play in the management of soil fertility in shifting cultivation contexts, which has recently attracted attention (Lambert & Arnason 1986, 1989; Saxena & Ramakrishnan 1983, 1984; Swamy & Ramakrishnan 1988).

Plate 13.2 Using a digging stick to uproot coarse grass in a fallowed garden.

Soil Mounds

The cultivation of sweet potato on mounds of soil in which plant residues are incorporated as compost is a characteristic feature of subsistence agriculture throughout the central highlands of Papua New Guinea (Waddell 1972). The practice is central to the feasibility of near continuous sweet potato production. The plano-convex mounds of soil, called locally *mond* by the Wola, vary in size, between two to three metres or so in diameter. The plant residues incorporated into the centre of them varies from garden to garden, depending largely on the time that elapses between cultivations. If women, the builders of mounds, rework gardens while they still support some crops, the residues will include herbaceous and grassy weeds[2] that have colonised the site together with uprooted crop remains, notably sweet potato vines. If they leave gardens fallow for many months, such that they grass over, the principal vegetation incorporated will be coarse grasses.[3]

The procedure of mound building starts with women pulling up the vegetation on a site, somtimes with men's assistance. They use small, one metre

Plate 13.3 Burning sun-dried fallow grass vegetation before mounding commences.

---

[2] The weeds will include such species as: *Arthraxon hispidus, Setaria pallita-fusca, Setaria sphacelata, Commelina diffusa, Polygonum nepalense, Erigeron sumatrensis, Crassocephalum crepidioides, Erechtites valerianifolia, Leersia hexandra, Paspalum conjugatum, Digitaria violascens, Biden pilosus, Desmodium repandum, Pletranthus scutellarioides, Impatiens* sp., *Selaginella* sp., among others (see Chapter 8).
[3] Notably *Ischaemum polystachum*.

long, digging sticks, or small spades today, to loosen the soil, pulling up plants with roots attached; they sometimes use bush knives to clear heavy growth, chopping against log cutting blocks. They call the clearance *ez aesay* (lit. *ez* spear) when rooting out weeds and crop remains, and *bol payay* (lit. *Ischaemum polystachum* cut) when clearing coarse grass from a longer fallowed site. They leave the pulled up vegetation strewn across the surface, usually for several days to dry out (although sometimes only a day or so); the mat of material protects the otherwise exposed and loosened topsoil from rainfall erosion (see Chapter 6).

When they are ready to mound a garden, women work systematically across it, preparing small areas at a time and heaping up the soil. They break up the earth into a fine crumb with their digging sticks or spades to start, and then prepare square depressions bounded by four-sided ridges of soil, called *aend saway* (lit. house earth-oven). The central depression will be the middle of the completed mound, and into it they heap some of the vegetation as compost, called *imbil mond aend wiy* (lit. garden-refuse mound house is); occasionally they burn the material, if dry enough. Next they scoop up the soil from the surrounding ridge over the compost, to start the mound, and continue building it up to the required size by digging up soil around its perimeter, breaking it into a fine tilth and heaping it up on top of the mound, called *mond kolay* (lit. mound heap). Finally, they plant the mound by pushing several sweet potato slips into its surface.

Plate 13.4 Scraping soil into an *aend saway* square to accommodate compost for the centre of a mound.

The heaping up of soil invariably improves sweet potato yields. At Aiyura in the Eastern Highlands Province for example, planting straight into the soil was inferior compared to mounds in a series of agricultural trials, and traditional mounds gave higher yields than ridges of corresponding size (Kimber 1970, 1971).[4] The potential benefits of mounding are manifold. According to the local population it ensures that the soil is friable, the compost giving mounds a soft centre into which tubers can readily expand, swelling into long and straight, regular shaped roots. The building of mounds also increases effective topsoil depth. Although this is not an essential consideration for mounding, because women heap mounds where good topsoil depths occur, it is nonetheless noteworthy in a region where they commonly cultivate steep slopes having inadequate topsoil depths for good sweet potato growth, local farmers observing that tuberisation is poor on plants that have to root into the denser, clayey subsoil.

The breaking up of the soil that attends mound building certainly ensures that bulk densities remain in ranges favourable to root penetration and tuber enlargement, and that structural deterioration and compaction when the soil is exposed under cultivation are unlikely to prompt site abandonment (see Chapter 2). Figure 13.1 demonstrates how effective the tillage that accompanies mounding is at maintaining soils round about the same bulk density regardless of the time they are kept under cultivation. While some increase in bulk density is evident up to three cultivations, it is not large, and the stan-

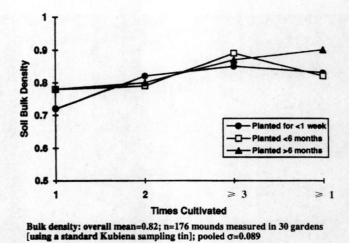

Bulk density: overall mean=0.82; n=176 mounds measured in 30 gardens [using a standard Kubiena sampling tin]; pooled σ=0.089

Figure 13.1 Mean soil bulk densities in gardens cultivated for different periods of time, according to time for which sweet potato mounds planted.

---

[4] At Keravat on the island of New Britain, planting in large ridges gave yields of 37.7 t/ha compared to 11.3 t/ha in traditional smaller ridges, although in the Tari basin there was no significant yield difference on a light soil between small and large mounds and ridges (Bourke 1985).

dard deviation indicates a considerable overlap of values between the different cultivation time classes. The variation between mounds built for different periods of time is not consistent either, the most recently planted not always having the lowest densities, which again suggests that bulk densities tend to vary at random about the overall mean in all gardens, regardless of age.

The extensive soil cultivation that attends mounding not only facilitates tuber physical expansion, it also promotes soil aeration and drainage by encouraging a loose friable structure, which is significant where rainfall is usually high. The down-side is that during occasional extended dry periods excessive drainage may be detrimental to crop growth, especially on soils containing volcanic ash which have inherently low available water capacities (see Chapter 4). On the other hand, the incorporation of large amounts of organic matter as compost is likely to improve water holding capacity. (In an investigation of tillage effects on crop yields on grass-fallowed soils in the New Guinea highlands, Clarke & Street (1967) maintain that increases in aeration

Plate 13.5 Starting to throw turned soil over compost in mound centre.

and drainage were central to the marked yield increase observed (irrespective of fertiliser treatment applied), subscribing to the suggestion (after Nye & Greenland 1960) that the induced acceleration in humus mineralisation and nitrification relieved the nitrogen deficiencies characteristic of soils under grass in the humid montane tropics.[5]) Other micro-climatic changes induced by mounding involve temperature modification. The microbial decomposition of the compost concentrated in the centre of the mound might be expected to increase the temperature appreciably, encouraging vigorous root growth. And according to one commentator (Waddell 1972), a significant climatic consequence of mounding is that it reduces the likelihood of plant frost damage by facilitating the drainage of cold air away from the crop planted over the upper surface of the mound (although this is only of real importance in some frost prone regions of the highlands, above altitudes of about 2000 m).

Another benefit of mounding cited by local people is that it reduces the incidence of disease and rotten tubers. In Enga Province in the central highlands, where people also say compost helps prevent tubers from rotting, black rot (caused by *Ceratocystis fibriata*) was observed four to five times (2.5% to 13.6%) less often on tubers from composted mounds than from uncomposted ones (Preston 1990). But care is required, especially when using crop residues in compost, not to increase the chances of disease; for example, caution is needed where severe weevil infestation exists if using sweet potato vines as composting material (Leng 1982). The incorporation of plant material in mounds also probably increases the total microbial population, increasing decomposition rates (Sanchez 1976). And a further beneficial feature of composted mounds is that they control weed growth, by burying surface germinating weed residues and associated seeds deep in the heart of mounds farmers slow down their proliferation, allowing sweet potato a head start in the competition for light.

It is the role played by composted mounds in the maintenance of soil fertility that is of particular interest. The practice permits the incorporation of nutrients stored in plant residues into the soil during cultivation (Siebert 1987[6]). The vegetation soon rots down into a soft compost, the generic Wola term for which is *paenpaen* (e.g. *iysh shor paenpaen*, literally 'tree leaf bedding-accumulated', for tree leaf compost, *gaimb paenpaen*, literally 'cane-grass bedding-accumulated', for cane grass compost, and so on). Organic matter plays a central role in maintaining soil fertility, it is generally recognised, notably on soils containing volcanic ash. Adding crop residues or straw is successful on a range of soils under shifting cultivation, giving yield responses as good, or higher than fertiliser or manure applications (Sanchez

---

[5] Clarke and Street go considerably further and speculate that tillage-improved-aeration corrects high reducing potentials and associated toxic concentrations of iron and manganese, which they think the low pH, calcium levels and properties of the surface organic layer indicate develop under grassy as opposed to forest vegetation.

[6] Siebert 1987 surmises that increases observed in the Philippines in soil nutrients, pH and organic matter in swiddens are due to the incorporation of weed and crop residues in the soil with the mounding that attends sweet potato cultivation.

Mounds of Soils: Grounds for Sound Land Resource Management 383

Plate 13.6 A mound starts to take shape as the compost is buried.

1976; Abu-Zeid 1973; Ofori 1973). And in sweet potato cultivation, "the addition of organic materials nearly always results in yield increases ... in addition to supplying plant nutrients, [they are] likely to have other beneficial effects on soil structure, water holding capacity, cation exchange capacity, and nutrient release rate" (Bouwkamp 1985:19). In twelve out of thirteen trials for example, conducted throughout Papua New Guinea, organic soil amendments increased sweet potato yields; these consistently positive responses have prompted suggestions that organic materials may offer more scope in this country for increasing food productivity than inorganic fertilisers (Bourke 1985; D'Souza & Bourke 1982; Leng 1982; Thiagalingam & Bourke 1982).

Compost

Early work on organic amendments established that the benefits of compost on sweet potato yield cannot be explained in additive terms alone of nutrient

Plate 13.7 Scooping up soil onto mound to produce friable bed.

supply or water holding capacity or whatever, but as a synergistic combination. According to Tsuno (1970), compost is especially effective as sweet potato manure because it results in more balanced nutrition, supplying nutrients in a slowly available form, besides improving soil moisture availability and soil aeration. The major impact of compost on volcanic ash soils is attributed to its augmentation of organic matter content, which by increasing cation exchange capacities reduces base losses, and by combining with active aluminium reduces phosphate fixation.

The incorporation of compost in mounds has been found to have a significant effect on sweet potato yields in trials throughout the New Guinea highlands. In a series of composting trials in the Southern Highlands Province for instance, on soils similar to those investigated here, compost treatments correlated significantly with increased tuber yields; the positive relationship between rate of compost application[7] and mean marketable tuber yield was linear (Floyd *et al.* 1987a, 1988). The observed response to the improvements in soil fertility was an increase in tuber initiation over tuber bulking, with more tubers per plant, although considerably more tubers also reached marketable size (>100 g). The increased yields achieved with compost amendments were of a similar order to those obtained with equivalent inorganic fertiliser applications. But with compost, the potassium and phosphorus added were used more efficiently, perhaps because grass

---

[7] At rates up to five times greater (at 100 t/ha) than those used by local subsistence farmers.

compost supplies a better balance and wider range of nutrients than fertilizers (such as magnesium and sulphur, together with micronutrients), plus differences between the organic and inorganic sources in the availability of the added nutrients, and also their different effects on biological activity and hence nutrient release.

The composting method employed in the highlands may result in nutrient uptake largely occurring directly from the decomposing vegetation concentrated at the centre of the mound, as roots grow through it, rather than from the soil complex (Floyd et al. 1987a).[8] The delay in the release of nutrients from the composted vegetation doubtless permits more opportunity for uptake, allowing time for initial root growth, and more intimate contact between roots and nutrients as they enter solution. Nutrient availability may be superior too from organic rather than mineral complexes. There are also

Plate 13.8 A sweet potato mound nears completion.

---

[8] Floyd et al. (1987a) also take the absence of any residual composting effect on the next crop as further evidence that it is direct nutrient uptake from decomposing compost that makes this manuring method particularly effective. When these workers cultivated a second crop without further composting, they dispersed the compost remaining widely throughout mounds, reducing the opportunity they argue for direct nutrient uptake. Under traditional cultivation however, women add new compost to mounds every time they recultivate a garden. Although there may be no evident residual effects on yield, the addition of organic matter in composting almost certainly makes a long-term contribution to the maintenance of soil fertility — carbon content decline being one of the few significant effects evident under on-going cultivation — by contributing to soil structure, promoting biological activity and boosting long-term availability of nutrients, notably phosphate and potassium.

differences probably in the stimulation of biological activity, as mentioned, resulting in disparities in the release of nutrients already held in the soil between compost and inorganic fertilisers. Whatever, in soils containing volcanic ash, added nutrients are readily lost, notably through leaching and fixation, and any manuring technique that reduces the extent of these losses might be anticipated to increase the efficiency of nutrient uptake.

It appears that composted mounds are particularly suitable for both sweet potato production and the management of soils containing volcanic ash, the mechanism of nutrient uptake which they afford being especially effective at overcoming phosphate fixation and poor base saturation. Regarding site history, gardens evidencing the highest response to compost are those that have previously been under grass fallow, whereas those under continuous cultivation give the lowest response. This may, in part, be due to disease and pest build-up under continuous cultivation, notably of root-knot nematodes (*Meliodogyne* spp.), any increased infestation by these being associated with reduced yields, curtailing a crop's ability to respond to compost. Fallowing also probably has some effect on the soil system, influencing the manner in which it responds to mound incorporated compost. Perhaps biological activity is higher in briefly fallowed soils than continuously cultivated ones, which would be advantageous, enhancing the breakdown of compost and release of nutrients, increasing yield. The flexibility afforded by the grass fallow option, which may vary in duration from only a few weeks to a year or more under grass, suggests that the traditional farming system has evolved a sustainable equilibrium with local soil resources, so long as the appropriate fallow time elapses for any site on any occasion. When sites show signs of reduced yields, farmers leave them longer under grass to recover their fertility status, whereas when they crop well, they may immediately recultivate them.

A long-term soil exhaustion experiment at Aiyura Highlands Agricultural Experiment Station in the Eastern Highlands Province, first planted in November 1955 on a heavy soil,[9] demonstrates the effectiveness of grass fallows between crops in maintaining sweet potato yields (Kimber 1974[10]). Yields fluctuated up to the fifteenth experimental planting, but it was noticed that they corresponded with the length of fallow preceding recultivation, which varied when fortuitous delays occurred in replanting, due to the weather being too dry or labour unavailable. The longer the fallow, the higher the yield of the next crop. The experiment was redesigned: on one half of the plot continuous cropping was practised and on the other a six week fallow was allowed between cultivations. While yield declined under both, it held up better under

---

[9] A brownish grey loam topsoil, over a dark grey, orange mottled, gleyed clay subsoil, on which phosphorus availability is low.
[10] Details in the Department of Agriculture, Stock & Fisheries Annual Reports 1959–60 to 1967–69.

the fallow (Figure 13.2).[11] But the extent to which grass fallows can restore yields was dramatically demonstrated when a sixteen month fallow under grass was left over the entire site between the twenty-second and twenty-third plantings, after which sweet potato yields returned to their original levels.[12]

An increase in the supply of plant nutrients is a critical feature of the incorporation of compost in mounds, linear responses to it indicating severe soil nutrient limitations.[13] Rates of composting, and hence amounts of nutrients incorporated in mounds varies between gardens, depending on the amount and kind of vegetation growing on the site at the time of recultivation. On occasions, when a garden is kept under continuous cropping, the sweet potato vine yield of the previous crop will largely determine the composting rate, together with colonising weeds. The yield of fresh vegetation in gardens at the later *puw* stage of cultivation (see Chapter 8) with no grassy fallow interval and fair top growth ranged in the Was valley from 28 t/ha to 35 t/ha, which suggests possible compost rates of between 20 and 40 t/ha fresh weight

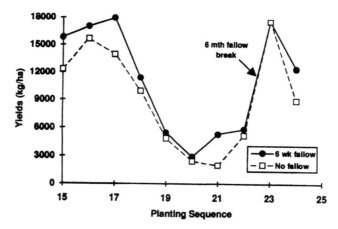

Figure 13.2 Sweet potato yields, Aiyura soil exhaustion experiment (from Kimber 1974).

---

[11] The long-term maintenance of soil fertility under appropriate management without outside amendments is not unusual. In the north-east of England, for instance, at Newcastle University's Cockle Park Farm the yield of hay from meadow receiving no manure or fertilisers has remained fairly constant over eighty-three years (Coleman, Shiel & Evans 1987). And in the south-east, the carbon and nitrogen contents of the soils on some of the Rothampstead long-term experimental control plots have remained reasonably constant for over one hundred years (Jenkinson & Rayner 1977).
[12] Experimental evidence indicates that a grass fallow is as, or more, beneficial than a semi-permanent legume; rotations which included a relatively long rest under a legume (e.g. pigeon pea) proved less productive than short grass fallow rotations (Kimber 1974).
[13] The increase in yield variation with larger compost applications may be interpreted as further evidence that soil fertility is a major constraint on yield because as the constraint is reduced with larger applications so other potentially limiting environmental factors start to become evident, causing yields to vary (Floyd *et al.* 1987a).

of vegetation.[14] The yield of green vegetation for compost from a garden at the early *mokombai* abandoned stage, under herbaceous regrowth largely and abandoned for six months or thereabouts, was of the same order, at 28 t/ha.[15] But when garden sites are left fallow under grassy *mokombai* regrowth for several months, the yield of compost increases considerably, composting rates varying from garden to garden, depending on the length of time they are left under fallow and the consequent extent of their grass cover, notably the growth of *bol* grass (*Ishaemum polystachyum*, Figure 13.3). The yield of grassy compost from such gardens ranged from 32 t/ha for a nine month or so fallow to 85 t/ha for a two years plus fallow,[16] with a mean rate of 62 t/ha.[17]

The extent to which compost added at these rates augments soil fertility varies with the vegetation mix. There are notable differences to be borne in mind between species in their uptake, and subsequent decomposition-release of nutrients.[18] Nevertheless, the Wola do not actively manage fallow sites to promote the growth of certain species above others. They use whatever vegetation comes naturally available (see Chapter 8). The evidence suggests that immediate recultivation of sites at the *puw* stage, incorporating crop residues and pioneer weeds as compost, or recultivation of them after short-term *mokombai* fallow intervals, when herbaceous recolonisation is more advanced, will result, if repeatedly used, in inadequate returns of nutrients to ensure tolerable sweet potato yields. Whereas, it only requires a longer fallow interval dominated by coarse grasses to intervene between cultivations to sustain available nutrient supplies indefinitely at levels sufficient to assure adequate yields of tubers (Table 13.1).

---

[14] This is considerably higher than the rates cited by Floyd *et al*. 1987a of between 10 and 20 t/ha, based on a mean fresh vine yield of 16 t/ha from a Mendi site under a traditional grass compost treatment.

[15] The vegetation weighed had been pulled up by the roots, and some soil still adhered to them, even after vigorous shaking, increasing the weights a little. The *puw* stage vegetation weighed included *Ipomoea batatas*, *Paspalum conjugatum*, *Arthraxon hispidus*, *Bidens pilosus*, *Cynoglossum javanicum*, *Setaria palmifolia*, *Paspalum* sp., *Isachne arfakensis*, *Polygonum nepalense*, *Cyperus* sp., *Crassocephalum crepidioides*, *Setaria sphacelata*, *Oenanthe javanica*, *Plectrantus scutellariodes*, *Selaginella* sp., *Brassica oleracea*, and a *Trema orientalis* sapling. The early *mokombai* stage vegetation included *Erigeron sumatrensis*, *Bidens pilosus*, *Adenostemma lavenia*, *Ipomoea batatas*, *Arthraxon hispidus*, *Crassocephalum crepidioides*, *Paspalum conjugatum*, *Cyclosorus* sp., and *Axonopus affinis*.

[16] There was a marked increase in the amount of decomposing grass litter the longer sites were under fallow, adding considerably to their compost yields. The dry weight of the vegetation was about one-third of its green weight.

[17] Again these yields are considerably higher than those reported by Floyd *et al*. 1987a, who maintain that rates higher than about 20 t/ha fresh weight are uncommon. Comparative data on composting rates in subsistence gardens are limited; according to Waddell (1972:159) they use 11–13 t/ha dry[?] weight at Wapenamanda, and according to Wohlt (1986) they use 40 t/ha fresh weight at Kandep, both in Enga Province.

[18] In Guatemala, for example, Tergas & Popenoe (1971) found that one fallow grass species accumulates about three times more phosphorus than average; in N.E. India, Ramakrishnan (1989) reports that one exotic weed species, together with bamboo, conserve and make available considerably increased amounts of potassium over other plants; and in Belize, Lambert & Arnason (1989) found that weeds accumulate a larger percentage of available nitrogen, phosphorus and potassium than crops, although they report that mulching was unsuccessful in increasing yields.

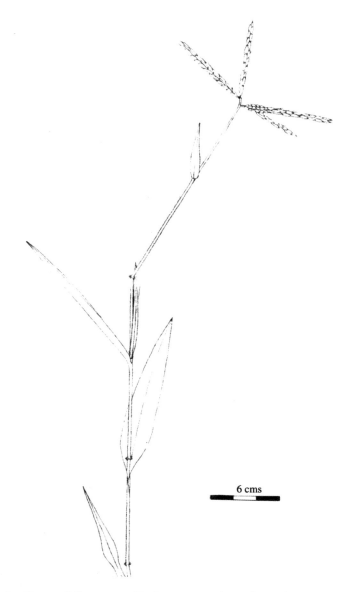

Figure 13.3 Coarse fallow grass *(Ischaemum polystachyum)*.

The supply of those nutrients identified as the probable major nutritional constraints on yield — namely potassium, phosphate and possibly nitrogen — is more or less sufficient from mature *bol* grassland at mean yield levels to meet the demands of sweet potato for optimum tuber yields. Only phosphorus requirements are apparently not met adequately, and boron of the micronutrients. Nonetheless, considering the limited availability of these nutrients in the soil, the compost represents a substantial boost in their supply to plants. The evidence suggests that when tuber yields decline below

Table 13.1

Chemical composition and nutrients supplied by vegetation successions used as compost in the Wola region[4]

NUTRIENT	EM PUW AEZHAE GARDEN COMPOST		MOKOMBAI SHORT FALLOW GARDEN COMPOST		EM BOL LONGER FALLOW GARDEN COMPOST[1]		SWEET POTATO OPTIMUM TUBER YIELD REQUIRE-MENTS[2]
	Dry Matter Fraction (%/ppm) [d.m. = 19% fresh wt]	Nutrient Levels (kg.ha^{-1}/20t compost ha^{-1})	Dry Matter Fraction (%/ppm) [d.m. = 28% fresh wt]	Nutrient Levels (kg.ha^{-1}/20t compost ha^{-1})	Dry Matter Fraction (%/ppm) [d.m. = 33% fresh wt]	Nutrient Levels (kg.ha^{-1}/60t compost ha^{-1})	(kg ha^{-1} for yield c.20t ha^{-1})[3]
N	1.80	68.4	1.45	81.2	1.03	203.94	100
P	0.22	8.36	0.11	6.16	0.09	17.82	26
K	1.68	63.84	0.97	54.32	0.63	124.74	>150
Ca	0.84	31.92	0.51	28.56	0.17	33.66	60
Mg	0.25	9.5	0.40	22.40	0.15	29.70	18

[1] *Ischaemum polystachyum* dominated.
[2] After Hackett 1985, and Scott & Bouwkamp 1974.
[3] A high tuber yield for the region, the mean is nearer to 12t ha^{-1}.
[4] In the Mendi region, equivalent levels of nutrients (kg/ha) supplied by different compost materials were as follows: *Ipomoea batatas* vines (rate of 20t/ha) = N: 55.8, P: 5.0, K: 100.4, Ca: 14.0, Mg: 10.0, S: 5.6, N: 58.6, P: 5.3, K: 72.5, Ca: 10.9, Mg: 4.7, S: 3.9, Mn: 0.28, Zn: 0.12; *Setaria sphacelata* (rate of 20t/ha) = N: 150.6, P: 20.1, K: 150.6, Ca: 68.7, Mg: 45.3, S: 26.7, Mn: 3.12, Zn: 0.78, Na: 1.17, Mn: 1.72, Zn: 0.17; and *Ischaemum polystachyum* (rate of 60t/ha) = N: 150.6, P: 20.1, K: 150.6, Ca: 68.7, Mg: 45.3, S: 26.7, Mn: 3.12, Zn: 0.78, Na: 1.17, Fe: 2.52, Cu: 0.12, B: 0.12 (Floyd *et al.* 1987:7,40; 1988:7); and at Taluma in Enga Province, equivalent levels of nutrients supplied by compost vegetation *Pennisetum clandestinum* (rate of 20t fresh weight/ha) were = N: 33.9, P: 5.97, K: 20.0; and at Kandep *Ischaemum polystachyum* (rate 60 t fresh weight/ha) = N: 96.3, P: 18.9, K: 167.4 (Preston 1990).

some acceptable point, ruling out both immediate recultivation or brief fallow rest because neither can return in their herbaceous growth nutrients sufficient to make good soil deficits, then a longer rest under coarse grasses subsequently returned as compost will boost them sufficiently to increase tuber yields again to tolerable levels.

While phosphorus availability may remain below that required by sweet potato for optimal yields, experimental evidence suggests that this may not have a marked effect given the crop's ability to manage on low phosphorus supplies (see Chapter 12). It is this aspect of sweet potato nutrition that particularly suits it to highlands soils. The interaction reported between compost and phosphate fertiliser applications, and possibly VA mycorrhizal infection (Floyd et al. 1987b), also suggest that the phosphate input of traditional compost is only having a limited impact. It is the boost in available potassium that is probably central to the success of composted mounds in sustaining the region's semi-permanent sweet potato cultivation. Grasses are well known for their high potassium contents. The superior sweet potato yields obtained in an experiment using traditional scarcely decomposed grass compost, compared with a fully rotted compost, support this conclusion, the yield decline under the fully decomposed manure being attributable to the substantial potassium losses, plus some overall nitrogen losses, which occurred with the leaching, volatilisation and biological incorporation that took place while the compost stood rotting down (Floyd et al. 1987a). There may also have been some unfavourable displacement of the potassium:nitrogen ratio contributing to the fall in tuberisation, for relatively fresh vegetation may reduce available nitrogen contents, microbial populations exploding to decompose the material, so immobilising nitrogen and maintaining a favourable ratio with the potassium boost. The changes in soil phosphate and potassium levels under cultivation further support this interpretation of the nutritional impact of composting on sweet potato yields (see Chapter 12). There is no time-related variation in extractable phosphate levels, they are continuously low from the time of first clearance, whereas available potassium levels show a significant decline over time. The grass fallow, and associated composted mounds, which feature in the cultivation of older gardens where soil potassium levels have fallen, keep them boosted to tolerable levels for sweet potato production.

The position with other crops is somewhat different. They have a problem coping with the high phosphate fixing capacities of the soils and consequent low extractable levels. In a series of experiments conducted in the Southern Highlands Province on the effects of various compost and phosphate fertiliser applications (up to 400 kg P/ha) on the yields of beans (*Phaseolus vulgaris*), cabbage (*Brassica oleracea*) and maize (*Zea mays*), the major responses observed were to added phosphate (Floyd et al. 1987a). While phosphate, potassium and nitrogen may all sometimes occur at sub-optimal levels for these crops, it is principally the low availability of phosphate that limits yield. When initially cleared, garden sites have their small reserves of biomass available phosphorus released, sufficient to support a wide variety of crops, but plant uptake and soil fixing processes soon deplete these boosted supplies. The subsequent cultivation of many of these crops cannot be further

Table 13.2
The average nutrient status of soils on abandoned house sites
(n = 4 topsoils, 4 subsoils)

	pH	P (ppm)	K (me/100g)	Ca (me/100g)	Mg (me/100g)	Na (me/100g)	CEC (me/100g)	C (%)	N (%)	C:N ratio
topsoil	6.28 (0.97)	33.45 (25.98)	2.99 (2.36)	64.23 (39.83)	9.56 (6.33)	0.85 (0.61)	33.63 (10.89)	19.83 (14.58)	1.05 (0.60)	18 (2.99)
subsoil	6.63 (0.88)	2.68 (4.53)	1.39 (1.21)	10.5 (12.56)	2.75 (2.94)	1.13 (1.08)	23.48 (3.40)	4.60 (2.12)	0.22 (0.22)	31 (13.1)

sustained by compost incorporation. The potential of green manure for supporting the yield of these more phosphorus demanding crops is limited; increased production will require some inorganic phosphate additions.

Another noteworthy source of nutrients for small cultivations near to homesteads, which tend to support a variety of crops, is the sweepings of everyday refuse from inside houses (*dowhuwniy*) and pig manure (*showmay iy hiym*), which people regularly toss onto adjacent gardens, acknowledging that such waste promotes fertility. The small enriched areas support a wide range of high yielding crops. People exploit this source of nutrients when they abandon houses too, using the sites for small mixed vegetable gardens after structures have rotted or their remains are burnt. These sites can initially prove particularly fertile (Table 13.2). Acidity approaches neutral, which favours phosphate availability, exchangeable cation availabilities are high, and CEC and nitrogen levels are substantial. All in all chemically very fertile sites, although topsoils are usually very thin and nutrient supplies are not sustained for long at these levels.

Traditional methods of organic manuring and soil management deserve close attention. There is little chance however, that local farmers can significantly increase their rates of composting, sufficient to exploit the possible linear increases in sweet potato yield reported in some trials (in which compost was added at rates way beyond those current in subsistence gardens). The major factors restricting composting rates are 1) the limited availability of suitable grassy and herbaceous composting material, and 2) the time and hard work required to collect it. Under the traditional agricultural regime only the vegetation uprooted from a site is incorporated in mounds there; if people collected more material from elsewhere this would deprive other fallowed garden sites of compost material, in all probability undermining the long-term maintenance of their fertility. Even if compostable vegetation was available from elsewhere, it would place an intolerable burden on women, the soil-tillers and mound-builders, to have them collect it and incorporate it into mounds. They produce sufficient food anyway for their subsistence needs under the traditional cultivation regime, in an ecologically sound and sustainable manner: why should they strive to achieve more?

CHAPTER 14

RACING WITH TIME: DEGRADATION OR CONSERVATION?

The Wola are keenly aware of their natural environment, its limitations and opportunities. The have a profound knowledge of their region's natural resources, albeit embodied more in a lived than a verbalised tradition. And they have evolved a subtle relationship with their environment, notwithstanding it having possibly undergone a revolution a few generations ago with the arrival of sweet potato. While they are attuned to their homeland's climate, land resources, vegetational successions, and so on, it would be stretching a point to suggest that they are conservationists. On the other hand, it would be untrue to depict them as agents of degradation. Their relationship with their environment is more equivocal, less easy to pigeon-hole by the black-or-white oppositions common to our ideological discourse. They do not agonise about protecting nature, nor do they heedlessly destroy her — they hover intriguingly somewhere in between.

One of the questions with which this book opened was: how is it that Wola farming practices, involving the cultivation of areas repeatedly without moving on, have not resulted in the degradation of land resources, as reported from other tropical regions following the continuous use of land without allowing adequate intervals for substantial natural vegetation to regenerate? The question no longer seems so bizarre, nor the Wola horticultural regime appear so improbable, given the nature of their soil resources, their management strategies, the crops they grow and their approach to their cultivation. The traditional subsistence regime of the Wola is, first impressions notwithstanding, broadly sustainable regarding its use of natural resources. But others disagree, maintaining that it is more appropriate to talk about Highlanders' degrading their land resources than to speak about them managing them (Wood 1982, 1984, 1991; Allen & Crittenden 1987; Manner 1982).

Degradation and Intensity of Cultivation

Regarding the reciprocal impacts of natural environment and agricultural practices on one another, it has been argued for the Tari basin, which adjoins the Kerewa-Giluwe region, that the near-continuous cultivation of 'dryland' sites, in a manner similar to that of the Wola, has brought the Huli people who live there to the brink of an ecological crisis (Wood 1982, 1984, 1991 — Crittenden 1982 subscribes in passing to a similar interpretation for the Nembi Plateau). Crop yields are reported to decline rapidly in the basin after forest clearance, correlating with a fall in available soil nutrients, notably potassium and phosphorus, in compliance with the classic assessment of the impact of shifting cultivation on the environment. The Huli continue semi-continuous cultivation until crop yields, notably of sweet potato, decline unacceptably, when they fallow the land for a long period, until secondary vegetation indicates an improvement in soil fertility. But forest infrequently re-establishes itself during the fallow break, the secondary succession tending to stabilise with cane grass, tree ferns and softwoods. The result is a permanent reduction in biomass and, it is argued, a decline in site yield potential, for the fall in soil nutrients under cropping is reportedly more severe on plots cleared from fallow vegetation than on new plots cleared from forest.

Plate 14.1 A mosaic of gardens, fallow grassland and cane regrowth with cloud shrouded primary forest on ridge crest.

Forest regeneration is recommended to maintain the productivity of the basin's volcanic ash soils, which are prone to degradation and rapid fertility decline with falling organic matter levels and topsoil loss in erosion. But the Huli are unable to act on this recommendation with their rapidly expanding population and the reported slow rates of soil nutrient recovery and organic matter build-up under natural vegetation, compared to declines under cultivation. Their inability to adopt this afforestation strategy has resulted in soil degradation, such that sustained cropping is not feasible in the long-term. But this interpretation fits uneasily with the evidence. The Huli cultivate some of their gardens near continuously for tens, even hundreds of years according to genealogical evidence. It can clearly take a long time for soil fertility to decline and yields to fall to unacceptable levels. The curves plotted from regression equations, to show changes in soil properties and crop yields over time, evidence not catastrophic falls but relatively smooth and shallow declines towards new equilibrium levels (re Chapter 12), albeit lower than at pre-clearance but not too low to sustain long-term sweet potato cultivation, together with some other crops, with appropriate soil management. Perhaps degradation is too pessimistic a view, and local management techniques are able to maintain a broadly sustainable agricultural regime, for "it is difficult to suggest improved techniques as Huli agricultural techniques are generally well adjusted to the conservation of soil fertility and prevention of erosion" (Wood 1991:13).

A modified version of the argument that land degradation occurs with intensification of land use under increasing demographic pressure has recently gained favour to account for the changes reported from regions like the New Guinea Highlands (Blaikie & Brookfield 1987). It innovates to include both human and pig populations, making it applicable to regions where human demographic growth, although strong, is not sufficient to account for intensive cultivation of land. While human population growth has contributed to land use intensification, and the consequent degradation of land resources, it is argued that the keeping of ever larger pig herds, fed on garden produce (notably sweet potato), has had a markedly more deleterious impact on the land. There is constant pressure to increase the size of herds because pigs feature in ceremonial exchanges, those competitive events in which men vie for status (even power in some regions according to certain reports). The up-shot is that demands on arable land resources increase inexorably as people struggle to grow evermore fodder to feed increasing numbers of animals, pressures on men to keep evermore pigs driving the expansion of herds, with consequent intensification of land use and its degradation (Brookfield 1972; Allen & Crittenden 1987).

This inventive argument sits uncomfortably with some aspects of Wola life. It overlooks the implications of pig herding arrangements (see Chapter 11). While men transact pigs and compete for social standing in ceremonial exchanges, it is women who largely herd and feed these animals. It is they, as the principal pig keepers, who control the size of herds through their labour

Plate 14.2 A pig, item of wealth and agent of soil erosion, standing in an area of soil that it has rooted over.

inputs. While some ambitious men may try to cajole the women in their households, and relatives elsewhere, into keeping a few more animals on occasion — usually for relatively brief periods before some planned exchange event — they are unable to oblige them to herd more animals than customarily thought reasonable. Furthermore, even though some women are more able than others and can handle larger herds, there is a physical limit to the number of pigs any of them can manage. A constant increase in pig herd size, driven by social pressures of exchange or whatever, sufficient to lead to a marked intensification of land use, would require a commensurate increase in the size of the adult female population to herd them, in the absence of some technical innovation increasing the productivity of their labour. In other words, increases in pig populations will correlate with increases in human populations, both are likely to remain in the same ratio over any time period.[1]

In some respects it is possible to reverse the porcine intensification argument and argue that pigs may reduce, not increase, the intensity with which people exploit their land resources. Pigs are agronomically useful because they convert into food what human-beings would otherwise waste, like

---

[1] Allowing for the growth and fall in pig populations with pig kill cycles. The data that I have on pig populations over two decades support this proposition.

animals in many other agricultural systems.² They are an integral part of the farming system, efficiently using resources otherwise of no use to humans — consuming such garden produce and waste as small stringy tubers, bean shuckings, the outer leaves off highland *pitpit* and cabbages, and so on — and turn them into meat for human consumption. They could reduce the need to farm some part of the area that people would otherwise need to cultivate to supply their needs, being a source of food — highly valued meat — raised on what humans would otherwise not use. On the down side, the Wola practice of keeping pigs way beyond maturity, sometimes for several years as adults, considerably reduces their food providing efficiency; their need of adult pigs to transact in exchanges increases the demand for cultivable land above possible optimal levels, but does not necessarily lead to such intensification of land use that degradation occurs.

The feeding of pigs on garden produce which human-beings would not otherwise consume, further undermines somewhat the exchange-pressure-on-pigs argument regarding land use intensification. The Wola grow their crops primarily to feed themselves, and fodder their pigs on that produce which they reject; in effect what they feed to animals is the by-product of what they cultivate to feed themselves. It is unusual to see people feeding large tubers suitable for human consumption to pigs. When women return from their gardens, they sort out the small and stringy tubers in their loads to feed the animals in their care, tubers called *showmay hokay* (lit. pig sweet-potato) not considered fit for humans to eat (Sillitoe 1983). If people increased the intensity with which they farmed their land specifically to grow more tubers to feed to increasing numbers of pigs, we might expect the supply of first-class crop produce to exceed human requirements and people regularly to feed human-consumption-standard tubers to pigs. This is not so, which implies that keeping increased numbers of pigs is matched by the simultaneous cultivation of more food for human consumption; that is, any increase in the porcine population is parallelled by an increase in the human population too, demanding more food. We should anticipate that any documented intensification in land use will relate to an increase in human numbers, which will augment pig feeding capacity by increasing agricultural output and waste production. The keeping of more pigs is a consequence not a contributor to these changes.

These arguments assume that intensity of land use relates to the size of human and pig populations, increasing as they grow, putting more demands on available land resources. Evidence from the Wola region questions this assumption. Large uncultivated areas exist in Wolaland, parts of which people could bring under cultivation if they wished. Instead they chose to

---

² In England for example, sheep and cattle convert grass into edible products, contributing significantly to the efficiency of farming, even its viability in some regions. Pigs likewise may consume waste, and in bygone days households kept them for this purpose.

farm some sites intensively time and again. My observations and their comments suggest that the intensive use of some sites has nothing to do with pressure on available land resources. Rather, they prefer some sites above others and cultivate them repeatedly because they are convenient to their homesteads, they are relatively easy to enclose effectively, they are near other gardens, they develop favourable agronomic properties, and so on. The evidence presented here further suggests that the intensive long-term use of sites need not result in degradation with appropriate land management. The reverse may even pertain, intensity of land use correlating positively with increases in staple crop yields. It may benefit humans and pigs alike to cultivate land intensively.

It is perhaps just as well that intensive land use has positive aspects, the Wola flying in the face of population-resource-pressure arguments further judging by other contemporary events. They currently evince intensification of land use with a decline in their resident population. Over the last decade, with increased labour migration out of their region — initially to the plantations of the Western Highlands and more recently elsewhere — the number of gardens under cultivation for extended periods of time appears to be on the increase. The migrants are predominantly men, those responsible for clearing and enclosing new gardens, consequently fewer new cultivations are being established. This obliges women and children left behind to cultivate for longer some garden sites which they might otherwise have abandoned. Furthermore, my impression is that absent men are placing less store by pigs and more by cash for transaction in exchanges. This turns the social-prestige-through-pigs-promotes-intensification argument on its head, for the pressures on females regarding pig herding may be less, but they are cultivating their garden sites more intensively.

Conservation and Environmentalism

The Wola cultivate some land intensively, whatever the reasons, perhaps increasingly so, and appear on first acquaintance to be reckless in their use of natural resources. They practice no soil conservation measures; they cultivate sites repeatedly with no outside inputs; they show little concern for the secondary regrowth that colonises sites when abandoned; they clear virgin rainforest on occasion; they assert that an abundant soil fauna is undesireable in an arable soil; they pay scant attention to the vegetation community on a site or its species composition before clearing it; they make little assessment of soils before cultivating them; and so on. All of these considerations appear to be contrary to what we think of as careful environmental management and suggest that the Wola are not good custodians of their natural resources. But a careful review of the evidence suggests otherwise.

Substantial areas of montane forest remain across their region, and across the Highlands generally. What seems extraordinary is not that swidden clearance by Highlanders, which extends back into antiquity (Golson &

Hughes 1976), has been responsible for much of the deforestation seen today, but that it has not resulted in greater loss of forest, particularly given their considerable population densities. It implies some regard perhaps for maintaining forested areas, certainly not the pillaging of natural resources. In the light of the archaeological evidence that the Highlands of New Guinea was among the first places on earth where human-beings tended plants (albeit perhaps in a mixed hunting-foraging context initially), the conclusion that the Wola practice a sustainable, not degrading, agricultural regime is perhaps not egregious. They have inherited a long-lived farming tradition, lessons learned over millenia passing down to them. It is not that traditional horticultural practices are degrading land resources, but that arguments to the contrary are degrading to the local population, implying that after countless generations they are unable to manage their land effectively.

These conclusions apply to the contemporary farming system. The time frame is relatively short, particularly in the context of soil processes. Any assessment of the longer term environmental impact of human activity requires an archaeological, not ethnographic perspective. The necessary prehistoric and palynological evidence is scarce for the Southern Highlands currently and non-existent for the Wola region. Nonetheless speculation is possible given evidence elsewhere, notably in the Western Highlands (Bayliss-Smith & Golson 1992; Hope & Golston 1995). It is widely assumed that the sweet potato arrived relatively recently in the Pacific, probably reaching the

Plate 14.3 The prelude to recultivation of a garden: pulling up coarse fallow grasses (newly planted sweet potato mounds on right).

New Guinea highlands some 300 years or less ago. Some postulate an 'Ipomoean Revolution' involving population growth, colonisation of higher altitudes and increased deforestation, some of the palynological evidence suggesting increased forest and soil disturbance (Oldfield *et al.* 1980, 1985a; Walker & Flenley 1979).

Perhaps today's land management strategies originated elsewhere with crops other than sweet potato and were introduced by small numbers of colonists arriving in Wolaland ten generations or so ago, with initially relatively little impact on forests and soils. Earlier horticulture may have centred on taro swidden cultivation with forest fallows, as west of the Strickland River today in the Ok region. Alternatively current day agricultural practices may descend from yam cultivation in mounds, although this crop is now uncommon in the high, wet Wola region (Sillitoe 1983). A combination of both is possible. Whatever, the agricultural regime of today, while dominated by a newly introduced crop, descends from an ancient horticultural tradition, inheriting and modifying upon its environmental knowledge and land use practices.

Land use has intensified under the current day sweet potato regime, reflecting the newly arrived tuber's suitability to the region, able to support a considerably increased population. It is difficult to assess the sustainability of practices in antiquity, but if impact on land resources is any indication they were not markedly degrading of the environment, people seem previously to have exploited their place leaving a minimal impression — although we should not confuse environmental impact with sustainability, humans changing their natural surroundings through their activities, as apparent contemporarily, yet their food production systems remaining sustainable. Today's agricultural regime descends from one that has perhaps been sustainable for millenia. It is difficult to assess the sustainability or otherwise of land use in the future with the rapid technological and social changes currently affecting the Highlands with increased interaction with the outside world.

It would be erroneous to go to the other extreme and depict the Wola as archetypal conservationists, as culturally conditioned environmentalists who can be relied upon for instance to use new technologies and innovations in a naturally sound way. There is no suggestion that they have an Amerindian-like conservation philosophy, featuring some sort of ethereal connection with nature, seeing themselves as custodians of the environment; indeed almost the reverse. They evidence little interest in environmentalism in their use of their land resources, displaying no apparent cultural recognition of responsibility for the natural world nor act as if they have a duty to conserve it. It will look after itself under their farming regime. The casual attitude they evince towards soil assessment reflects their off-hand attitude towards these issues. Unable to foresee its agricultural potential, they acknowledge little control over or responsibility for their soil resources. The soil is there to be exploited, to the full, even if that exploitation appears somewhat hit-and-miss. They push the soil to its fertility limits, until costs oblige them to relent. They follow their shifting agricultural strategy and abandon garden sites to natural vegetation and regeneration, not they say to protect the environment, but because the sites inadequately repay the labour they put into their cultivation, crop yields declining below tolerable levels (for any one of several possible reasons, not just because of fertility problems). The suggestion is not that they invite land degradation if they can avoid it, but that they talk about it in terms of wasted time and labour, not the environmental costs. If a soil develops favourable agricultural attributes under cultivation they are likely to crop it repeatedly, adopting cultivation strategies, like short-term grass fallow and green composting, to extend its useful life for so long as it suits them.

There is no notion of protecting the natural environment from the potentially harmful effects of human activities, but of exploiting it to the full within the constraints of available technology. The Wola have abundant land resources and can move to new sites when old ones become unusable; with adequate potential arable land available they are not exploiting their resources near to the margins given their current technology, when we might expect environmentalist-like concerns perhaps to become apparent. According to their traditional perspective they are in no danger whatsoever of destroying nor using up their region's natural resources. The rainforest is so vast for example, that its destruction is virtually unimaginable, demon guardians and all, it taking days to walk through it in some directions before reaching other human settlements. They think the environment has a infinite buffer capacity, rather as we thought of the sea until recently. Perhaps environmentalist concerns only become apparent when people realise that their environments have a limited capacity to accommodate interference, which in turn may relate partly to increasing demographic pressures.

The Wola are unlikely to have their conservation consciousness raised in the near future by perceiving that there is a threat of running out of fertile land with their sustainable exploitation of land resources. When necessary, they can leave abandoned sites to recover under natural vegetation. This is an inevitable natural process, not one dependent upon human agency, although with appropriate management many sites can remain in production for years, even decades. It is their agricultural technology, coupled with a modest population density, rather than any voiced cultural ideology, that protects the Wola environment in the long-term. It is necessary in this context to bear in mind the practical status of some their knowledge. They act out their understanding, living their balanced and sustainable use of the environment without any need to articulate it (and certainly no wish to preach it!). They are as oblivious to aspects of their stable horticultural regime as they are to facets of their acephalous political constitution (where lack of awareness, I have argued, is critical regarding the isolation of power beyond the reach of individuals or groups).

Some soils progressively improve the Wola maintain the longer they are under cultivation, and the only way to find out is to garden them. This is the exact reverse of the current image of soils under shifting subsistence regimes, as soon run down and forcing a change of site. The cultivation practices of these New Guinea highlanders appear to contradict accepted wisdom regarding low-input agriculture. But the understanding that we have of such tropical subsistence agriculture derives in considerable part from work done in regions on ancient land surfaces with old soils, like Africa and South America with their Oxisols and Ultisols, and there is a tendency to generalise from it to the equatorial tropics as a whole, whereas we should not expect people living throughout this region of the world necessarily to follow similar practices. Commentators, even some agronomists, have been inclined to consider humid tropical soils as homogenous, with common properties and chemistry: synonymous with highly weathered acidic red earths of low fertility, often featuring laterite. Even where they acknowledge that large differences exist between soil resources, some nevertheless tend to treat problems regarding crop production as universal, whether under traditional subsistence or mechanised cash crop agriculture.

While many of the soils of the humid tropics are acid, and some are red, and a few feature laterite — characteristics which relate to a range of other variables cited as limitations to crop production, including lack of nutrient reserves in weatherable minerals, low organic matter contents, poor response to liming, and low cation exchange capacities — it is mistaken to think these have pan-tropical relevance. The evidence suggests that we should not lump the subsistence cultivators of New Guinea, nor those elsewhere cultivating on relatively young soils (e.g. on other Pacific islands and parts of S.E. Asia), with shifting cultivators living on ancient land surfaces. Soils of the Inceptisol order show relatively little variation compared to older orders because they have not existed long enough to bear the imprint of local environmental variations and diverge markedly in productive potential. The geologically recent volcanic rejuvination of the soils in some of these regions, including Wolaland (see Chapter 5), has further contributed to their uniformity and youth. Nor should we consider those farming at higher altitudes to be similar to those farming in lowland forests. Farmers in the tropics clearly do not all face the same constraints, given the wide variation in the land resources available to them.

We have a long way to go before we understand the variety and dynamics of these tropical agricultural systems. Clearly, those who have lived by them for millennia can only help further our knowledge (Richards 1985; Inglis 1993). The ethno-environmental approach offers other cultures' perspectives on issues concerning our shared natural world, for which as user-custodians we all bear some responsibility to each other and future generations; although the scale of our relative impacts differ, for example the volume of hydrocarbons people like the Wola release in burning vegetation and its impact on the global climate is small compared to that of our massive per capita consumption of fossil fuels. We may well have something to learn from them about moderating our behaviour in relation to our natural surroundings, about attitude and approach, so as to reduce adverse environmental changes, instead of ethnocentrically foisting on them our culture's contemporary anxieties about environmental degradation.

The technological impact on the environment of the West, often detrimental, is frequently the unintended consequence or cost of other activities; the excessive spreading of inorganic fertilisers to optomise crop yields for example polluting ground water supplies. We need to learn perhaps to accept more and meddle less, to heed well-tried practices and trust less to fashionable overly intellectual panaceas, at least until we have a better understanding through experience of the consequences of our actions. Maybe we think too much and feel too little for Nature. We depend perhaps too heavily on vogue academic discourse and too little on unspoken practical knowledge, as rapid technological change transforms our relationship with Nature; many 'Greens' for instance warn that we are in a 'race against time' to achieve a sustainable balance in our use of natural resources, or degradation of place, entropy before time is unavoidable. But change is inevitable, it is pace and direction that we debate. Regarding land resources, of the Wola region or wherever, time will inevitably win place.

No matter how technically primitive people may appear, we should not allow this to fool us into thinking that their understanding of their environment is deficient in some regards, nor that their knowledge of the world as they experience it is underdeveloped and simplistic, even erroneous. Why is it that we are inclined to assume that generations of applied knowledge is inadequate and inappropriate because it comes from a different cultural tradition, why are we prone to think for example that our soil science is the more insightful because it debates analytically the management of natural processes? Their understanding might derive from a tradition that more lives its management of natural resources than moralizes about it, but this need not reduce its effectiveness, although it makes its excogition in a study like this alien and difficult. It is disquietening to contemplate our conceit in assuming that we know better given the environmental chaos and destruction that our nostrums for agricultural development have inflicted on various parts of the tropical world, foisting inappropriate programmes on people who for millennia have successfully managed their soils and natural environments to ensure their livelihoods. We have, until recently, assumed that we have to go out to technologically lesser developed nations like Papua New Guinea and further their 'development', whereas they are already fully developed along their own cultural lines. If this environmental ethnography achieves nothing beyond demonstrating that local lore and practices can withstand close scientific investigation, which cannot better them, so lending support to the tenets of sustainable agriculture and local community participation now gaining ground, then it has accomplished something. It is high time that experts respected local knowledge and consulted it closely, before they try to improve on it (Chambers *et al.* 1989; Warren 1991).

APPENDIX

GLOSSARY OF WOLA ENVIRONMENTAL TERMS

aena or *ae*	direction marker: the S–E quadrant
aend saway	one of the stages in mounding: preparation of square depression bounded by four-sided ridges of soil
aenk sem	pandans
aeray	chert
aiben or *mumung* or *haelboi*	axe blade stone (metamorphic rock)
aiben pay	very large level area of land
aluw	small hill
amuw	direction marker: the N–W quadrant
araytol	chert covered by a dull accretion of lime carbonate
baen biy	fine, dry, sunny weather; drought conditions
baen nat	sun heralding a fine day
baeray aez hae	steep mountain slope negotiable on foot
bol	coarse grass regrowth
bol payay	clearing coarse grass from a fallow site before mounding
bulhenjip	Southern hemisphere winter, S.E. season (April to September)
buriy	strong
buwbiyaeb	lightening
buwkbuwk suw	very large level area of land
chay	rain
chay kung	wet, overcast period; low cloud-mist
chay nat	weak sun before rain
chay onda	rainstorm
chay pokot	drizzle, fine rain
chay suw	rain while sun shines
chay taendak	hailstorm
day ebay or *dayngel*	hungry period, famine
den sem	grasses, herbaceous plants
deraen	outside; wild
deret biy	fretting of leaves due to grazing of insects
dez or *poborgaim*	mineral oil
dimbuw	watery, softening rot of tubers
domay	rotten
dorpind	disaster falling from the sky
dowhuwniy	sweepings from house; friable and granular structure
duwlbaeray	earthquake or tremor
ebenjip	Southern hemisphere summer, N.W. season (October to May)
elelbiy	insects (invertebrates and reptiles)

eltaekis	gas seep
em	garden
em bort bway	crops; plants cultivated in gardens
em gaimb uw imb shay	virgin, uncultivated cane grassland
em g^emb	small mixed vegetable garden, usually near houses
em hul	long-term garden
em imb	garden cleared from virgin vegetation
em puw	garden being harvested for second or subsequent time
em waeniy	new garden being harvested for first time
ez aesay	pulling up weeds and crop remains before mounding
gaimb	cane grass or cane grass regrowth
gaimb kolowmon	thick litter layer of rotten cane grass stems
gaip say	rotting of standing plant with browning and drying
g^enk	small
gogay	worm
gogay iy	worm casts
goiz sem	palm trees
haen	rock, large stones
haen abaenda	rock overhang
haen bagiy	rockface
haen h^egben	rounded water-borne rocks
haen hok	calcareous mudstones, siltstones and rotten limestone; arrested soil (Rendoll, Rendzina)
haen kolkol	friable stone fragments cemented together, conglomerate
haen kombay	avalanche
haen naenk	sandstone rock or boulders
haen paenj	soil-like accumulation of weathered limestone; regolith
haen tay	base of a rock face or overhang
haen nek	pot hole
haenbora	rocky vegetation; skeletal soils
haeret hezay	to be lost, have no sense of direction
haeriy	brown rot marks in tuber flesh
haez	axe blade stone (glaucophane schist)
hakuwl kay	thunder
hat	mountain
hat bagiy	near vertical or vertical slope unnegotiable on foot
hat chem or hat maenda or hat nay	mountain summit
hat shuwol	sinkhole
hat suwl	poorly drained depressions
hat tongowlow	large valley
hathaen	limestone
h^egben	dry gully; steep sided stream gulley; ravine
hezaembuwl	brown rot marks in tuber flesh
hiy	short
hokay em	sweet potato garden
hokay haerok	black scab on sweet potato tubers
howma	ceremonial ground, settlement clearings
hundbiy	orange-brown

Appendix

hunduwm menay	curling of leaves due to insect attack
huwbiyp	dark coloured, extrusive and intrusive rocks (largely basalts)
huwguwk	mottles in gleyed soils
huwguwmb say	mildew and bacterial moulds
huwpiy	cool weather, chilly conditions
hwiybtowgow	ritual following forest demon attack
ibil	higher altitude regions
iybtit	forest spirit or demon
imbil mond kolay	heaping soil up over compost/ash into mound
iyb	river, water
iyb bombok	water flowing into pot hole or water flowing through arch
iyb day	river in spate
iyb dor tilai	recent alluvial soils (Fluvent; Fluvisol)
iyb hondai	water flowing into pothole
iyb kaen	steep sided ditch; ravine
iyb komb deray	water flowing out of a pothole
iyb liy	wild water
iyb ma	upstream
iyb mael	calm water; pond
iyb maeray or *iyb solow*	river bank
iyb muw	unconsolidated sand or gravel; sandy soils (Psamment; Arenosol)
iyb muw tongom	pale grey sand
iyb shoba or *iyb taip*	confluence, river fork
iyb sond	waterfall
iyb tay	downstream
iyb uw damiy	peat soils (Saprist; Histosols)
iysh	trees, woody plants
iysh shor paenpaen	litter layer of relatively unrotted plant material
iysh tam ebay	large landslide that uproots trees
iyshabuw	montane forest
iyshpondamahenday	ritual performed in hard times, famine ritual
izuw	a place at the same altitude as the speaker not separated by a valley
kaena	channels between sweet potato mounds
kas	bright reddish brown to orange clay; cracks in tubers
keshond	crysalis
kol	brown rot marks in tuber flesh
kolbatindiy	subsoil with reddish tuff and pyroclastic cinders
kolbatindiy haen	unbreakable tuff
kolbatindiy kolkol	friable tuff
komay	small length of ritual fence
kond lay	leaf spot necroses, galls and rusts
kopenaend	cocoon around a pupa
kunguwp	constricted tubers; fleshy roots
kuwng	ridge top
kwimb sem	mosses and liverworts

ma em	taro garden
ma hok	black scab on taro corms
maend bay	stem rot; rot infected plant; rotten fruit or tuber
mobo	somewhere nearby at the same altitude as the speaker, separated by a valley
mogo	somewhere distant at the same altitude as the speaker, separated by a valley
mokombai	recently abandoned garden; grassy/herbaceous regrowth vegetation
mond	mound of earth
mondow	soft sticky liquid mud
mor	somewhere at the same altitude as the speaker, separated by a valley
moray nais	community representatives in famine ritual
munk shor sem	large leaved shrubs
muwlol	cloud, mist
muwlol chay maenmaen biy	heavy black storm clouds
muwlol munga payay	cloud-mist filling valleys
muwtiy	small hill
naen	direction
nat day	sunny, sunshine
nat taendabiy	intense sun (midday)
nat turiybiy	warm sun (morning)
ngoltombae	ritual performed in hard times; famine ritual
ningiylbap	rainbow
nobow	somewhere nearby but lower in altitude than speaker
nuw piriy haeray	procedure to stop rain
obael	secondary woodland
oliy	husband; one of a pair
omb	bright reddish brown to orange clay
omb hul	small laterite-like concretions found in clay subsoils
onyin or *onyuw*	down; direction marker: the S–W quadrant; somewhere distant and lower in altitude than speaker
pa	wetland vegetation
pa tongom	gley soils (Aquent; Gleysols)
paenpaen	compost of unrotted vegetation
paerai mbay	sticky (of soils etc.)
pay sem	hollow stemmed plants (e.g. bamboos)
payhonez	greyish brown silty clay
paym	axe blade stone
pendet pay	movement of material downslope; stone rolling downhill
pibiy	dew
pibiyhond	frost
pitpit	Pidgin term for cane grass
pletbok	thicket of dense vegetation
pombray	black
popo	breeze, light wind
popo hae	gritty, sandy texture

popo onda or *posaebsuw*	strong wind, gale
pung biy	rotten, stinking
saem	forest spirit or demon
saem aenda	forest demon ritual house
saezuwp sem	ferns
sem	family or group of related persons/phenomena (e.g. plants)
semg^enk	small family or kin group
semonda	large family or kin group
sez sem	fungi
shor baebray	yellowing of leaves
showmay	pig
showmay hokay	pigs' sweet potato
showmay iy hiym	pig manure
showmay suw pen kayor	transport of soil by rooting pigs
shuwol	low-lying region
sinjbilay	watery softening rot of tubers
sol	tall, high
sol tukay	cumulus cloud
suw	a land area, territory; soils
suw aevril	cracks that appear in soil before a landslide
suw araytol	stony soil
suw engay may	landslide
suw haeriy	poor soil state; hard, cloddy soil
suw hemem	excellent soil state; good depth of friable, dark topsoil
suw hombshor	resting a garden area under grass fallow
suw hul or *suw kompae*	large blocky soil aggregates
suw hundbiy	bright-brown clay subsoil; clayey soil (Tropept; Cambisol)
suw huwniy	friable soil state, good for sweet potato
suw iyba or *hobor*	soil 'blood' or 'grease'; organic matter
suw iybom tok pay	transport of soil by water; water erosion
suw ka	soil under long-standing natural vegetation; newly cleared soil
suw kabiy	exhausted, tired soil
suw kolkol bay	friable, crumbly soil
suw mundit may	landslide
suw ombo	small sized soil aggregates, crumbs
suw ora	father or custodian of the land
suw pa	wetland, bog or swamp; waterlogged soil
suw pombray	dark brown to black topsoil
suw saway	heavy cloud, overcast; chilly weather
suw silsil pay	movement of soil particles downslope; erosion
suw taebowgiy	poor soil state; hard cloddy soil
taengbiyp	early herbaceous/grassy phase of regrowth
tag day	over ripe fruit
tiyptiyp	ashy agglomerate; volcanic ash soils (Andepts; Andosols)
tiyptiyp haen	unbreakable ashy agglomerate
tiyptiyp kolkol	friable ashy agglomerate

tol	dirt; concretional covering on rock
tombow nae hae	hunger
tomiy	weak, soft
tongom	white or grey clay
turuwpiy	warm weather, hot
tuw	male; non-cone-bearing casuarinas
tuwguwn	in between, middle
waibsuw	lower altitude regions
waip	litter layer; peat soils (Folist, Fibrist; Histosols)
way	female; cone-bearing casuarinas
weray	wife; one of a pair
wezow	shadow; life force
ya sem	vines or climbers
yin or *yuw*	up; direction marker: the N–E quadrant
yin pay	steep mountain slope negiotiable on foot
yom	alpine vegetation
yuw	somewhere distant and higher in altitude than the speaker
yuwbuw	somewhere nearby and higher in altitude than the speaker

REFERENCES

Abu-Zeid, M.O. 1973 Continuous cropping in areas of shifting cultivation in Southern Sudan. *Tropical Agriculture* **50**:285–90.
Acock, M.C. & Garner, J.O. 1984 Effect of fertiliser and watering methods on growth and yields of pot-grown sweet potato genotypes. *Hortscience*, **19**(5):687–689.
Ahmad, N. & Brechner, E. 1974 Soil erosion on three Tobago soils. *Tropical Agriculture*, **51**:313–324.
Allbrook, R.F. & Radcliffe, D.J. 1987 Some physical properties of Andepts from the Southern Highlands, Papua New Guinea. *Geoderma*, **41**:107–121.
Allan, W. 1967 *The African husbandman*. London: Oliver & Boyd.
Allen, B.J. 1982 Yam gardens and fallows in the Torricelli foothills, Dreikiki District, East Sepik. In R.M. Bourke & V. Kesavan (eds.) *Proceedings of the Second Papua New Guinea Food Crops Conference*. Konedobu: Department of Primary Industry.
Allen, B.J. (ed.) 1984 Agricultural and nutritional studies on the Nembi Plateau, Southern Highlands. University of Papua New Guinea, Geography Department Occasional Paper (new series) No. 4.
Allen, B. & Brookfield, H.C. 1989 The future: questions for policy and research. *Mountain Research & Development*, **9**:329–334.
Allen, B., Brookfield, H.C. & Byron, Y. 1989 Frost and drought through time and space. II: the written, oral and proxy records and their meaning. *Mountain Research & Development*, **9**(3):279–305.
Allen, B. & Crittenden, R. 1987 Degradation and a pre-capitalist political economy: the case of the New Guinea highlands. In P. Blaikie & H. Brookfield (eds.) *op. cit.* (pp. 143–156).
Arnason, T., Lambert, J.D.H., Gale, J., Cal, J. & Vernon, H. 1982 Decline of soil fertility due to intensification of land use by shifting agriculturalists in Belize, Central America. *Agro-Ecosystems*, **8**:27–37.
Ash, J. 1982 The *Nothofagus* Blume (Fagaceae) of New Guinea. In J.L. Gressitt (ed.) *Biogeography and ecology of New Guinea*. The Hague: W. Junk *Monographiae Biologicae*, Vol. **42**:355–380.
Austin, M.E., Aung, L.H. & Graves, B. 1970 Some observations on the growth and development of sweet potato (*Ipomoea batatas*). *Journal of Horticultural Science*, **45**:257–64.
Aweto, A.O. 1981 Secondary succession and soil fertility restoration in south-western Nigeria. 2. Soil fertility restoration. *Journal of Ecology*, **69**:609–14.
Bain, J.H.C. 1973 A summary of the main structural elements of Papua New Guinea. In P.J. Coleman (ed.) *The Western Pacific island arc, marginal seas, geochemistry*. Perth: U.W.A. Press (pp. 147–61).
Barrow, C.J. 1991 *Land degradation: development and breakdown of terrestrial environments*. Cambrigde University Press.
Barry, R.G. 1978 Aspects of the precipitation characteristics of the New Guinea mountains. *Jour. Tropical Geog.*, **47**:13–30.
Bartholomew, V., Meyer, J. & Laudelout, H. 1953 Mineral nutrient immobilization under forest and grass fallow in the Yangambi (Belgian Congo) region. *Institut National pour l'Etude Agronomique du Congo Ser. Science*, No. **57**:1–27.
Bayliss-Smith, T. & Golson, J. 1990 A Colocasian revolution in the New Guinea Highlands? Insights from Phase 4 at Kuk. *Archaeology in Oceania*, **27**:1–21.
Beckerman, S. 1983 Does the swidden ape the jungle? *Human Ecology*, **11**:1–12.
Behrens, C.A. 1989 The scientific basis for Shipibo soil classification and land use: changes in soil-plant associations with cash cropping. *American Anthropologist*, **91**:83–100.

Berlin, B. 1991 The chicken and the egg-head revisited: further evidence for the intellectualist bases of ethnobiological classification. In A.Pawley (ed.) *Man and a half: essays in Pacific anthropology and ethnobiology in honour of Ralph Bulmer*. Auckland: Polynesian Society Memoir No. 48 (pp. 57–66).

Berlin, B. 1992 *Ethnobiological classification: principles of categorization of plants and animals in traditional societies*. Princeton University Press.

Bernhard–Reversat, F. 1975 Nutrients in throughfall and their quantitative importance in rain forest mineral cycles. In F.B. Golley & E. Medina (eds.) *Tropical ecological systems*. Berlin: Springer (pp. 153–159).

Biersack, A. (ed.) 1995 *Papuan borderlands: Huli, Duna, and Ipili perspectives on the Papua New Guinea highlands*. Ann Arbor: Michigan University Press.

Bik, M.J.J. 1967 Structural geomorphology and morphoclimatic zonation in the Central Highlands, Australian New Guinea. In J.N. Jennings & J.A. Mabbutt (eds.) *Landform studies from Australia and New Guinea*. Canberra: Australian National University Press (pp. 26–47).

Blaikie, P. & H. Brookfield (eds.) 1987 *Land degradation and society*. London: Methuen.

Bleeker, P. 1983 *Soils of Papua New Guinea*. Canberra: CSIRO & Australian National University Press.

Bleeker, P. & Healy P.A. 1980 *Analytical data of Papua New Guinea soils* (2 vols.). CSIRO Division of Land Use Research Technical Paper No. 40.

Blong, R.J. 1982 *The time of darkness: local legends and volcanic reality in Papua New Guinea*. Canberra: A.N.U. Press.

BOSTID 1984 *Casuarinas: nitrogen-fixing trees for adverse sites*. Washington, D.C.: National Academy Press (National Research Council Board on Science and Technology for International Development 'Innovations in Tropical Reforestation' Series).

Bourke, R.M. 1977 Sweet potato (*Ipomoea batatas*) fertilizer trials on the Gazelle Peninsula of New Britain: 1954–76. *Papua New Guinea Agriculture Journal*, 28 (2,3,4):73–95.

Bourke, R.M. 1984 Growth analysis of four sweet potato (*Ipomoea batatas*) cultivars in Papua New Guinea. *Tropical Agriculture*, 61 (3):177–81.

Bourke, R.M. 1985 Sweet potato (*Ipomoea batatas*) production and research in Papua New Guinea. *Papua New Guinea Journal of Agriculture, Forestry & Fisheries*, 33 (3,4):89–10.

Bourke, R.M. 1988 *Taim hungre:* variation in subsistence food supply in the Papua New Guinea Highlands. Ph.D. thesis, Australian National University, Canberra.

Bourke, R.M. 1989 Influence of soil moisture extremes on sweet potato yield in the Papua New Guinea Highlands. *Mountain Research & Development*, 9:322–328.

Bouwkamp, J.C. 1983 Growth and partitioning in sweet potatoes. *Ann. Tropical Res.*, 5:53–60.

Bouwkamp, J.C. & Hassam, M.N.M. 1988 Source–sink relationships in sweet potato. *Journal of the American Society of Horticultural Science*, 113 (4):627–29.

Bouwkamp J.C., 1985 Production requirements. In J.C. Bouwkamp (ed.) *Sweet potato products: a natural resource for the tropics*. Boca Raton (Florida): CRC Press (pp. 10–33).

Bradbury, J.H. & Holloway, W.D. 1988 *Chemistry of tropical root crops: significance for nutrition and agriculture in the Pacific*. Canberra: Australian Centre for International Agricultural Research Monograph No. 6.

Brams, E.A. 1971 Continuous cultivation of west African soils:organic matter diminution and effects of applied lime and phosphorus. *Plant and Soil*, 35:401–414.

Bridge, J. & Page, S.L.J. 1982 Plant nematodes of Papua New Guinea: their importance as crop pests. Report of a survey conducted by the Commonwealth Institute of Parasitology, St. Albans.

Brinkman, L.F. & Nascimento, J.C. de 1973 The effect of slashand burn agriculture on plant nutrients in the tertiary region Central Amazonia. *Turrialba*, 23:284–90.

Brookfield, H.C. 1972 Intensification and disintensification in Pacific agriculture: a theoretical approach. *Pacific Viewpoint*, 13:30–48.

Brookfield, H.C. & Allen, B. 1989 High altitude occupation and environment. *Mountain Research and Development* (spec. ed. 'Frost and drought in the highlands of Papua New Guinea' eds. B. Allen and others), 9 (3):201–209.

Brookfield, H.C. & Brown, P. 1963 *Struggle for land: agriculture and group territories among the Chimbu of the New Guinea Highlands*. Melbourne: A.N.U. and Oxford University Press.

Brown, C.H. 1984 *Language and living things: uniformities in folk classification and naming*. New Brunswick: Rutgers University Press.

Brown, C.M. & Robinson, G.P. 1977 Explanatory notes on the Lake Kutubu geological sheet. Port Moresby: Dept. Lands, Surveys and Mines, Geological Survey of Papua New Guinea Report No. 77/12.

Brown, M. & Powell, J.M. 1974 Frost and drought in the highlands of Papua New Guinea. *Jour. Tropical Geography*, **38**:1–6.

Brown, P. & Podolefsky, A. 1976 Population density, agricultural intensity, land tenure and group size in the New Guinea Highlands. *Ethnology*, **40**:211–238.

Bruijnzeel, L.A. 1989 Nutrient content of bulk precipitation in south-central Java, Indonesia. *Journal of Tropical Ecology*, **5**:187–202.

Bruijnzeel, L.A. 1991 Nutrient input-output budgets of tropical forest ecosystems: a review. *Journal of Tropical Ecology*, **7**:1–24.

Brunton, R. 1980 Misconstrued order in Melanesian religion. *Man*, **15**:112–128.

Bulmer, R.N.H. 1968 "Worms that croak & other mysteries of Kalam natural history" *Mankind*, **6**(12):621–39.

Bulmer, R.N.H. 1974 Folk biology in the New Guinea Highlands. *Social Science Information*, **13**:9–28.

Burnett, R.M. 1963 Some cultural practices observed in the Simbai Administrative Area, Madang District, Papua New Guinea. *Papua New Guinea Agriculture Journal*, **16**:79–85.

Chambers, R., Pacey, A. & Thrupp, L.A. (eds) 1989 *Farmer first: farmer innovation and agricultural research* London: Intermediate Technology Publications.

Chang, J-H. 1968 Rainfall in the tropical S.. Pacific. *Geog. Review*, **58**:142–44.

Chapman, S.B. (ed.) 1973 *Methods in plant ecology*. Oxford: Blackwell.

Charles, A.E. 1976 Shifting cultivation and food crop production. In K. Wilson & R.M. Bourke (eds.) *Proceedings of the Papua New Guinea Food Crops Conference*. Kondeobu: Department of Primary Industry (pp. 75–78).

Chartres, C.J. & Pain, C.F. 1984 A climosequence of soils of late Quaternary volcanic ash in the Highlands of Papua New Guinea. *Geoderma*, **32**:131–155.

Clarke, W.C. 1971 *Place and people: an ecology of a New Guinean community*. Berkeley: University of California Press.

Clarke, W.C. & Street, J.M. 1967 Soil fertility and cultivation practices in New Guinea. *Journal of Tropical Geography*, **24**:7–11.

Coleman, S.Y., Shiel, R.S. & Evans, D.A. 1987 The effects of weather and nutrition on the yield of hay from Palace Leas meadow hay plots, at Cockle Park Experimental Farm, over the period from 1897 to 1980. *Grass and Forage Science*, **42**:353–358.

Conklin, H.C. 1957 Hanunoo agriculture: a report on an integral system of shifting cultivation in the Philippines. *F.A.O. Forestry Development Paper* No. 12. Rome: F.A.O.

Constantin, R.J., Jones, L.G. & Hernandez, T.P. 1977 Effects of potassium and phosphorus fertilization on quality of sweet potatoes. *Journal of the American Society of Horticultural Science*, **102**:779–81.

Crittenden, R. 1982 Sustenance, seasonality and social cycles on the Nembi Plateau, Papua New Guinea. Unpub. Ph.D. thesis, Australian National University, Canberra.

Cunningham, R.K. 1963 The effect of clearing a tropical forest soil. *Journal of Soil Science*, **14**:334–45.

D'Souza, E.J. & Bourke, R.M. 1982 Compost increases sweet potato yields in the highlands. *Harvest*, **8**(4):71–75.

D'Souza, E.J. & Bourke, R.M. 1986 Intensification of subsistence agriculture on the Nembi Plateau, Papua New Guinea (Parts 1,2,& 3). *Papua New Guinea Journal of Agriculture, Forestry & Fisheries*, **34**:19–48.

Dantas, M. & Phillipson, J. 1989 Litterfall and litter nutrient content in primary and secondary 'terra firme' rain forest. *Journal of Tropical Ecology*, **5**:27–36.

Davies, H.L. & Smith, I.E. 1971 Geology of east Papua. *Bulletin of the Geological Society of America*, **82**:3299–312.

Denham, D. 1969 Distribution of earthquakes in the New Guinea — Solomon Islands region. *Journal of Geophysical Research*, **74**:4290–99.

Diem, H.G. & Dommergues, Y.R. 1990 Current and potential uses and management of Casuarinaceae in the tropis and subtropics. In C.R. Schwintzer & J.D. Tjepkema (eds.) *The bilogy of Frankia and actinorhizal plants*. London: Academic Press (pp. 317–342).

Djazuli, M. & Tadano, T. 1990 Comparison of tolerance to low phosphorus soils between sweet potato and potato. *Journal of Faculty of Agriculture Hokkaido University*, **64**(3):190–200.

Dokuchaev, V.V. 1883 *Russian chernozem*. St. Petersburg.

Dove, M.R. 1985 *Swidden agriculture in Indonesia: the subsistence strategies of the Kalimantan Kantu*. Berlin: Mouton.

Driessen, P.M., Buurman, P. & Permadhy 1976 The influence of shifting cultivation on a "podzolic" soil from Central Kalimantan. *Soil Research Institute Bulletin* Bogor, Indonesia, **3**:95–115.

Drover, D.P. 1973 Chemical & physical properties of surface peats in the Wahgi valley, Western Highlands. *Science in New Guinea*, **1**(2):8–10.

Duchafour, P. 1982 *Pedology: pedogenesis & classification*. Allen & Unwin.

Dvorak, K.A. 1988 *Indigenous soil classification in semi-arid tropical India* Patancheru (Andhra Pradesh): International Crops Research Institute for the Semi-Arid Tropics.

Edwards, P.J. 1977 Studies of mineral cycling in a montane rain forest in New Guinea: II. The production and disappearance of litter. *Journal of Ecology*, **65**:971–992.

Edwards, P.J. 1982 Studies of mineral cycling in a montane rain forest in New Guinea: V. Rates of cycling in throughfall and litter fall. *Journal of Ecology*, **70**:807–827.

Edwards, P.J. & Grubb, P.J. 1977 Studies of mineral cycling in a montane rain forest in New Guinea: I. The distribution of organic matter in the vegetation and soil. *Journal of Ecology*, **65**:943–69.

Edwards, P.J. & Grubb, P.J. 1982 Studies of mineral cycling in a montane rain forest in New Guinea. IV. Soil characteristics and the division of mineral elements between the vegetation and soil. *Journal of Ecology*, **70**:649–666.

Ellen, R.F. 1986 What Black Elk left unsaid: on the illusory images of Green primitivism. *Anthropology Today*, **2**:8–12.

Ellen, R.F. 1993 Rhetoric, practice and incentive in the face of changing times: a case study in Nuaulu attitudes to conservation and deforestation. In Milton, K. (ed.) *op. cit.* (pp. 126–143).

Elwell, H.A. 1981 A soil loss estimation technique for southern Africa. In R.P.C. Morgan (ed.) *Soil conservation: problems and prospects*. Chichester: Wiley (pp. 281–292).

Enyi, B.A.C. 1977 Analysis of growth and tuber yield in sweet potato *(Ipomoea batatas)* cultivars. *Journal of Agricultural Science*, **88**:421–430

Ewel, J.J., Berish, C., Brown, B., Price, N. & Raich, J. 1981 Slash and burn impacts on a Costa Rican wet forest site. *Ecology*, **62**:816–29.

FAO 1957 Shifting cultivation. *Tropical Agriculture*, **34**:159–64

FAO 1978 Report on the FAO/UNEP expert consultation on methodology for assessing soil degradation. Rome: FAO Project No. 1106–75–05

FAO-SIDA 1974 *Shifting cultivation and soil conservation in Africa*. Rome: FAO Soils Bulletin 24.

FAO-Unesco 1974 *Soil map of the world. Vol. 1 Legend*. Paris:Unesco.

FitzPatrick, E.A. 1983 *Soils: their formation, classification and distribution*. Harlow: Longman

Fitzpatrick, E.A. 1965 Climate of the Wabag-Tari area. In C.S.I.R.O. *Lands of the Wabag-Tari area, Papua New Guinea*. In Perry R.A. et al. General report on lands of the Wabag-Tari area, Territory of Papua New Guinea, 1960–61. Melbourne: C.S.I.R.O. Land Research Series No. 15 (pp. 56–69).

Fitzpatrick, E.A., Hart, D., & Brookfield, H.C. 1966 Rainfall seasonality in the tropical southwest Pacific. *Erdkunde*, **20**:181–94.

Flenley, J.R. 1969 The vegetation of the Wabag region, New Guinea Highlands: a numerical study. *Journal of Ecology*, **57**:465–490.

Floyd, C.N., D'Souza, E.J. & LeFroy, R.D.B. 1987a Composting and crop production on volcanic ash soils in the Southern Highlands of Papua New Guinea. Port Moresby: Department of Primary Industry Technical Report 87/6.

Floyd, C.N., D'Souza, E.J. & Lefroy, R.D.B. 1987b Phosphate and potash fertilisation of sweet potato on volcanic ash soils in the Southern Highlands of Papua New Guinea. Konedobu: Department of Primary Industry Technical Report 87/7.

Floyd, C.N., Lefroy, R.D.B. & D'Souza, E.J. 1988 Soil fertility and sweet potato production on volcanic ash soils in the highlands of Papua New Guinea. *Field Crops Research*, **19**:1–25.

Fox, R.L., Hashimoto, R.K., Thompson, J.R. & de la Pena, R.S 1974 Comparative external phosphorus requirements of plants growing in tropical soils. *Tenth International Congress of Soil Science* (Moscow), **4**:232–39.

Fujise, K. & Tsuno, Y. 1967 The effect of potassium on dry matter production of sweet potato. *Proceedings of the International Symposium on Tropical Root Crops*, **1**:20–33.

Furbee, L. 1989 A folk expert system: soil classification in the Colca Valley, Peru. *Anthropological Quarterly*, **62**:83–102.

Gagné, W.C. 1977 Entomological investigations of agrosilviculture using the composted contour systems in Papua New Guinea. *Science in New Guinea*, **5**:85–101.

Gagné, W.C. 1982 Staple crops in subsistence agriculture: their major insect pests, with emphasis on biogeographical and ecological aspects. In J.L. Gressitt (ed.) *Biogeography and ecology of New Guinea* vol. 1. The Hague: W.Junk Pubs. *Monographiae Biologicae* Series vol. 42 (pp. 229–259).

Gebhardt, H. & Coleman, N.T. 1974 Anion adsorption by allophanic tropical soils. III. Phosphate adsorption. *Proceedings of the Soil Science Society of America*, **38**:255–61.

Gillison, A.N. 1969 Plant succession in an irregularly fired grassland area — Doma Peaks region, Papua. *Journal of Ecology*, **57**:415–429.

Gillison, A.N. 1970 Structure and floristics of a montane grassland/forest transition, Doma Peaks region, Papua. *Blumea*, **18**:71–86.

Glasse, R.M. 1963 Bingi at Tari. *Journal of the Polynesian Society*, **72**:270–71.

Godelier, M. & Strathern M. (eds.) 1991 *Big men and great men: personifications of power in Melanesia*. Cambridge: Cambridge University Press.

Gollifer, D.E. 1980 A time of planting trial with sweet potatoes. *Tropical Agriculture*, **57**:363–367.

Golson, J. 1982 The Ipomoean revolution revisited: society and the sweet potato in the upper Wahgi valley. In A.J. Strathern (ed.) *Inequality in New Guinea Highlands societies*. Cambridge: Cambridge University Press (pp. 109–136).

Golson, J. & Gardner, D.S. 1990 Agriculture and sociopolitical organization in New Guinea Highlands prehistory. *Annual Review in Anthropology*, **19**:395–417.

Golson, J. & Hughes, P.J. 1976 The appearance of plant and animal domestication in New Guinea. In J. Garanger (ed.) *La préhistoire océanienne*. Nice: IXe Congrès de l'Union Internationale des Sciences Préhistoriques et Protohistoriques (pp. 88–100).

Goodbody, S. & Humphreys, G.S. 1986 Soil chemical status and the prediction of sweet potato yields. *Tropical Agriculture*, **63**(2):209–11.

Greenland, D.J. 1977 Soil structure and erosion hazard. In D.J. Greenland & R. Lal *Soil conservation and management in the humid tropics*. Chichester: Wiley (pp. 17–23).

Greenland, D.J. & Nye, P.H. 1959 Increase in the carbon and nitrogen contents of tropical soils under natural fallows. *Journal of Soil Science*, **10**:284–99.

Greenland, D.J. & Okigbo, B.D. 1983 Crop production under shifting cultivation and the maintenance on soil fertility. *International Rice Research Institute Symposium on Potential Productivity of Field Crops under Different Environments*, Los Banos, Philippines 1980:505–24.

Grubb, P.J. 1977 Control of forest growth and distribution on wet tropical mountains with special reference to mineral nutrition. *Annual Review of Ecological Systems*, **8**:83-107.

Grubb, P.J. & Edwards, P.J. 1982 Studies of mineral cycling in a montane rain forest in New Guinea. III. The distribution of mineral elements in the above-ground material. *Journal of Ecology*, **70**:623–648.

Guillet, D. 1989 A knowledge-based-systems model of native soil management. *Anthropological Quarterly*, **62**:59–67.

Guillet, D.W., Furbee, L., Sandor, J. & Benfer, R. 1995 The Lari soils project in Peru — a methodology for combining cognitive and behavioural research. In D.M. Warren, L.J. Slikkerveer & D. Brokensha (eds.) *The cultural dimension of development*. London: Intermediate Technology Publications (pp. 71–81).

Haantjens, H.A. & Bleeker, P. 1970 Tropical weathering in the Territory of Papua and New Guinea. *Australian Journal of Soil Research*, **8**:157–77.

Haantjens, H.A. & Rutherford, G.K. 1967 Soil zonality and parent rock in a very wet tropical region. *Proceedings of the 8th. International Congress of Soil Science, Bucharest 1964*, **5**:493–500.

Haberle, S. 1991 Ethnobotany of the Tari basin, Southern Highlands Province, Papua New Guinea. Canberra: Australian National University Department of Biogeography & Geomorphology.

Hackett, C. 1985 *National relationships between plants and the environment: an aid to broad-scale evaluation of land for plant production. 1. Sweet potato* (Ipomoea batatas). Canberra: CSIRO Institute of Biological Resources Technical Memorandum 85/5.

Hahn, S.K. 1977 Sweet potato. In P.T. Alvim & T.T. Kozlowski (eds.) *Ecophysiology of tropical crops*. New York: Academic Press (pp. 237–248).

Hammett, L.K., Constantin, R.J., Jones, L.G. & Hernandez, T.P. 1982 The effect of phosphorus and soil moisture levels on yield and processing quality of 'Centennial' sweet potatoes. *Journal of the American Society of Horticultural Science*, **107**:119–122.

Harrison, S. 1993 The commerce of cultures in Melanesia. *Man*, **28**:139–158.

Hays, T.E. 1979 Plant classification and nomenclature in Ndumba, Papua New Guinea Highlands. *Ethnology*, **18**:253–270.

Hays, T.E. 1982 Utilitarian/adaptionist explantations of folk biological classifications: some cautionary notes. *Journal of Ethnobiology*, **2**:89–94.

Hays, T.E. 1991 Interest, use, and interest in uses in folk biology. In A.Pawley (ed.) *op. cit.* (pp. 109–114).

Healey, C.J. 1979 Taxonomic rigidity in biological folk classification: some examples from the Maring of New Guinea. *Ethnomedizin*, **5**:361–383.

Henty, E.E. (ed.) 1981 *Handbooks of the flora of Papua New Guinea*, vol. 2. Melbourne: University Press.

Henty, E.E. 1982 Grasslands and grassland succession in New Guinea. In J.L. Gressitt (ed.) *op. cit.* (pp. 459–473).

Henty, E.E. & Pritchard, G.S. 1973 *Weeds of New Guinea and their control*. Lae: Dept. Forests, Botany Bulletin No. 7.

Hide, R. *et al.* 1979 A checklist of some plants in the territory of the Sinasina Nimai (Simbu Province, Papua New Guinea), with notes Auckland: Anthropology Dept., Auckland University: Working Papers in anthropology, archaeology, linguistics, Maori Studies 54.

Hide, R.L., Goodbody, S. & Gertru, G. 1984 Agriculture. In R.L. Hide (ed.) *South Simbu: studies in demography, nutrition and subsistence. Research report of the Simbu land use project Vol. VI*. Waigani: Institute of Applied Social & Economic Research (pp.207–289).

Hill, K.C. 1991 Structure of the Papuan Fold Belt. *Bulletin of the American Association of Petroleum Geologists*, **75**:857–872.

Hill, W.A., Dodo, H., Hahn, S.K., Mulongoy, K. & Adeyeye, S.O. 1990 Sweet potato root and biomass production with and without nitrogen fertilization. *Agronomy Journal*, **82**:1120–22.

Hillel, D.J. 1992 *Out of the earth: civilization and the life of the soil*. Berkeley: University of California Press.

Hirsch, E. & O'Hanlon, M. (eds.) 1995 *The anthropology of landscape*. Oxford: Clarendon Press.

Hodgson, J.M. (ed.) 1976 *Soil survey field handbook: describing and sampling soil profiles*. Harpenden: Soil Survey Technical Monograph No. 5.

Hope, G. & Golson, J. 1995 Late Quaternary changes in the mountains of New Guinea. *Antiquity*, **69**:818–830.

Hozyo, Y. 1970 Growth and development of tuberous root in sweet potato. *Proceedings of the 2nd. International Symposium on Tropical Root and Tuber Crops, Hawaii* **1**:24.

Hudson, N.W. 1981 *Soil conservation*. London: Batsford.

Humphreys, G.S. 1984 The environment and soils of Chimbu Province, Papua New Guinea with particular reference to soil erosion. Port Moresby: Department of Primary Industry Research Bulletin No. 35.

Humphreys, G.S. 1990 Soil maps of Papua New Guinea: a review. *Science in New Guinea*, **17**(2):77–96.

Hunn, E.S. 1982 The utilitarian factor in folk biological classification. *American Anthropologist*, **84**:830–847.

Huypers, H., A. Mollema & E. Topper, 1987 *Erosion control in the tropics.* Wageningen: Agromisa (Agrodok 11).

Hyndman, D. 1991 The Kam Basin homeland of the Wopkaimin: a sense of place. In A. Pawley (ed.) *op. cit.* (pp. 256–265).

Hynes, R.A. 1974 Altitudinal zonation in New Guinea *Nothofagus* forest. In J.R. Flenley (ed.) *Altitudinal zonation in Malesia.* Hull: Hull University Geography Department Miscellaneous Series, 16, (pp. 75–109).

Inglis, J.T. (ed.) 1993 *Traditional ecological knowledge: concepts and cases.* Ottawa: International Development Research Centre.

Jaiyebo, E.O. & Moore, A.W. 1964 Soil fertility and nutrient storage in different soil-vegetation systems in a tropical rainforest environment. *Tropical Agriculture*, **41**:129–39.

Jenkinson, D.S. & Rayner, J.H. 1977 The turnover of soil organic matter in some of the Rothampstead classical experiments. *Soil science*, **123**:298–305.

Jennings, J.N. & Bik, M.J.J. 1962 Karst morphology in Australian New Guinea. *Nature*, **194**:1036–38.

Jenny, H. 1941 *Factors of soil formation.* New York: McGraw-Hill.

Jenny, H., Bingham, F.T. & Padilla-Saravia, B. 1948 Nitrogen and organic matter contents of equatorial soils of Colombia, South America. *Soil Science*, **66**:173–186.

Johns, R.J. 1976 A provisional classification of montane vegetation of Papua New Guinea. *Science in New Guinea*, **4**:105–127.

Johns, R.J. 1980 Notes on the forest types of Papua New Guinea. Part I: The *Dacrydium* swamp forest of the Southern Highlands. *Klinki*, **1**:3–9.

Johns, R.J. 1982 Plant zonation. In J.L. Gressitt (ed.) *op. cit.* (pp.309–330).

Johnson, R.W. 1976 Late Cainozoic volcanism and plate tectonics at the southern margin of the Bismarck Sea, Papua New Guinea. In Johnson, R.W. (ed) *Volcanism in Australasia.* Amsterdam: Elsevier. pp. 101–16.

Jordan, C.F. 1985 *Nutrient cycling in tropical forest ecosystems.* New York: Wiley.

Jordan, C.F. 1989 *An Amazonian rain forest: the structure and function of a nutrient stressed ecosystem and the impact of slash-and-burn agriculture.* Paris: UNESCO Press.

Juo, A.S.R. & Lal, R. 1977 Effects of fallow and continuous cultivation on chemical and physical properties of an Alfisol in southern Nigeria. *Plant & Soil*, **47**:567–84.

Juo, A.S.R. & Kang, B.T. 1989 Nutrient effects of modification of shifting cultivation in west Africa. In J. Proctor (ed.) *Mineral nutrients in tropical forest and savanna ecosystems.* Oxford: Blackwell (pp. 289–300).

Kalkman, C. & Vink, W. 1970 Botanical exploration in the Doma peaks region, New Guinea. *Blumea*, **18**:88–135.

Kalma, J.D. 1972 Solar radiation over New Guinea and adjacent islands. *Australian Meteorological Magazine*, **20**:116–27.

Kang, B.T. & Juo, A.S.R. 1986 Effects of forest clearing on soil chemical properties and crop performance. In R. Lal, P. Sanchez & R.W. Cummings (eds.) *Land clearing and development in the tropics.* Rotterdam: A.A. Balkema (pp. 383–94).

Kanua, M.B. & Floyd, C.N. 1985 A study of site x variety interactions for sweet potato in the highlands of Papua New Guinea. Konedobu: Department of Primary Industry Technical Report 85/17.

Kays, S.J. 1985 The physiology of yield in the sweet potato. In. J.C. Bouwkamp (ed.) *op. cit.* (pp. 80–132).

Keig, G., Fleming, P.M. & McAlpine J.R. 1979 Evaporation in Papua New Guinea. *Jour. Tropical Geog.*, **48**:19–30.

Kellman, M.C. 1969 Some environmental components of shifting cultivation in upland Mindanao. *Journal of Tropical Geography*, **28**:40–56.

Kellogg, C.E. & Pendleton, R.L. 1948 Soil conservation. *FAO Agriculture Studies*, No. 4.

Kerven, C., Dolva, H. & Renna, R. 1995 Indigenous soil classification systems in Northern Zambia. In D.M. Warren, L.J. Slikkerveer & D. Brokensha (eds.) *The cultural dimension of development*. London: Intermediate Technology Publications (pp. 82–87).

Kimber, A.J. 1970 Some cultivation techniques affecting yield response in sweet potato. In D.L. Plucknett (ed.) *Proceedings of the Second International Symposium on Tropical Root and Tuber Crops*. Honolulu: University of Hawaii (pp. 32–36).

Kimber, A.J., 1971 Cultivation practices with sweet potato. *Harvest*, 1(1):31–33.

Kimber, A.J., 1974 Crop rotations, legumes and more productive arable farming in the highlands of Papua New Guinea. *Science in New Guinea*, 2(1):70–79.

King, G.A. 1985 The effect of time of planting on yield of six varieties of sweet potato *(Ipomoea batatas)* in the southern coastal lowlands of Papua New Guinea. *Tropical Agriculture*, 62:225–228.

Kinnel, P.I.A. 1981 Rainfall intensity: kinetic energy relationships for soil loss prediction. *Soil Science Society of America Proceedings*, 45:153–155.

Knight, J. 1992 Anthropology and environmentalism: the 1992 ASA conference. *Anthropology Today*, 8(3):15–17.

Kocher Schmid, C. 1991 *Of people and plants: a botanical ethnography of Nokopo village, Madang and Morobe Provinces, Papua New Guinea*. Basel: Basler Beiträge zur Ethnologie, Band 33.

Krebs, J.E. 1975 A comparison of soil under agriculture and forests in San Carlos, Costa Rica. In F. Golley & E. Medina (eds.) *Tropical ecological systems*. New York: Springer (pp. 381–90).

Lal, R. 1974 Soil erosion and shifting cultivation. In FAO *Shifting cultivation and soil conservation in Africa*. Rome: FAO Regional Seminar Ibadan (pp. 48–71).

Lal, R. 1977 Analysis of factors affecting rainfall erosivity and soil erodibility. In D.J. Greenland & R. Lal (eds.) *Soil conservation and management in the humid tropics*. New York: Wiley (pp. 49–56).

Lal, R. 1990 *Soil erosion in the tropics: principles and management*. New York: McGraw-Hill.

Lambert, J.D.H. & Arnason, J.T. 1986 Nutrient dynamics in milpa agriculture and the role of weeds in initial stages of secondary succession in Belize, Central America. *Plant & Soil*, 93:303–22.

Lambert, J.D.H. & Arnason, J.T. 1989 Role of weeds in nutrient cycling in the cropping phase of milpa agriculture in Belize, Central America. In J. Proctor (ed.) *op. cit.* (pp. 301–13).

Landon, J.R. (ed.) 1991 *Booker tropical soil manual*. Harlow: Longman Scientific & Technical.

Landsberg, J. & Gillieson, D.S. 1980 *Toksave bilong graun*: common sense or empiricism in folk soil knowledge from Papua New Guinea. *Capricornia*, 8:13–23.

Laudelot, H. 1961 *Dynamics of tropical soils in relation to their fallowing techniques*. Rome: FAO Paper 11266/E.

Lavelle, P. 1974 Les vers de terre de la savane de Lamto. In *Analyse d'un écosystemè tropical humide: la savane de Lamto*. Bull. de Liason des Chercheurs de Lamto. No. spéc., 5:133–166.

Lederman, R. 1986 *What gifts engender: social relations and politics in Mendi, Highland Papua New Guinea*. Cambridge.

Lee, K.E. 1967 Microrelief features in a humid tropical lowland area, New Guinea, and their relation to earthworm activity. *Australian Journal of Soil Research*, 5:263–274.

Lee, K.E. 1983 Earthworms of tropical regions — some aspects of their ecology and relationships with soils. In J. E. Satchell (ed.) *Earthworm ecology: from Darwin to vermiculture*. London: Chapman & Hall (pp. 179–193).

Leng, A.S. 1982 Maintaining fertility by putting compost into sweet potato mounds. *Harvest*, 8(2):83–84.

Lewis, G. 1980 *Day of shining red: an essay on understanding ritual*. Cambridge University Press.

Löffler, E. 1977 *Geomorphology of Papua New Guinea*. Canberra: CSIRO & Australian National University Press.

Manner, H.I. 1981 Ecological succession in new and old swiddens of montane Papua New Guinea. *Human Ecology*, 9:359–77.

Manner, H.I. 1982 Ecological perspectives on the intensification of subsistence agricultural systems: with special reference to Papua New Guinea. In R.M. Bourke & V. Kesavan (eds.) *op. cit.* (pp. 208–17).
Mawe, T. 1982 *Mendi culture and tradition: a recent survey*. Waigani: National Museum and Art Gallery of P.N.G.
McAlpine, J.R., Keig, G. & Short, K. 1975 *Climatic tables for Papua New Guinea*. CSIRO, Division of Land Use Research Technical Paper No. 37.
McAlpine, J.R., Keig, G. & Falls, R. 1983 *Climate of Papua New Guinea*. Canberra: CSIRO & Australian National University Press.
Meggitt, M.J. 1958 Mae Enga time-reckoning and calendar, New Guinea. *Man*, **58**:74–77.
Merlan, F. & Rumsey, A. 1991 *Ku Waru: language and segmentary politics in the western Nebilyer valley, Papua New Guinea*. Cambridge: University Press.
Midgley, S.J., Turnbull, J.W. & Johnston, R.D. 1983 *Casuarina ecology, management and utilization*. Melbourne: CSIRO.
Milton, K. (ed.) 1993 *Environmentalism: the view from anthropology*. London: Routledge.
Mishra, B.K. & Ramakrishnan, P.S. 1983a Secondary succession subsequent to slash and burn agriculture at higher elevations of north-east India. I. Species diversity, biomass and litter production. *Acta Oecologica-Oecologia Applicata*, **4**:95–107.
Mishra, B.K. & Ramakrishnan, P.S. 1983b Secondary succession subsequent to slash and burn agriculture at higher elevations of north-east India. II. Nutrient cycling. *Acta-Oecologica-Oecologia Applicata*, **4**:237–45.
Mishra, B.K. & Ramakrishnan, P.S. 1983c Slash and burn agriculture at higher elevations in north-eastern India. I. Sediment, water and nutrient losses. *Agriculture, Ecosystem and Environment*, **9**:69–82.
Mishra, B.K. & Ramakrishnan, P.S. 1983d Slash and burn agriculture at higher elevations in north-eastern India. II. Soil fertility changes. *Agriculture, Ecosystem and Environment*, **9**:83–96.
Modjeska, C.N. 1982 Production and inequality: perspectives from Central New Guinea. In A.J. Strathern (ed.) *op. cit.* (pp. 50–108).
Mohr, E.C.J., van Baren, F.A. & van Shuylenborgh, J. 1972 *Tropical soils: a comprehensive study of their genesis*. The Hague: Mouton.
Morauta, L., J. Pernetta & W. Heaney (eds.) 1982 *Traditional conservation in Papua New Guinea: implications for today*. Waigani: IASER (pp. 93–114).
Morellato, L.P.C. 1992 Nutrient cycling in two south-east Brazilian forests. I Litterfall and litter standing crop. *Journal of Tropical Ecology*, **8**:205–215.
Morgan, R.P.C. 1986 *Soil erosion and conservation*. Harlow: Longman.
Mueller-Harvey, I., Juo, A.S.R. & Wild, A. 1989 Mineralization of nutrients after forest clearance and their uptake during cropping. In J. Proctor (ed.) *op. cit.* (pp. 315–24).
Mueller-Harvey, I., Juo, A.S.R. & Wild, A. 1985 Soil organic C, N, S and P after forest clearance in Nigeria: mineralization rates and spatial variability. *Journal of Soil Science*, **36**:585–91.
Mulongoy, K., Callens, A. & Okogun, J.A. 1988 Differnces in mycorrhizal infection and P uptake of sweet potato cultivars (*Ipomoea batatas* L.) during their early growth in three soils. *Biology & Fertility of Soils*, **7**:7–10.
Nakano, K. 1978 An ecological study of swidden agriculture in Northern Thailand. *South East Asian Studies*, **16**:411–46.
Nakano, K. & Syahbuddin, 1989 Nutrient dynamics in forest fallows in South-East Asia. In J. Proctor (ed.) *op. cit.* (pp. 325–36).
Newton, K. 1960 Shifting cultivation and crop rotation in the tropics. *Papua New Guinea Agriculture Journal*, **13**:81–118.
Ngeve, J.M. & Roncadori, R.W., 1985 The interaction of vesicular-arbuscular mycorrhizae and soil phosphorus fertility on growth of sweet potato (*Ipomoea batatas*). *Field Crops Research*, **12**:181–85.
Nicholaides, J.J., Chancy, H.F., Mascagni, H.J., Wilson, L.G. & Eaddy, D., 1985 Sweet potato response to K and P fertilization. *Agronomy Journal*, **77**:466–70.

Nye, P.H. 1961 Organic matter and nutrient cycle under moist tropical forest. *Plant & Soil*, 13:333–346.

Nye, P. H. & Greenland, D.J. 1960 *The soil under shifting cultivation.* Harpenden: Commonwealth Bureau of Soils Technical Bulletin No. 51.

Ofori, C.S. 1973 Decline in fertility status in a tropical ochrosol under continuous cropping. *Experimental Agriculture*, 9:15–22.

Oldfield, F., Appleby, P.G. & Thompson, R. 1980 Palaeoecological studies of lakes in the highlands of Papua New Guinea. *Journal of Ecology*, **68**:457–477.

Oldfield, F., Worsley, A.T. & Baron, A.F. 1985a Lake sediments and evidence for agricultural intensification: a case study form the highlands of Papua New Guinea. In I.S. Farington (ed.) *Prehistoric intensive agriculture in the tropics.* BAR International Series S232 (pp; 385–391).

Oldfield, F., Appleby, P.G. & Worsley, A.T. 1985b Evidence from lake sediments for recent erosion rates in the highlands of Papua New Guinea. In I. Douglas & E. Spencer (eds.) *Environmental change & tropical geomorphology.* London: Allen & Unwin (pp. 185–195).

Ollier, C.D., Drover, D.P. & Godelier, M. 1971 Soil knowledge amongst the Baruya of Wonenara, New Guinea. *Oceania*, **42**:33–41.

Paglau, M. 1982 Conservation of soil, water and forest in Upper Simbu valley (trans. A. Goie). In L. Morauta, J. Pernetta & W. Heaney (eds.) *op. cit.* (pp. 115–119).

Paijmans, K. (ed.) 1976 *New Guinea vegetation.* Canberra: ANU Press & CSIRO.

Paijmans, K. 1970 An analysis of four tropical rain forest sites in New Guinea. *Journal of Ecology*, **58**:77–101.

Pain, C.F. & Blong, R.J. 1976 Late Quaternary tephras around Mt. Hagen & Mt. Giluwe, Papua New Guinea. In R.W. Johnson (ed.) *op. cit.* (pp. 239–51).

Papua New Guinea Dept. Agriculture, Stock & Fisheries 1959/60–1967/69. *Annual Reports.* Konedobu: Govt. Printer.

Parfitt, R.L. 1976 Shifting cultivation: how it affects the soil environment. *Harvest*, 3:2.

Parfitt, R.L. & Mavo, B. 1975 Phosphate fixation in some Papua New Guinea soils. *Science in New Guinea*, 3:179–90.

Pearson, M.N. 1982 A review of virus and mycoplasma diseases of food crops in Papua New Guinea. In R.M. Bourke & V. Kesavan (eds.) *op. cit.* (pp. 448–57).

Perry, R.A. 1965 Outline of the geology and geomorphology of the Wabag-Tari area. In R.A. Perry et al. *op. cit.* (pp. 70–84).

Philips-Howard, K.D. & Kidd, A.D. 1991 Knowledge and management of soil fertility among dry season farmers on the Jos Plateau, Nigeria. Durham University Geography Department Jos Plateau Environmental Programme Report No. 25.

Pieters, P.E. 1982 Geology of New Guinea. In J.L. Gressitt (ed.) *op. cit.*

Pivello, V.R. & Couthino, L.M. 1992 Transfer of macro-nutrients to the atmosphere during experimental burning in an open *cerrado* (Bravilian savanna). *Journal of Tropical Ecology*, 8:487–497.

Popenoe, H.L. 1959 The influence of the shifting cultivation cycle on soil peoperties in Cental America. *Proceedings of the Ninth Pacific Science Congress, Bangkok*, 1:72–77.

Powell, J.M. 1976a "Ethnobotany" in K. Paijmans (ed.) *op. cit.* (pp. 106–183).

Powell, J.M.1976b "Some useful wild and domesticated plants of the Huli of Papua" *Science in New Guinea*, 4(3):173–201.

Preston, S.R. 1990 Investigation of compost * fertilizer interactions in sweet potato grown on volcanic ash soils in the highlands of Papua New Guinea. *Tropical Agriculture*, 67(3):239–42.

Quigley, D. 1993 Raymond Firth on social anthropology (an interview). *Social Anthropology*, 1(2):207–222.

Radcliffe, D.J. 1985a The land resources of Kuma experimental farm. Konedobu: Dept. Primary Industry Tech. Report No. 85/11.

Radcliffe, D.J. 1985b The land resources of Kiburu experimental farm. Konedobu: Dept. Primary Industry Tech. Report No. 85/9.

Radcliffe, D.J. 1986 The land resources of Upper Mendi. (2 vols) Konedobu: Dept. Primary Industry Research Bulletin No. 37.

Radcliffe, D.J. & Gillman, G.P. 1985 Surface charge characteristics of volcanic ash soils from the Southern Highlands of Papua New Guinea. In E. Fernandez Caldas & D.H. Yaalon (eds.) *Volcanic soils: weathering and landscape relationships of soils on tephra and basalt*. Cremlington: Catena Supplement 7 (pp. 35–46).

Ramakrishnan, P.S. 1984 The sciences behind rotational bush fallow agriculture systems (*jhum*). *Proceedings of the Indian Academy of Sciences (Plant Sciences)*, **93**:379–400.

Ramakrishnan, P.S. 1989 Nutrient cycling in forest fallows in north-eastern India. In J. Proctor (ed.) *op. cit.* (pp. 337–52).

Ramakrishnan, P.S. & Toky, O.P. 1981 Soil nutrient status of hill agro-ecosystems and recovery pattern after slash and burn agriculture (*jhum*) in north-eastern India. *Plant & Soil*, **60**:41–64.

Rambali, P. 1993 *It's all true: in the cities and jungle of Brazil*. London: Heinemann.

Rappaport, R.A. 1968 *Pigs for the ancestors: ritual in the ecology of a New Guinea people*. New Haven: Yale University Press.

Rappaport, R.A. 1972 The flow of energy in an agricultural society. In J.G.Jorgensen (ed.) *Biology and culture in modern perspective* (Readings from *Scientific American*). San Francisco: Freeman (pp. 345–356).

Raumolin, J. (ed.) 1987 Special issue on swidden cultivation. *Suomen Anthropologi*, **12** (4).

Reay, M. 1961 Mushroom madness in New Guinea. *Oceania*, **31**:137–139.

Reynders, J.J. 1961 Some remarks about shifting cultivation in Netherlands New Guinea. *Netherlands Journal of Agricultural Science*, **9**:36–40.

Richards, P. 1985 *Indigenous agricultural revolution*. London: Hutchinson.

Robbins, R.G. 1960 The anthropogenic grasslands of Papua and New Guinea. In UNESCO Symposium on the impact of man on humid tropics vegetation, Goroka (pp. 313–329).

Robbins, R.G. & Pullen, R. 1965 Vegetation of the Wabag–Tari area. In R.A. Perry *et al. op. cit.* (pp. 100–115).

Roberts, G. 1982 Food shortages, Tombudu valley, Southern Highlands. In R.M. Bourke & V. Kesavan (eds.) *op. cit.* (pp. 279–81).

Robison, D.M. & McKean, S.J. 1992 *Shifting cultivation and alternatives: an annotated bibliography 1972–89*. Wallingford: C.A.B International (with CIAT).

Rose, C.J. & Wood, A.W. 1980 Some environmental factors affecting earthworm populations and sweet potato production in the Tari Basin, Papua New Guinea Highlands. *Papua New Guinea Agricultural Journal*, **31**:1–13.

Rutherford, G.K. & Haantjens, H.A. 1965 Soils of the Wabag-Tari area. In R.A. Perry *et al. op. cit.* (pp. 85–99).

Ruxton, B.P. 1967 Slopewash under mature primary rainforest in northern Papua. In J.N. Jennings & J.A. Marbutt (eds.) *Landform studies from Australia and New Guinea*. Canberra: Australian National University Press (pp. 85–94).

Ryan, D'A. 1959 Clan formation in the Mendi valley. *Oceania*, **29**:257–90.

Ryan D'A. 1961 Gift exchange in the Mendi valley. Ph.D. Thesis, Sydney.

Sabhasri, S. 1978 Effects of forest fallow cultivation on forest production and soil. In P. Kunstadter, E.C. Chapman & S. Sabhasri (eds.) *Farmers in the forest: economic developments and marginal agriculture in Northern Thailand*. Honolulu: Univ. Hawaii Press (pp. 160–84).

Sajjapongse, A. & Wu, M.H. 1989 Effect of some climatic factors on sweet potato yield. *Kasetsart Journal (Natural Science)*, **23**:86–92.

Sanchez, P.A. 1973 Soil management under shifting cultivation. In P.A. Sanchez (ed.) *A Review of soils research in tropical Latin America*. North Carolina Agriculture Experiment Station Technical Bulletin No. 219.

Sanchez, P.A. 1976 *Properties & management of soils in the tropics* New York: Wiley.

Saxena, K.G. & Ramakrishnan, P.S. 1984 Hebaceous vegetation development and weed potential in slash and burn agriculture (*jhum*) in N.E. India. *Weed Research*, **24**:135–42.

Saxena, K.G. & Ramakrishnan, P.S. 1986 Nitrification during slash and burn agriculture (*jhum*) in north-eastern India. *Acta Oecologica-Oecologia Plantarum*, **7**:319–31.

Scott, D.A., Proctor, J. & Thompson, J. 1992 Ecological studies on a lowland evergreen rain forest on Maraca Island, Roraima, Brazil. II Litter and nutrient cycling. *Journal of Ecology*, **80**:705–717.

Scott, L.E. & Bouwkamp, J.C. 1974 Seasonal mineral accumulation by the sweet potato. *HortScience,* **9**(3):233–35.
Shaw, D.E. 1984 Microorganisms in Papua New Guinea. Port Moresby: Dept. of Primary Industry Research Bulletin No. 33.
Siebert, S. 1987 Land use intensification in tropical uplands: effects on vegetation, soil fertility and erosion. *Forest Ecology and Management,* **21**:37–56.
Sillitoe, P. 1979 *Give & take: exchange in Wola society.* New York: St. Martin's.
Sillitoe, P. 1981 Dance with the cassowaries. *The Geographical Magazine,* **53**:534–38.
Sillitoe, P. 1983 *Roots of the earth: crops in the highlands of Papua New Guinea.* Manchester: Manchester University Press.
Sillitoe, P. 1987 Sorcery divination among the Wola. In M. Stephens (ed.) *Sorcerer and witch in Melanesia.* Rutgers University Press (pp. 121–46).
Sillitoe, P. 1988 *Made in Niugini: technology in the highlands of Papua New Guinea.* London: British Museum Publications.
Sillitoe, P. 1994 *The Bogaia of the Muller Ranges, Papua New Guinea: land use, agriculture and society of a vulnerable population.* Sydney: Oceania Monographs No. 44.
Skeldon, R. 1977 Volcanic ash, hailstones and crops: oral history from the Eastern Highlands of Papua New Guinea. *Journal of the Polynesian Society,* **86**:403–9.
Smith, J.M.B. 1975 Mountain grasslands of New Guinea. *Jour. Biogeography,* **2**:27–44.
Smith, E.S.C. & Thistleton, B.M. 1982 Some common pests of vegetables and fruit in Papua New Guinea and their control. In R.M. Bourke & V. Kesavan (eds.) *op. cit.* (pp. 476–82).
Sowden, F.J., Griffith, S.M. & Schnitzer, M. 1976 The distribution of nitrogen in some highly organic tropical volcanic soils. *Soil Biology & Biochemistry,* **8**:55–60.
Spenceley, A.P. 1980 Garden and fallow plants of the Nembi plateau. *Science in New Guinea,* **7**:47–56.
Stamps, D.J., Shaw, D.E. & Cartledge, E.G. 1972 Species of Phytophthora and Pythium in Papua New Guinea. *Papua New Guinea Journal of Agriculture,* **23**:41–45.
Steensberg, A. 1980 *New Guinea gardens.* London: Academic Press.
Sterly, J. 1974 "Useful plants of the Chimbu" *Ethnomedizin,* **3**(3/4):353–90.
Sterly, J. 1977 Research work on traditional plantlore and agriculture in the Uper Chimbu Region, Papua New Guinea. *Bulletin of the International Committee on Urgent Anthropological and Ethnological Research,* **19**:95–114.
Straatmans, W. 1967 Ethnobotany of New Guinea in its ecological perspective. *Journal d'Agriculture Tropicale,* **14**:1–20.
Strathern, A.J. 1969 Finance and production: two strategies in New Guinea Highlands exchange systems. *Oceania,* **40**:42–67.
Strathern, A.J. 1970 The female and male spirit cults in Mount Hagen. *Man,* **5**:571–85.
Strathern, A.M. 1972 *Women in between: female roles in a male world.* London: Seminar Press.
Strathern, A.M. 1988 *The gender of the gift: problems with women and problems with society in Melanesia.* Berkeley: University of California Press.
Strauss, M. & Tischner, H. 1962 *Die Mi-Kultur der Hagenberg-stamme.* Hamburg: Cram, de Gruyter & Co.
Swamy, P.S. & Ramakrishnan, P.S. 1988 Ecological implications of traditional weeding and other imposed weeding regimes under slash and burn agriculture (*jhum*) in north-eastern India. *Weed Research,* **28**:127–36.
Taylor, G.A.M. 1971 An investigation of volcanic activity at Doma Peaks. *Aust. Bur. Miner. Resour. Records* 1971: 137 (pp. 15).
Tergas, L.E. & Popenoe, H.L. 1971 Young secondary vegetation and soil interactions in Izabel, Guatemala. *Plant & Soil,* **34**:675–90.
Thiagalingam, K. 1983 Role of casuarina in agroforestry. In S.J. Midgley, J.W. Turnbull & R.D. Johnson (eds.) *Casuarina ecology, management and utilization.* Melbourne: CSIRO.
Thiagalingam, K. & Bourke R.M. 1982 Utilization of organic wastes in crop production. In R.M. Bourke & V. Kesavan (eds.) *op. cit.* (pp. 218–26).

Thiagalingam, K. & Famy, F.N. 1981 The role of *Casuarina* under shifting cultivation: a preliminary study. In R. Wetselaar, J.R. Simpson & T. Rosswall (eds.) *Nitrogen cycling in South-East Asian wet monsoonal ecosystems*. Canberra: Australian Academy of Science (pp. 154–156).

Thistleton, B.M. & Masandur, R.T. 1985 *Surveys of insects associated with food crops in three study areas in the Southern Highlands Province*. Konedobu: Department of Primary Industry Research Bulletin No. 36.

Thurston, H.D. 1991 *Sustainable practices for plant disease management in traditional farming systems*. Boulder: Westview Press.

Togari, Y. 1950 A study on the tuberous root formation of sweet potato. *Bulletin of the National Agriculture Experimental Station* (Tokyo), **68**:1–96.

Toky, O.P. & Ramakrishnan, P.S. 1981 Run-off and infiltration losses related to shifting cultivation (*jhum*) in north-eastern India. *Environmental Conservation*, **8**:313–21.

Toky, O.P. & Ramakrishnan, P.S. 1983a Secondary succession following slash and burn agriculture in north-eastern India. I. Biomass, litterfall and productivity. *Journal of Ecology*, **71**:735–45.

Toky, O.P. & Ramakrishnan, P.S. 1983b Secondary succession following slash and burn agriculture in north-eastern India. II. Nutrient cycling. *Journal of Ecology*, **71**:747–57.

Trenbath, B.R. 1989 The use of mathematical models in the development of shifting cultivation systems. In J. Proctor (ed.) *op. cit.* (pp. 353–69).

Tsuno, Y. 1970 *Sweet potato: nutrient physiology and cultivation*. Berne: International Potash Institute.

Tuneera-Bhadauria, T., Ramakrishnan, P.S., & Bhadauria, T. 1991 Population dynamics of earthworms and their activity in forest ecosystems of north-east India. *Journal of Tropical Ecology*, **7**:305–318.

Turvey, N.D. 1974 Water in the nutrient cycle of a Papuan rain forest. *Nature*, **251**:414–415.

Uhl, C. 1987 Factors controlling succession following slash-and-burn agriculture in Amazonia. *Journal of Ecology*, **75**:377–407.

Ungemach, H. 1969 Chemical rain water studies in the Amazon basin. In J.M. Idrobo (ed.) *Il simposio y foro de biologica tropical Amazonica, simposio del associacion pro Biologia Tropical*. Bogotá: Pax (pp. 354–358).

USDA. 1975 *Soil taxonomy: a basic system of soil classification for making and interpreting soil surveys*. Washington: U.S. Soil Conservation Service Agriculture Handbook No. 436.

Utomo, W.H. & Mahmud, N. 1984 The possibility of using the Universal Soil Loss Equation in mountainous areas of east Java with humus-rich andosols. *Proceedings of the Fifth ASEAN Soil Conference* E5.1–E5.13.

Vicedom, G.F. & Tischner, H. 1943–48 *Die Mbowamb*. Hamburg: Friederichsen, de Gruyter & Co.

Vitousek, P.M. & Sanford, R.L. 1986 Nutrient cycling in moist tropical forest. *Annual Review of Ecology & Systematics*, **17**:137–69.

Vine, H. 1954 Is the lack of fertility of tropical African soils exaggerated? *Proceedings of the 2nd Inter-Africa Soils Conference*, **1**:389–412.

Wada, K. 1977 Allophane and imogolite. In J.B. Dixon & S.B. Weed (eds.) *Minerals in soil environments*. Wisconsin: Soil Science Society of America (pp. 603–638).

Waddell, E. 1972 *The mound builders: agricultural practices, environment, and society in the Central Highlands of New Guinea*. Seattle: University of Washington Press.

Waddell, E. 1975 How the Enga cope with frost: responses to climatic perturbations in the central highlands of New Guinea. *Human ecology*, **3**:249–73.

Walker, D. 1966 Vegetation of the Lake Ipea region, New Guinea Highlands. I Forest, grassland and 'garden'. *Journal of Ecology*, **57**:503–533.

Walker, D. 1970 The changing vegetation of the Montane tropics. *Search*, **1**:217–221.

Walker, D. & Flenley, J.K. 1979 Late quaternary vegetational history of the Enga Province of upland Papua New Guinea. *Philosophical Transactions of the Royal Society of London*, **286**:265–344.

Wallace, K.B. 1971 Residual soils of the continually wet highlands of Papua New Guinea: a basic study of their occurrence and geotechnical properties. *Geotechnique*, **23**:203–218.

Waller, J.M. 1984 Diseases of small holder food crops in the Southern Highlands Province of Papua New Guinea. Results of survey conducted by the Commonwealth Mycological Institute for the World Bank Southern Highlands Development Project (AFTSEMU).

Warren D.M. 1991 Using indigenous knowledge in agricultural development. Washington: World Bank Discussion Paper No. 127.

Watson, J.B. 1963 Krakatoa's echo? *Journal of the Polynesian Society*, **72**:152–55.

Watters, R.F. & Bascones, L. 1971 The influence of shifting cultivation on soil properties at Altamira-Calderas, Venezuelan Andes. *FAO Forestry Development Paper*, **17**:291–99.

Watters, R.F. 1971 *Shifting cultivation in Latin America*. Rome: FAO Forestry Development Paper No. 17.

Webster R. 1977 *Quantitative and numerical methods in soil classification and survey* Oxford: Clarendon Press.

Wild, A. 1972 Mineralization of soil nitrogen at a savanna site in Nigeria. *Experimental Agriculture*, **8**:91–97.

Williams, A.R. 1982 Biological consevation techniques to help control soil erosion. In R.M. Bourke & V. Kesavan (eds.) *op. cit.* (pp. 227–36).

Williams, P.W. 1972 Morphometric analysis of polygonal karst in New Guinea. *Bulletin of the Geological Society of America*, **83**:761–96.

Wilson, L.A. 1982 Tuberization in sweet potato *(Ipomoea batatas)*. In R.L. Villareal & T.D. Griggs *Sweet potato: proceedings of the 1st international symposium*. Tainan (Taiwan): Asian Vegetable Research & Development Institute (pp. 79–94).

Wilson, L.A. & Lowe, S.B. 1973 Quantitative morphogenesis of root types in the sweet potato *(Ipomoea batatas* L.) root system during early growth from stem cuttings. *Tropical Agriculture*, **50**(4):343–45.

Wischmeier, W.H., Johnson, C.B. & Cross, B.V. 1971 A soil erodibility nomograph for farmland and construction sites. *Journal of Soil & Water Conservation*, **26**:189–93.

Wischmeier, W.H. & Smith, D.D. 1978 *Predicting rainfall erosion losses: a guide to conservation planning*. USDA Agriculture Handbook No. 537.

Wohlt, P.B. 1986 *Subsistence systems of Enga Province*. Division of Primary Industry, Department of Enga Province, Technical Bulletin no. 3.

Womersley, J.S. (ed.) 1978 *Handbooks of the flora of Papua New Guinea. vol. 1*. Melbourne: University Press.

Wood, A.W. 1979 The effects of shifting cultivation on soil properties: an example from the Kirimui & Borrai Plateaux, Simbu Province, Papua New Guinea. *Papua New Guinea Agricultural Journal*, **30**:1–9.

Wood, A.W. 1982 Food cropping systems in the Tari Basin. In R.M. Bourke & V. Kesavan (eds.) *op. cit.* (pp. 256–67).

Wood, A.W. 1984 Land for tomorrow: subsistence agriculture, soil fertility and ecosystem stability in the New Guinea Highlands. Ph.D. thesis, University of Papua New Guinea.

Wood, A.W. 1987 The humic brown soils of the Papua New Guinea Highlands: a reinterpretation. *Mountain Research and Development*, **7**:145–156.

Wood, A.W. 1991 Responses to land degradation by the Huli. Paper presented at New perspectives on the Papua New Guinea highlands; an interdisciplinary conference on the Duna, Huli and Ipili peoples, Canberra.

Wood, A.W. & Humphreys, G.S. 1982 Traditional soil conservation in Papua New Guinea. In L. Morauta, J. Pernetta & W. Heaney (eds.) *op. cit.* (pp. 93–114).

Wrightson, J. & Newsham, J.C. 1919 *Agriculture*. London: Crosby, Lockwood & Son.

Yen, D.E. 1974 *The sweet potato and Oceania: an essay in ethnobotany*. Honolulu: B.P. Bishop Museum Bulletin No. 236.

Young, A. 1976 *Tropical soils and soil survey*. Cambridge University Press.

Zinke, P.J., Sabhasri, S. & Kunstadter, P. 1978 Soil fertility aspects of the Lua' forest fallow system of shifting cultivation. In P. Kunstadter, E.C. Chapman & S. Sabhasri (eds.) *op. cit.*

INDEX

Abu-Zeid, M.O.   383
Acalypha sp.   177, 187, 224, 191
acephalous politics (*see also* stateless political order)   334–336
Aceratium tomentosum   57
Acock, M.C. and Garner, J.O.   365n
Acorus calamus   172
Acrisols   279–283
Adenostemma lavenia   192, 220
Africa   23, 37, 215, 401
AFTSEMU   69, 285, 289, 359
ages, men's   333–334
agriculture and prehistory   164, 398–400
Ahmad, N. and Breckner, E.   155
Aiyura   38, 84, 86, 364, 380, 386–387
Albizia fulva   249
Alfisols   19, 31, 37–38
Allan, W.   315
Allbrook, R.F. and Radcliffe, D.J.   148, 150
Allen, B.   77, 298, 315,
Allen, B. and Brookfield, H.C.   79
Allen, B. and Crittenden, R.   14, 139, 332, 392, 395
allophane   287, 353–354
alluvial soils   289–290, 361
Alocasia sp.   174, 211
alpine vegetation   188–190
aluminium (Al)   348–349
Alyxia sp.   184
Amynthas spp.   257
ancestor spirits   24, 215
Andepts   38, 283–289, 355
Andisols   271, 283–289
animal decomposers   253–256
anthropology, problems and distortions   13
anthropology and natural science   10–11, 13
Aquent   292–293
Aquept   292–293
Araucaria sp.   191, 251
archaeology and agriculture   164, 398–400

Arenosol   291
Arnason, T. *et al.*   36
Arrhenotus sp.   252
Arthraxon sp.   172, 193, 220
Artocarpus sp.   222
Ash, J.   201–203
ash, *see* volcanic ash
ash, nutrients   36
aspect, *see* site aspect
assimilate partition   366–368
Astelia sp.   188
Austin, M.E. *et al.*   47
Aweto, A.O.   34, 37

bacteria   249
Bain, J.H.C.   123
Barrow, C.J.   140
Barry, R.G.   60
Bartholomew, V. *et al.*   35
base pumping   239
Bayliss-Smith, T. and Golson, J.   79, 399
Bazzania sp.   183
beans   43
Beckerman, S.   29
Behrens, C.A.   265
Belize   388
Bena Bena   161
Berlin, B.   8, 171
Bernhard-Reversat, F.   234
Bidens pilosus   192, 220
Biersack, A.   23
big men *see also ol howma*   20–21
big women (*ten howma*)   335
Bik, M.J.J.   156
biomass   240–242
biota and soil   6
bird scares   252
Bismarck Range   29, 34, 230, 364
Bismarck Sea   131, 133
Blaikie, P. and Brookfield, H.   395
Bleeker, P.   142–143, 151, 154–155, 159–161, 163, 240, 279, 282, 287, 290, 293–294
Bleeker, P. and Healy, P.A.   148, 151, 287

Blong, R.J.   131–132
Blumea arnakidophora   178
Boletus erythropus   249
"bone gardens"*see also* semi-permanent cultivation   4, 39
BOSTID   225
botanical classification   167–179
Bourke, R.M.   46–47, 49, 76, 84–85, 366, 368, 372, 380, 383
Bouwkamp, J.C.   49, 366, 368, 383
Bouwkamp, J.C. and Hassam, M.N.M.   49
Brachyposium sp.   188
Bradbury, J.H. and Holloway, W.D.   45
Brams, E.A.   32, 312
Brazil   8, 158
Bridge, J. and Page, S.L.J.   252–253, 262, 339, 375
Brinkman, L.F. and Nascimento, J.C. de   30, 358
Brookfield, H.C.   395
Brookfield, H.C. & Brown, P.   37, 115, 127, 273
Brookfield, H.C. and Allen, B.   73
Brown, C.H.   254
Brown, C.M. and Robinson, J.P.   123
Brown, M. and Powell, J.M.   73, 75
Brown, P. and Podolefsky, A.   204
Brunton, R.   96
Bubbia sp.   94, 210
Bulmer, R.N.H.   172, 175, 256
Burnett, R.M.   161

Cacatua galerita   252
Cambisol   279–283
Campnosperma brevipetiolata   91
cane grass, *see Micanthus* spp.
cane grassland (*gaimb*)   194–198, 245, 362
carbon (C)   239, 283, 288, 290, 294, 354–355
carbon:nitrogen ration (C:N)   33–34, 356–358
Carex sp.   188
cassowary, *see Casuarius bennetti*
Castanopsis sp.   183, 221, 248, 280
Casuarina sp.   187, 191, 196, 223–228
   botany   224–226
   cultivation   228
   roots, symbionts   225–227
Casuarius bennetti   57
cation exchange capacity (CEC)   31–32, 239, 283, 288, 290–292, 294, 353–354, 358–359

Cayratia sp.   184
Cenococcum sp.   225
Central America   38
Ceratocystis fimbriata   46, 382
ceremonial exchange   20–21, 91–92, 94, 315, 335, 395
Chambers, R. *et al.*   12, 162, 403
Chang, J.H.   58, 72
Charles, A.E.   36
Charmosyna papou   88
Chartres, C.J. and Pain, C.F.   150–151, 266
chert   128, 280, 301–302
Chisocheton ceramicum   94, 210
Christianity   24
Clarke, W.C.   37, 55, 72, 110, 135, 143, 190, 344
Clarke, W.C. and Street, J.M.   29, 298, 320, 364, 381–382
classification   8–9
   garden sites   303–304
   in pairs   176–177
   plants   169–179
   rocks   127–131
   soils   265–296
   vegetation communities   179–200
clayey soils   279–289, 345
climate   15–17, 53–78
   change   14
   environmental processes   53
   food shortages   76
   soil   5, 101–102
   sweet potato   76, 79, 81–86
   S.W. Pacific   54
   type, lower montane humid   54
   see also frost, rain, temperature, cloud, wind
cloud   63–66
   cover daily   65
   as smoke   65–66
   types   65
Cockle Park   387
Coix sp.   187, 196, 199
Coleman S.Y., Shiel, R.S. and Evans, D.A.   387
Collybia sp.   249
Colocasia sp.   173
Cominisia sp.   191
Commelina sp.   191
communities, human *see also* sem   22
Compositae   187, 196, 199
compost   373, 377–379, 382–392
   application rates   387–388
   nutrients   385–386, 388–391

rate limits   392
trials   384–387
Comprosoma sp.   188
Conklin, H.C.   4, 315
conservation   14, 398–403
Constantin, R.J. *et al.*   366, 368
Coprosma sp.   189
Cordyline sp.   178, 191, 196, 223
Coscinocera anteus   256
Cracticus quoyi   57
Crassocephalum sp.   172, 193, 220
Crittenden, R.   75, 79, 392
crops   41, 43, 190–193, 362, 364, 391
    cover   157–159
    pathogens   250
cropping cycle   159
Cryptocarya sp.   196, 198, 220–221, 236, 249, 254
CSIRO   266, 271, 285, 289
curcurbits   43, 173, 190
cultivation *see also* shifting cultivation
    intensity of   393–398
    regimes   4
Cunningham, R.K.   30
Cupaniopsis sp.   213
curing rites   209–214
Cyathea spp.   169, 183, 188, 196, 198, 211, 252, 295
Cyclosorus sp.   183, 187
Cynoglossum javanicum   192, 220
Cyperus sp.   177, 187
Cypholophus sp.   177
Cyrtandra sp.   183

D'Souza, E.J. and Bourke, R.M.   383
Dacrycarpus sp.   188
Dacrydium sp.   187
Dantas, M. and Phillipson, J.   236
Danthonia sp.   188
Daphniphyllum sp.   183
Davies, H.L. and Smith, I.E.   123
decomposition   229, 247–262
demographic pressure   395–398
demography and hungry times   79
den, see plant families
Dendrobium sp.   184
deposition   290
depositional features   135
descent   23
Deschampsia sp.   188
Desmodium sp.   196, 199
destructive plate margin   133–134
development and indigenous knowledge   12, 403

Dicaeum geelvinkianum   184
Dicranoloma sp.   183
Dicranopteris sp.   174, 189, 196, 200, 280, 295
Diem, H.G. and Dommergues, Y.R.   224
Dimorphanthera sp.   184, 189, 207
diseases   249–251
    spread   251
    naming of   249–250
Diski, J.   9
Djazuli, M. and Tadano, T.   368–370
Dobsonia moluccensis   252
Dodonea sp.   187, 196, 198, 220–221
Dokuchaev, V.V.   5
Doma Peaks   133
Dorcopsis vanheurni   91
dorpind myth   103, 132–133
Dove, M.R.   39
Driessen, P.M. *et al.*   38
drizzle   61
drought and crop stress   77
drought and crop yields   76–77
drought and soil water content   77
drought magic   73
Duchafour, P.   270
Dvorak, K.A.   265
Dystrandept   283

earth mounds *see* soil mounds
earth oven   120
earth tremor   105
    folk saying   105
earthquakes   134
earthworms   256–262
    exotic introduced   259
    fear of   256–257
    and soil fertility   257–259
Eastern Highlands Province   38, 84, 194, 357, 364, 380, 386
Edwards, P.J.   35, 155, 163, 230, 232–237, 246, 351
Edwards, P.J. and Grubb, P.J.   183, 230, 234, 239–240, 246, 262, 352, 357
Elaeocarpus spp.   57, 183, 191
Elatostema sp.   176
Eleocharis sp.   177
Ellen, R.F.   203
Elwell, H.A.   142
emic view   266
Enga people   55, 87, 95, 132
Enga Province   129, 388, 390
Entisols   265
environment   4–7

environmental anthropology 14
environmental knowledge and lore
    7–15
  informality of 9
environmentalism 14, 398–403
Enyi, B.A.C. 46–48, 84, 364
Equisetum 187
Erigeron sumatrensis 220
erosion 29–30, 106, 135, 139,
    140–142, 227, 311–312
  conservation practices (P) 160–161
  control measures 161–163
  local terminology 141–142
  pigs 157, 396
  protection, vegetation cover (C)
    155–160
  slope 152–155
  soil losses 161–164
  water 139, 140–142
erosivity index (R) 143–147
erodibility index (K) 147–151
ethno-environmental perspective 11
ethnobotany 167–200
  methods 178
ethnogeoscience 105
ethnometeorology 53–78
ethnopedology 4, 265–266, 271–276,
    276–296
ethnoscience, environment 7–15
ethnoscience classification 8
Euphorbia sp. 191, 198, 224
Europe 42–43
Ewel, J.J. *et al.* 29–30
exchangeable cations 31–32, 236–237,
    242–243, 245–246, 279, 288–291,
    293, 295, 350–353, 358
experiential knowledge 9, 401

fallow *see* secondary vegetation
  brief grassy 38
families (*sem*) 22–23, 172–173,
    175–176, 247, 253, 298, 326–332
famine foods 79–80
FAO 26, 36, 161–162, 217, 276,
    281–282, 353
FAO-SIDA 36
fertilisers 359, 364–366, 370
fertility depletion 38
Festucia sp. 188
Fibrist 295–296
Ficus spp. 183, 196, 198, 221–223, 284
FitzPatrick, E.A. 5–6
Fitzpatrick, E.A. 53, 60
Flenley, J.R. 185, 216

Floyd, C.N. *et al.* 42, 156, 349, 359,
    367, 370, 375, 377, 384–385,
    387–388, 390–391
Fluvent 289–290
Fluvisol 289–290
Foi people 113, 305
Folist 295–296
food shortages 77–78, 85–86
  rituals 86–99
  and geological events 105
forest (*iyshabuw*) 182–186, 201–203,
    245, 362
  attitudes towards 203–204, 215–217
  conservation 215–216
  destruction 216
  fear of 204, 215
  floristic composition 183–184
  vegetation, transition between 200
forest demons 204–214
  iyb tit 207–209
  rites 209–214
  saem 205–207
Fox, R.L. *et al.* 368
Frankia sp. 224–225
Freycinetia sp. 183
frost 73–75, 77
  and altitude 74
  and crop damage 74–75
  and seasons 75
  and topography 74
  and vegetation 74
frost damage celebration 74–75
Frullania sp. 183, 187
fungal black rot 46
fungi 209, 225, 247–249
  fruiting cycle of 248
  naming of 249
Furbee, L. 265

Gagné, W.C. 163, 252
gaimb see cane grassland
Galbulimima sp. 183
Gallicolumba spp. 57
Garcinia sp. 57, 183
gardens
  abandonment 364
  times cultivated 339–344
garden sites, *see also* site
  burning off *see also* vegetation
    burning 40, 42
  clearing 40
  enclosing 40
  planting 41
  selection 20, 315, 334

garden types   19–20, 336–337, 339–344
    sweet potato   339–344
    taro   187, 339–344
    mixed vegetable   339–344
garden vegetation see also crops, vegetation   190–192, 220
Gardenia sp.   184
gas, mineral   130–131
Gebhardt, H. and Coleman, N.T.   349
gender   335
Gentiana sp.   188
geographical directions   113–115
geology   120–137
    geological eras   123–126
    geological features   121–122
geomorphological processes   134–137
Ghana   31
Gigaspora sp.   370–371
Gillison, A.N.   190, 273
Giluwe, Mount   66, 109, 131, 145, 162
Glasse, R.M.   132–133
Gleichenia sp.   188
gley soils   292–293, 302, 361
Gleysols   292–293
Glochidion sp.   187
Glomus sp.   225, 370–371
Godelier, M. and Strathern, A.M.   20
Goldie's loritkeet   57
Gollifer, D.E.   82, 84
Golson, J.   164
Golson, J. and Gardner, D.S.   79, 164, 204
Golson, J. and Hughes, P.J.   164, 398
Goodbody, S. and Humphreys, G.S.   370, 376–377
Goroka   364
gound doves   57
Gramineae see grasses
Graptophyllum sp.   191, 223
grasses, see also *Miscanthus*, cane grassland, secondary vegetation   187, 193–198
grassland   245, 362
Greenland, D.J.   147
Greenland, D.J. and Nye, P.H.   35
Greenland, D.J. and Okigbo, B.D.   37
Grifola frondosa   174, 209, 249
Grubb, P.J.   232
Gryllotalpa sp.   252, 257
Gryllus bimaculatus   257
Guatemala   38–39, 388
Guillet, D.   265
Gulubia sp.   177
Gymnogryllus angustus   252, 257

Haantjens, H.A. and Bleeker, P.   135
Haantjens, H.A. and Rutherford, G.K.   240, 266
Haberle, S.   167, 176
Hackett, C.   83–84, 390
Haelaelinja   59, 63, 70–71, 142–143, 181
haen paenj, see weathering product
Hagen people   61, 87, 206
Hahn, S.K.   45, 48–49, 82, 85, 86
hailstorm   61
Hammett, L.K. et al.   85, 368, 369
hard-time rituals   86–99
Harmsiopanax ingens   56
Harrison, S.   95
Hays, T.E.   8, 167
Healey, C.J.   174
Helicia oreadum   57
Hemist   293–295
henk, see plant families
Henty, E.E.   194, 197, 204
Henty, E.E. and Pritchard, G.S.   220
Heterodontonyx bicolor   252
Hibiscus manihot   178
Hide, R. et al.   84, 168
Hill, K.C.   124
Hill, W.A. et al.   366
Hirsch, E. and O'Hanlon, M.   115
Histosols   265, 293–296
Hoboga Makwes   213
Hodgson, J.M.   299, 316
hokay - see sweet potato
Homolanthus sp.   187
Hope, G. and Golson, J.   399
Horwar Saliyn (spirit being)   51, 86–95, 99, 131
house sites, abandoned   392
household waste, as manure   392
houseyard vegetation   191, 223–224
Hozyo, Y.   48
Hudson, N.W.   144–145, 152, 154
Huli people   113, 132, 153, 158, 393–395
human-beings and soil   7
humid tropics, soils of   402
humidity   70–71
Humphreys, G.S.   135, 140–146, 151, 153–154, 157, 159, 162–163, 253, 258–259, 262, 268, 270
Humult   279–283
humus see also organic matter, litter   282
    under fallow   35
hunger and pigs   81
hungry times   79–81

Hunn, E.S.   8, 254
Huypers, H. *et al.*   139, 163
Hydrandept   283
Hymenogaster sp.   225
Hyndman, D.   112
Hynes, R.A.   202

igneous rocks (*huwbiyp*)   128–129
Impatiens sp.   199, 220
Imperata sp.   28, 156, 187–188, 197, 221
in-/out-field   42
Inceptisols   265, 316, 402
India   28, 388
indigenous knowledge   12, 403
   distortions   177–179
Inglis, J.T.   12, 402
Inocybe sp.   209
insect pests   46, 252–256
"intellectualists" (cognition and classification)   8
inter-tropical convergeance zone (ITCZ)   66
intercropping   29, 41, 158, 364
invertebrates   253–256
   metamorphosis   254–256
Ipomea batatas see sweet potato
Isachne sp.   178, 187, 193, 196, 220
Ischaemum sp.   172, 187, 193, 194, 196, 199, 220–221, 379, 388–389
iysh see plant families
iyshabuw see montane rainforest
iyshpondamahenday ritual   51, 89–95

Jaiyebo, E.O. and Moore, A.W.   38
Jenkinson, D.S. and Rayner, J.H.   387
Jennings, J.N. and Bik, M.J.J.   119
Jenny, H.   5–6, 101, 240, 357
Johns, R.J.   179
Jordan, C.F.   25, 30, 235–236, 240, 243
Junus effusus   176–177
Juo, A.S.R. and Kang, B.T.   25–26, 37–38, 359
Juo, A.S.R. and Lal, R.   29

Kalam people   256
Kalkman, C. and Vink, W.   201, 203
Kalma, J.D.   73
Kaluli people   132
Kanua, M.B. and Floyd, C.N.   369, 375
Karimui people   25
Kays, S.J.   82, 84–86, 370
Keig, G., Fleming, P.M. and McAlpine, J.R.   61
Kellman, M.C.   36

Kellogg, C.E. and Pendleton, R.L.   26
Keravat   366, 380
Kerewa, Mount   109, 207
Kerewa-Giluwe region *see also* Wola region   119–120, 276
   climate   54–55
   geology   123–126
   soils   295
   topography   109
   vegetation   179
Kerven, C. *et al.*   265
Kewa people   132
Kimber, A.J.   38, 364, 380, 386–387
King, G.A.   84
Kinnel, P.I.A.   145
Knight, J.   14
Kocher Schmid, C.   167
Kombie swamp   69
Krebs, J.E.   30
Kuma   69–70
Kundiawa   143, 145
kwimb see plant families
Kyllinga sp.   177, 187

labour
   female   21
   migration   398
Lai valley   75
Lake Egari   109
Lake Kutubu   113, 130, 207–208, 311
Lal, R.   29, 144–145, 151–152, 154, 157–160
Lambert, J.D.H. and Arnason, J.T.   28, 38, 377, 388
Lamprima adolphinae   254, 257
land
   access to   21
   attitudes to   3
   degradation   393–398
   degradation in New Guinea   14–15
   resources   401
   rotation   4
land tenure   22–23, 326, 333–334
landforms   115–120
   alluvium   119
   areas covered   119–120
   folded   115–116
   low rounded hills   119
   polygonal karst   119
   steep sided mountain ridges   118–119
   volcanic   115–116
   volcanic ash plains   117–118
   volcanic mountains   117
Landon, J.R.   149, 322, 347, 358

Landsberg, J. and Gillieson, D.S. 265, 273
landscape *see also* landforms 106–109
  and geology 129
landslides 134–135, 162
Laportea sp. 191, 251
Latin America 34, 37, 357
Laudelot, H. 30
Lavelle, P. 258
law of the minimum 367
leaching nutrient losses 27, 351, 356
Lederman, R. 22, 120, 175, 207
Lee, K.E. 257, 259
Leersia sp. 186–187, 193, 220
Leng, A.S. 382–383
Lentinus spp. 249
Lepidoptera, as pig spirits 256
Lepidozia sp. 183
Lewis, G. 96
Liasis spp. 257
life-form
  and biomass 241
  classes 176
lightning 61
limestone (*hathaen*) 127–128, 136–137, 278–279
  fossils in 127
Linnean taxonomy 10
Lithocarpus spp. 183, 221–222, 249, 254
Lithosol 278
Litoria iris 62
litter *see also* humus, organic matter 165, 300
  decomposition 235–236
  fall 235–237
  mass 235
  nutrient status 236–237
local orography *see also* landforms 109–111
local soil assessment 310–312, 359–361
local topographical knowledge 112–115
locales and landscape 112
location terminology 114–115
Löffler, E. 119, 126, 131, 140, 156, 162
long-lerm shifting cultivation *see also* semi-permanent cultivation 4, 37–39
low input cultivation 49–50
Lycopodium sp. 188

Macaranga sp. 221
Maenget Kem 51, 57

Maenget Pes 165–167
Maesa sp. 187
Manner, H.I. 34, 392
Maring people 55, 135, 222
marriage 21, 335
marsupial offering *see also* pig offering 98
marxism 20
mass movements 162
Mawe, T. 87, 89
Mayka Sal 133, 373
Mbowamb people 61, 87, 206
McAlpine, J.R. *et al.* 53–54, 59–60, 63, 66, 69, 71, 75, 77, 143–144, 158, 320
Mecopodinae 252
Meggitt, M.J. 55–56, 75
Melastomaceae 183
Melidictes rufocrissalis 62
Meliodogyne sp. 386
Meliosma pinnata 56
Mendi 59–60, 63, 70–71, 75, 87, 113, 143, 145, 285, 316, 388
Merlan, F. and Rumsey, A. 23
metamorphic rocks 129
Metapheretima spp. 257
Meterorium sp. 183, 187
micronutrients 358, 371–372
Microsorium punctatum 174
Midgley, S.J. *et al.* 224
Milton, K. 14
mineral cycling, montane forest 230
mineral cycling, seconday woodland 231
mineral cycling, cane grassland 232
mineral cycling, coarse grasses 233
mineral oil, (*dez*) 130–311
mineralisation 240
mineralogy, soil 287–288, 292, 353–354
Miscanthus spp. 57, 72, 75, 169–170, 174, 186–187, 189, 194–198, 214, 220–221, 236
  botany of 194–195
Mishra, B.K. and Ramakrishnan, P.S. 29–30, 34
mist 61, 65, 71–72
Modjeska, N. 21
Mohr, E.C.J. *et al.* 101, 281, 286, 357
Mollisols 19, 38, 265
montane forest *see* forest
Morauta, L. *et al.* 203
Morellato, L.P.C. 235–236
Morgan, R.P.C. 143–145, 151–152, 159, 163
moss forest 188–189
mounding, of soil *see* soil mounds

mountain terrain *see also* landforms,
    landscape   106–107, 110
Mueller-Harvey, I. *et al.*   33, 350, 356
Mulongoy, K. *et al.*   369–370
multivariate analysis   299
munk shor see plant families
Musa sp.   191
mycorrhizae   225–227, 369–371, 391
myth
    of white-skinned spirit woman
        87–88
    of *dorpind*   103, 132–133

Nakano, K.   38–39
Nakano, K. and Syahbuddin,   39
Nastus sp.   171, 191
natural resources, access to   20–22
natural science and anthropology
    10–11, 13
nematodes   252–253
Nembi Plateau   371, 392
Neopsittacus musschenbroekii   252
net assimilation rate (NAR)   47, 367
New Britain   366, 380
New Guinea   37–38, 50,
    geology   122–123
    climate   53
Newton, K.   26, 36–37
Ngeve, J.M. and Roncadori, R.W.   369
ngoltombae ritual   95
Nicholaides, J.J. *et al.*   366–369
Nigeria   33, 37, 158
nitrogen (N)   33–34, 236–237,
    242–243, 245–246, 283, 288,
    290–291, 293–295, 355–356
nitrogen fixation   224–225
Norfolk-four-course   42
Nothofagus sp.   168, 174, 183, 185,
    201–203, 254
nutrient cycling   167, 229–246
    rainfall inputs   233–234
    throughfall inputs   233
nutrient ions   26–27
    released upon burning *see also*
        vegetation, burning of   243–245
    up-take under fallow   34
    vegetational distribution   242–244
nutrient replenishment under fallow
    36–37
nutrient shortages   246
Nye, P.H.   35, 234–235
Nye, P.H. and Greenland, D.J.   4,
    26–28, 30, 33, 217, 236, 312, 348,
    382

ochre   281
Oenanthe sp.   187, 191, 196, 199
Ofori, C.S.   30, 383
oil seeps   93
Ok region   400
ol howma (big men)   20–21, 315–323,
    330, 333–334, 337
Oldfield, F. *et al.*   164, 400
Olethrinus tyrannus   252
Ollier, C.D. *et al.*   265, 273
organic matter *see also* humus, litter
    150–151, 238–240, 245, 354–358
    breakdown   354, 357
    build up   240
    content   321–322
    nutrient status   238–240
Oxisols   19, 32, 37–38, 401
Oxya japonica   252

Paglau, M.   161
Paijmans, K.   179, 240
Pain, C.F. and Blong, R.J.   131
palynology   164
Pandanus spp.   57, 74, 170, 177, 185,
    191, 196, 207, 223, 249, 252, 254
Panicum sp.   187
Papuacedrus sp.   188
Papuana sp.   252
parent material, and soil *see also* soil
    formation   6
Parfitt, R.L.   225
Parfitt, R.L. and Mavo, B.   349
Paspalum sp.   192–193, 196, 199, 220
Pearson, M.N.   249
peaty soils   293–296
pebbles,   128, 135
Pennisetum macrostachyum   174, 196
Perry, R.A.   123
Peru   31
pests   251–253
pests and diseases   28
pH (acidity)   30, 238, 279, 282, 288,
    290–292, 294–295, 322, 326, 341,
    347–349, 359
Philippines   382
Philips-Howard, K.D. and Kidd, A.D.
    265
Pholiota sp.   249
phosphorus (P)   236–237, 242–243,
    245–246, 279, 283, 288–289,
    290–291, 293–295, 322, 326,
    349–350, 358, 391
    availability   32, 349
Phyllanthus sp.   183

Phytophthora sp. 203
Pieris sp. 77, 252
Pieters, P.E. 123
pig keeping, women's labour 395–397
pig offering *see also* marsupial offering 90–91, 94, 100
pigs 20–21, 315–316, 335, 395–397
  feeding 397
  and soil erosion 157, 396
Piperaceae 183, 187, 198
Pipturus sp. 172, 177, 183, 196, 198
Pisolithus sp. 225
Pittosporum sp. 184, 196
Pivello, V.R. and Couthino, L.M. 26, 233
plants
  classification of 169–179
  classification, disagreements over 173–175
  communities, Wola identification of 179–182
  communities, biomasses 240–241
  cultivated see also crops 173
  diseases 203
  life-form classes 170–172
  naming 173–175
plant families
  den 170, 172
  henk 169–170, 172
  iysh 168, 170, 172
  kwimb 170
  munk shor 170
  ya 170–171
plant kingdom 175
Platea excelsa 57
Plectranthus sp. 193, 196, 199,
Pleistocene 124–126, 131, 283
Pleutrotus djamor 249
Poa sp. 188
Podocarpus sp. 187–188
poison 60
Pol Kot 103, 133
Polygonum sp. 172, 192–193, 199, 220
polygyny 21
Polyporus arcularius 249
Popenoe, H.L. 32, 38, 240, 355
post-modern critiques 12–13
potholes 111, 120–121
Pouzolzia sp. 178
Powell, J.M. 168
prehistory 398–400
  farming 164
  food shortages 79
  vegetation 203

Preston, S.R. 382, 390
Psamment 291
Psittacella madaraszi 252
Psitteuteles goldiei 57
Pteridium sp. 187–188, 200
puddling of soil 312
Pycnoporus sanguineus 249
Pyralidae 252

Quigley, D. 13

Racemobambos sp. 183, 189
Radcliffe, D.J. 73, 75, 77, 148–151, 181, 197, 240, 266, 270–271, 273, 284–285, 287, 289, 316, 349, 354, 358
Radcliffe, D.J. and Gilman, G.P. 353
rainfall 194
  averages 58–59
  erosivity 142–147
  forecasting 62
  intensity 62, 142–147
  interception 234
  kinetic energy 143–146
  magic 62–63
  mean 61
  orographic effects 60
  rainbow 61
rainforest *see* forest
Ramakrishnan, P.S. 26–27, 29–30, 33, 388
Rambali, P. 8
Ranunculus sp. 188
Rapanea sp. 188, 207
Rappaport, R.A. 29, 37, 76, 222
rat scares 251
Rattan sp. 184
Rattus spp. 251
Raumolin, J. 43
Reay, M. 209
regowth vegetation, *see* secondary vegetation
Rendoll 278–279
Rendzina 278–279
Reynders, J.J. 37
Rhizobium spp. 225
Rhododendron spp. 188
Rhyticaryum sp. 187
Richards, P. 12, 162, 402
rills 141
rituals, *see iyshpondamahenday, ngoltombae*, forest demons
  borrowing 95–96
  consequences of 100–101

rituals, *see iyshpondamahenday,*
    ngoltombae, forest demons — *cont.*
  dance 91–92
  diet 90
  house 90, 94
  incantations 91, 98, 213–214
  and misfortune 100
  specialists 89
  variations in 87, 95–96, 99
  water pools 92–93
rivers 107–108, 111
Robbins, R.G. 179, 194, 197, 204
Robbins, R.G. and Pullen, R. 179, 201
Roberts, G. 76
Robison, D.M. and McKean, S.J. 26, 36–37
rocks
  classification 127–131
  local use of 120
  miscellaneous 130
rocky vegetation (*haenbora*) 187, 278
Rose, C.J. and Wood, A.W. 257–258, 262
Rothampstead 387
Rubus sp. 57, 193, 196, 200
Rungia sp. 190
Russula sp. 249
Rutherford, G.K. and Haantjens, H.A. 271, 273, 279, 284–285, 293
Ruxton, B.P. 156
Ryan, D'A 22, 120, 175

Sabhasri, S. 38–39
Sabkabyinten (spirit being) 51, 57, 61, 66, 73, 87, 99
Saccharum sp. 173, 187, 190, 196
Sacciolepsis sp. 196
Sajjapongse, A. and Wu, M.H. 82–83, 85–86
salinity, soil 358
Sanchez, P.A. 4, 25–29, 31–32, 34–35, 37–39, 312, 348, 350, 354, 356–357, 382
sand/gravel 136
sandstone (*naenk*) 128
sandy soils 291
Saprist 293–295
Saurauia sp. 188, 196, 198, 221
Saxena, K.J. and Ramakrishnan, P.S. 28, 33, 377
Schefflera sp. 56, 184, 188, 196, 198
Schuurmansia sp. 196
Scincella elegantoides 62
Scott, D.A. *et al.* 235

Scott, L.E. and Bouwkamp, J.C. 368, 371, 390
Scutellospora sp. 370
seasons 58, 60–61, 73
  and bird migration 57
  and clouds 65
  and crops 75
  and cultivation 75–76
  and deciduous trees 56–57
  and droughts 77
  and marsupials 57
  NW (*ebenjip*) 51, 53, 55–58, 60–62
  SE, (*bulhenjip*) 53, 55–58, 60–61
  SW Pacific 60
  and solar changes 56
secondary forest vegetation (*obael*) 198–200, 245, 362
secondary regrowth
  managing 222–224
  successions 217–222
  vegetation 34, 39, 192–194, 393
Selaginella sp. 196, 199
sem see families
sem groups
  demography and soil use 327
  semg^{e}nk 22, 175–176, 333–334, 337
  semonda 22, 175–176, 345
semi-permanent cultivation 4, 14–15, 19–20, 39–44, 49–50, 347–358, 364, 377, 395, 398
  reasons for 10–11
senselessness 208–209
Setaria spp. 172, 190, 193, 196, 199, 220
settlements 22
shade 29
shallow soils 278–279
Shaw, D.E. 251
shifting cultivation 4, 25–43, 339
  burning *see* vegetation, burning of
  cations 31–32
  CEC 32
  fallow period 34–36
  humus decline under 32–33
  inefficient land use 37
  nitrogen (N) decline 33
  nutrient decline 25
  pest and disease multiplication 28
  phosphorus (P) 32
  relativity of *see also* semi permanent cultivation 36–39
  short cropping periods 36
  soil erosion 29–30
  soil nutrient deterioration 30–34

soil organic matter 32–33
soil pH 30
strategy 25
weed proliferation 28–29
yield declines 28, 38
Siebert, S. 25, 30, 38, 154, 348, 350, 382
Sillitoe, P. 8, 18, 20–23, 41, 43–44, 57, 87, 94, 113, 120, 128–130, 158, 173, 175–176, 190, 204, 207–208, 215, 223, 250–251, 281, 284, 299, 323, 327, 332, 334, 339, 364, 397, 400
Simbai 161
Simbu Province 84, 86, 143–146, 153–154, 158–159, 161–162, 376
site
  abandonment, reasons for 28–34
  abandonment, social and cultural considerations 34
  altitude 320
  aspect 305, 317
  characteristics 323–326, 327–330, 341
  classification 303–304
  selection 20, 315, 334
Skeldon, R. 132
sky-beings see Sabkabyinten
slope, garden sites 317–320, 341
  length (L) 152–155
  steepness (S) 152–155
  topography 154–155
Smith, J.M.B. 72
Smith, E.S.G. and Thistleton, B.M. 252
social groups see also sem groups 22, 326–330
social standing see ol howma
  and crop yields 331–332
  and garden type 336–337
  and soil assessment 330–331
  and time garden cultivated 336–337
soil (suw) 4–7, 263–364
  agricultural potential 312–315
  and crops 315
  and environmental factors 4–7
  cultivation of (see also soil mounds) 151, 283, 289–291, 293–295, 297–298
  cultivation and gardeners' ages 323
  inspection of 297–298, 309–310, 314–315, 333
soil assessment 298–315, 340–341
  assessment classes 313–314
soil classification 265–296
  and parent material 266
  problems with 267–268, 270–271
  natural systems approach 268–269
  artificial systems approach 269–270
  flexibility 272–273
  and soil properties 272–273
  non-hierarchical schemes 273–276
soil conservation 30, 160–161, 224
soil drainage 293
soil erosion (see also under erosion) 29–30, 106, 227, 311–312
soil fauna and flora 256–262
  changes in 30
  macro fauna 253
  populations 259–262
  microbial population 30
soil fertility (see also chemical elements) 322, 326, 375–376
  management 224, 359
  and social status 332–336
  under cultivation 344–364
  under fallow 36–37
soil "grease" (suw iyba) 165, 167, 263, 276, 291, 295, 310–314, 322, 361
soil formation 137, 163
  and vegetation 167
  Rendzina 279
  Tropept 281–283
  Andepts 286–288
  Fluvent 290
  Aquent 292
  Hemist 294
  Folist 295
  geophysical resources 105
  and moisture 101
  and temperature 101–102
soil horizons 147, 321
  classification 273, 299–302
soil management 160–161, 168, 359, 362, 376–377
soil mineralogy 151, 287–288, 292, 353–354
soil moisture 61, 85
  available water capacity 149–150
  soil water state 311–312
soil mounds 41, 54, 75, 160, 163, 376, 373, 377, 378–383
  and disease 382
  soil structure 380–382
  fertility 382–383
soil nutrition see soil fertility 344
  nutrient cycles 35
  nutrient status under cultivation 347–358
soil organic matter 32–33, 150–151, 238–240

soil pH, see pH
soil physical condition, changes under
    fallow 35
  changes under cultivation 29
soil profiles, classification 302–303
  technical descriptions 279–280,
    284, 289, 291–293, 295
soil protection 155–161
soil resources, classification 304
  classes and indigenous soil classes
    306–308
  classes and land use 306–309
  and gardeners' ages 323–326, 331
  and kin group affliliation 326–331
  and social standing 315–337
soil samples, laboratory tests 345n
  field tests 316n
soil state classes, *suw hemem* 313
  suw huwniy 313–314
  suw ka 313
  suw pa 314
  suw taebowgiy 314
soil structure 227–228, 312–313
  bulk density 148–149, 380–381
  consistence 148
  porosity 149
  stoniness 310
  strength 310, 314
  structural degradation 29
  texture 147–148
soil survey 298–304, 316, 345
  classes 361–362
soil temperature 30, 69–70
soil types, *haenbora* skeletal soil 278
  haen hok arrested soils 278–279
  hundbiy clay soils 279–283
  iyb dor tilai alluvial soils 289–290
  iyb muw sandy soils 291
  iyb uw damiy peat soils 293–295
  pa tongom glay soils 292–293
  tiyptiyp volcanic ash soils 283–289
  waip peat soils 295–296
  nutrient status 279, 282–283,
    288–291, 293–296
soil, under cultivation 339–364
soil variability 314–315
Solanum sp. 191
solar radiation 72–73
Solomon Islands 82, 84
sorcery 24
S.E. Asia 38, 357, 402,
S.W. Pacific region 69
Southern Highlands Province 38, 130
Spathoglottis sp. 178, 187–188

spirits 24, 51, 57, 61, 66, 73, 86–95,
    99, 131, 204–214
  and potholes 120–121
Staich, E. 59, 143
stateless political order 20, 23, 93–94
  and productiom 334–336
statistical analyses 299, 316–317, 347
Steensberg, A. 4
Sterculia sp. 56, 58, 183, 254
Sterly, J. 168
stone axes 129–130
Straatmans, W. 167
Stratherm, A.M. 14, 23, 335
Strathern, A.J. 21, 61, 89, 94,
Strauss, M. and Tischner, H. 61, 87,
    89, 206
Strickland River 400
Strobilmyces velutipes 249
structuralism 9–10
Styphelia sp. 188
subsoil 276, 279, 280–281, 284–285,
    289–292, 301–302
  mottling 292
sunshine 71–73
sunshine and slope aspect 72
supernatural beliefs see also spirits 24
surface wash 140
suw see soil 3, 273–276
suw iyba see soil "grease"
swamp forest 187
swamp vegetation (*pa*) 186–187
Swamy, P.S. and Ramakrishnan, P.S.
    28, 377
sweet potato (*hokay*) 22, 41, 43–50,
    172, 190, 193, 336, 339, 376, 397,
    400
  botany 45
  cultivation see also soil mounds
    45–46
  productivity 46–49
  semi-permanent cultivation 44,
    347–358, 364, 395, 398
  shortages 76
sweet potato, environmental responses
    49–50
  and climate 76, 79, 81–86
  and rainfall 83–86
sweet potato, physiology 46–50
  dry matter production 46–49, 83
  growth stages 47
  leaf area 47, 49, 82, 366–368
  net assimilation rate (NAR) 47, 367
  partitioning 46
  source-sink relationships 48

storage root development 46–49,
    82–85, 366–368
  vegetative phase 47
sweet potato - yield 44
sweet potato and leaf area 82
sweet potato and weather 76
sweet potato dry matter production 83
sweet potato nutrition 364–372
  micronutrients 371–372
  mycorrhizae 369–371
  nitrogen 366–368
  pH 371
  phosphorus 368–371
  phosphorus nutrient experiment 369n
  potassium 366–368
sweet potato yields 38, 44, 46, 375
  increasing yields 14
  and solar radiation 82
  and temperature 82–83
sword grass see Miscanthus spp.
Symplocos sp. 183
Syzygium spp. 56–57, 187

Talegalla jobiensis 62
Tari Basin 38, 154, 158, 393
taro gardens 187, 336–337, 339–344
taxonomy see classification
Taylor, G.A.M. 131
tectonic activity 131–134
temperature 69–70, 82–83
temperature, daily 69
temperature and altitude 69
Tergas, L.E. and Popenoe, H.L. 33, 39, 353, 357, 388
Thelypteridaceae 196, 200
Themeda australis 156
Thiagalingam, K. 224
Thiagalingam, K. and Bourke, R.M. 383
Thiagalingam, K. and Famy, F.N. 225
thickets (*pletbok*) 196
Thistleton, B.M. and Masandur, R.T. 339, 375
thunder 61
Thurston, H.D. 251
Thylogale bruijni 91
time and soil 6–7
time of darkness legends 132–133
Timonius sp. 184
Togari, Y. 48–49
Toky, O.P. and Ramakrishnan, P.S. 27, 33–34, 37
Toona sureni 56

topography 106–120
topography and soil 5–6
  topographical features, vocabulary for 110–111
topsoil *see also* soil 276, 279–280, 282, 284, 289, 291–293, 300–301, 313, 320–322
  colour 311, 321, 341
  depth 310, 320–321, 341
  replenishment of 163
trees, cultivated 191, 223–228
  and garden boundaries 223
Trema sp. 196, 198, 220–221, 236, 249, 254
Trenbath, B.R. 27, 36
Trochocarpa sp. 188
Tropepts 279–283, 345
Troporthent 278
Tsuno, Y. 384
Tull, J. 272
Tuneera-Bhadauria, T. *et al.* 259
Turvey, N.D. 246

U.S.A. 144–145
Uhl, C. 36
Ultisol 19, 31, 35, 37–38, 265, 401
Ungemach, H. 233
universal soil loss equation (USLE) 142
Urticaceae 183
USDA 61, 70, 265, 267, 269, 271–272, 281–283, 353
USLE 144–145, 151, 154, 159
"utilitarianists" (utility and classification) 8
Utomo, W.H. and Mahmud, N. 151

Vaccinium sp. 189
valley winds 66
vegetables, green leafy 43
  tubers 43
vegetation 344
  burning, woody fraction undestroyed 245
  houseyard 191, 223–224
  nutrient cycling 229–246
  nutrient leaching 234
  and soil fertility 298
vegetation, burning 26–27, 40–42, 243–245, 350–352, 362
  nutritional effects 26–27
  physical effects 27
vegetation communities 179–200
  areas covered 181–182

Index

vegetation communities — *cont.*
  human interference 197–198
vegetation cover 320, 326
  changes in 157
Vertisols 37
Vicedom, G.F. and Tischner, H. 61, 87, 89
Vine, H. 37
vine bridges 108-109
Viola sp. 172, 178, 193, 196, 220
viral disease 46
Vitousek, , P.M. and Sanford, R.L. 25
volcanic ash (tephra) 19, 131–134, 273, 278, 293-289, 349, 353
  debris 129
volcanic activity 126–127

Wabag 216
Wada, K. 354
Waddell, E. 44, 54, 74, 76, 95, 332, 378, 382, 388
Waebis Homalow 263
Waigani 84
Walker, D. 193, 196–197, 201, 203
Walker, D. and Flenley, J.K. 197, 400
Wallace, K.B. 148, 150
Waller, J.M. 249–250, 339, 375
Warren, D.M. 12, 162, 403
Was (Wage) valley 60, 74–75, 86, 181, 248, 298, 316, 327, 345, 387
waterlogged soils (*pa*), *see also* gley soils 292–293, 314
Watson, J.B. 133
Watters, R.F. 26, 29, 36, 217
Watters, R.F. and Bascones, L. 350
weather *see also* climate 54
  control over 99–100
  and cultivation 75–78
  folklore sayings 60–61, 78
  forecating 72
weathering 135, 266, 279, 281–283, 285–287
  product (*haen paenj*) 278
Webster, R. 299
weeds/weeding 28–29, 38, 158, 191–193, 220–221
Wenja Neleb 1, 3, 130
West Africa 32, 38
Western Highlands Province 89, 129, 399
wet periods 76

white-skinned spirit women *see also* Horwar Saliyn and Sabkabyinten 86–88, 99
Wild, A. 33
Williams, A.R. 163
Williams, P.W. 119
Wilson, L.A. 47–48, 86
Wilson, L.A. and Lowe, S.B. 46, 48
wind 66–69
  mountain 66
  orographic effects 66, 68
  roses 67
  and seasons 66, 69
  types 66
Wischmeier, W.H. *et al.* 151
Wischmeier, W.H. and Smith, D.D. 142, 144–145, 154, 159
Wohlt, P.B. 388
Wola people 15–24
Wola region (Kerewa-Giluwe region), climate 15–17
  daily weather pattern 54
  geology 17–18, 123–127
  geomorphology 18
  local lithology 127–131
  original colonists 327
  orogeny 124–126
  rainfall 57–63
  seasons 55–57
  soils of 18–19, 276–296
  topography 17
  volcanic activity 126–127
  vegetation 179
Wood, A.W. 4, 14, 25, 37–38, 139, 153–154, 240, 266, 273, 298, 309, 344, 368, 377, 392, 395
Wood, A.W. and Humphreys, G.S. 161, 163

Xenobactrachus sp. 62

ya see plant families 170–171
Yen, D.E. 44
yields 14, 28, 38, 44, 46, 82–83, 375
Young, A. 268

Zaire 35
Zimbabwe 144–145
Zingerberaceae 183, 200
Zinke, P.J. *et al.* 27, 38, 312
zonal spol perspective 101, 266